Microeconomic Theory and Computation

Michael R. Hammock • J. Wilson Mixon

Microeconomic Theory and Computation

Applying the Maxima Open-Source Computer Algebra System

Springer

Michael R. Hammock
Department of Economics and Finance
Middle Tennessee State University
Murfreesboro, TN, USA

J. Wilson Mixon
Berry College
Mount Berry, GA, USA

ISBN 978-1-4614-9416-4 ISBN 978-1-4614-9417-1 (eBook)
DOI 10.1007/978-1-4614-9417-1
Springer New York Heidelberg Dordrecht London

Library of Congress Control Number: 2013954834

Printed on acid-free paper

Springer is part of Springer Science+Business Media (www.springer.com)

To Shelby: Some value is beyond computation.

Preface

The *Maxima* computer algebra system is an open-source tool for conducting mathematical analysis using a computer. The analyst can use it to manipulate symbols, to derive numerical computations, and to graph relationships. Also, it provides a powerful programming language that allows the analyst to extend *Maxima*'s capabilities to address specific questions. *Maxima*'s symbolic and computational abilities allow one to go farther into a given problem and develop a deeper understanding of microeconomic principles than would be possible without such an aid.

Reviewing and Extending Textbook Analysis

This book shows how one can examine many of the economic relationships that form the core of microeconomic theory using this computer algebra system. It lets readers see how to set up the relevant analysis and how symbolic expressions, numerical computations, and graphical representations can be derived and combined. Once this knowledge is available, readers are in a position to extend the analysis, either by changing selected parameter values or by making more fundamental changes in the models. In particular, it facilitates comparative statics analysis.

The material developed in this book provides readers a way to gain a deeper understanding of a set of fundamental economics principles. It has two additional goals: to provide a foundation for readers who will pursue additional coursework in computational economics, and to help readers develop their ability to use a computer algebra system for more general purposes.

Computational Economics

Students who advance to courses in computational economics will build on and extend the material in this book. They will consider topics that are more advanced than, but directly related to, the ones developed here. Likewise, they will move to powerful applications of *Maxima* that are briefly introduced in this book, specifically the use of its powerful and flexible programming language.

Beyond coursework, many students will find that mathematical analysis is part of their career. For these, *Maxima* can be an important part of their careers. It is possible, however, that institutions in which they are employed use *Mathematica* or other proprietary programs. In these cases, familiarity with *Maxima* makes mastery of the other programs a relatively direct matter.

Engaged Learning

The paragraphs above refer to students and readers, but students/readers (you) should be actively engaged. The web site http://www.wxmaximaecon.com/ contains a set of *wxMaxima* workbooks that allow recreation of all of the input and output that appears in the book. (*wxMaxima* is the user interface form *Maxima* that we use throughout this book.) You should open the relevant file and work with it as you read the section. The text contains a few suggested exercises, but further experimentation will yield rewards (and, occasionally, will be fun). You won't break anything that cannot be replaced by a fresh download.

Also, do not be reluctant to go all the way back to look at Chap. 1 when working through the material in later chapters. Chapter 1 explains many basic techniques that can become confusing when you are working through advanced examples. You may find it surprisingly helpful to look at them again in a simpler context.

In addition to the workbooks that underpin the book, the web site contains exercise sets that extend the material in the text, and it contains other applications that do not relate directly to the text. Also, it contains links to sites that pertain directly to *Maxima*. These sites provide the information required to download and install *Maxima* on your computer and tips on *Maxima*'s use.

Readers who are already familiar with proprietary software such as *Mathematica* or *Maple* will see the similarities between these programs and *Maxima*. This similarity points to another source of insights into using *Maxima*: books and web sites that offer instruction in the use of these other programs.

Key to Textbooks

We have tried to make the material self-contained by sketching the theory on which the *Maxima* applications are based. Space precludes a complete development of the theory, however, so we provide the following key to some microeconomic theory textbooks.

The columns identify the textbooks that are listed below the table. The first column, labeled HM, refers to this book. Thus, each row identifies a chapter in this book and the related chapter(s) in the other textbooks. Brackets below correspond to the column labels.

HM	BB	BZ	McA	ML	MWG	NS	P	V1	V2
2	1,2	1,2	2	2,6	–	1,2	2.3	–	1
3	3	3	5	12	1	3	4	3	3,4
4	4	3	5	12	2	4	4	3	5
5	5	4	5	13	3	5,6	5	3	6–8,14
6	6	7	4	9	5	9	6	1	18–20
7	7	8	4	9	5	10	7	1	7
8	8	9	4	10	10	11	8	2	22–23
9	9	9	4	10	10	11	8	2	22–23
10	10	–	–	–	10	12	9	–	–
11	16	–	5	14	–	13	10	5	31–32
12	11	11,13	6	15	12	14	11	2	25
13	12	12	–	–	–	–	12	–	–
14	13,14	13	7	17	12	15	13,14	2	27
15	–	17	4	11	20	17	16	–	10

BB Besanko D, Braeutigam R (2011) Microeconomics, 4th edn. Wiley, New York

BZ Browning EK, Zupan MA (2011) Microeconomics: theory and applications, 11th edn. Wiley, New York

McA McAfee RP (2009) Introduction to economic analysis. Available at www.mcafee.cc/Introecon/. The notes at the site indicate the differences between this text and the next one

ML McAfee RP, Lewis TR (2009) Introduction to economic analysis. Available at catalog.flatworldknowledge.com/catalog/editions/mcafee-introduction-to-economic-analysis-1-0

MWG Mas-Colell A, Whinston MD, Green JR (2005) Microeconomic theory. Oxford University Press, New York

NS Nicholson W, Snyder C (2008) Microeconomic theory: basic principles and extensions, 11th edn. South-Western, Mason

P Perloff JM (2012) Microeconomics with calculus, 2nd edn. Addison-Wesley, Boston

V1 Varian HR (1992) Microeconomic analysis, 3rd edn. Norton, New York

V2 Varian HR (2009) Intermediate microeconomics: a modern approach, 8th edn. Norton, New York

Murfreesboro, TN, USA Mike Hammock

Mount Berry, GA, USA J. Wilson Mixon

Acknowledgements

This book grew out of the authors' interest in using open-source software in the teaching and application of economic theory. Both authors are economists; neither is formally trained in the use of *Maxima*. Therefore, we made full use of the generosity of members of *Maxima*'s development team. Robert Dodier provided encouragement early in the project and helped at several points as the project progressed. Andrej Vodopivec, *wxMaxima*'s developer, and Mario Rodríguez Riotorto, who developed the *draw* package that links *Maxima* to the open-source graphing utility *gnuplot*, were most gracious and informative in responding to our questions about using these valuable tools.

As noted, we learned the elements of *Maxima* as we developed this project. This observation has three implications. First, the three experts whose inputs we much appreciate are not responsible for any infelicities that are due to our oversight of better ways to carry out some task. Second, one need not be an expert (or even familiar) with a computer algebra system to put it to use. Third, you will likely find better ways to carry out some of the tasks than we have; we hope so.

We also thank three economists for helping us in various ways. R. Preston McAfee helped directly by looking at material that we later incorporated into this text and offering insights and encouragement. He helped indirectly by writing *Introduction to Economic Analysis*, which is used at points throughout this book. Alan G. Isaac, who maintains "Software for Economists" (http://econpy.googlecode.com/svn/trunk/software4econ.xhtml), provided direction and encouragement early in the project. Hal R. Varian graciously agreed to allow us to translate *Mathematica* workbooks that accompany his *Microeconomic Analysis* to *Maxima*. These are posted at our web site (http://www.wxmaximaecon.com/).

Our wives Liz and Barbara provided encouragement, patience, and support (including Barbara's reading of the manuscript) without which this project could not have been completed.

Contents

Chapter 1
Introduction

Abstract The open-source software system *Maxima* can compute and manipulate data symbolically, numerically, and graphically. Systems like *Maxima* are called *Computer Algebra Systems*. *Maxima*'s capabilities can greatly aid the study of economics. Its symbolic abilities can reduce the amount of tedious algebra that can be required to solve problems. Also, its use reduces the chance of computational errors. More importantly, *Maxima*'s numeric and graphical abilities help analysts go further into a given problem and extract insights and interpretations that might otherwise be missed. This chapter introduces the salient aspects of *Maxima* as it applies to microeconomic analysis.

1.1 Overview

This book considers *Maxima* as it applies to the study of economics, but much of the material that is developed here applies beyond this context. One of the attractions of software like *Maxima* is that one need not become an instant expert. Rather, learning a few basics and then applying them to a set of problems facilitates expanding the range of inquiry and the associated range of *Maxima*'s capabilities.

This chapter introduces the aspects of *Maxima* that are used in the subsequent chapters. It is important to work through this chapter, trying out the examples in *Maxima*, before moving into the rest of the text. This practice provides the background for understanding the examples that appear throughout the remainder of the text.

This text does not detail the use of *Maxima* on any particular computer operating system. *Maxima* is developed to operate on a Linux platform, but is easily installed and used on Windows, Linux, and Macintosh operating systems.[1] The material developed here can be applied in any of these environments.

---

[1]*Maxima* can also be run in the Android environment but (at the time that this material was written) without full graphical capabilities.

M.R. Hammock and J.W. Mixon, *Microeconomic Theory and Computation: Applying the Maxima Open-Source Computer Algebra System*,
DOI 10.1007/978-1-4614-9417-1_1, © Springer Science+Business Media New York 2013

The remainder of this chapter consists of five sections. The first section introduces *Maxima* and the *Maxima* interface *wxMaxima*. (The *wxMaxima* interface is not the only one available to *Maxima* users, but it is the one used throughout this text. In fact, this text was written mostly within *wxMaxima*.) The second section addresses the symbolic analysis of equations and systems of equations. The third section introduces the rudiments of *Maxima*'s treatment of calculus. The fourth section considers the symbolic analysis of ordinary differential equations (ODEs). The fifth section introduces the basics of manipulating expressions, including factoring and expanding them. The final section discusses graphing.

1.2 Basics

This section introduces the mechanics of interacting with *Maxima*. It shows how to enter commands, either individually or as a group, and how the resulting output is expressed.

1.2.1 Interacting with Maxima

Maxima is primarily an interactive tool into which the user types and enters commands, causing the computer to respond to those commands immediately, though sets of commands may be gathered into a file and entered in a batch mode. Also, for the interface used in this book, *wxMaxima*, sets of commands may be entered as small batches, using *wxMaxima*'s input groups. When waiting for input, *Maxima* displays a prompt, usually an arrow:

```
-->
```

or an input prompt:

```
%i1
```

The commands that we type are typically algebraic expressions, assignments, or function definitions. *Maxima* is case sensitive; pay attention to this, especially when issuing *Maxima* commands. Terminate commands by a semicolon (;) or a dollar sign ($) to inform *Maxima* that it has reached the end of the command. A semicolon causes *Maxima* to print the result of executing the command, whereas a dollar sign suppresses that output but keeps the result in *Maxima*'s memory. A long command can be extended over several lines. The `Enter` key starts a new line. *Maxima* attempts to execute the command only when it encounters a `;` or a `$`. In the *wxMaxima* interface, an input cell may contain any number of commands.

These are implemented when the cursor is placed anywhere within the cell and either a `Shift Enter` or a `Control Enter` combination is entered. The following are simple examples of inputs and *Maxima* responses. Notice the number convention. The prompt for the first input is `(%i1)`, and the output that results from executing the first command is output (`%o1`). Until this session is ended, the output (`%o1`) remains in *Maxima*'s memory and can be recalled for use in further analysis.

```
(%i1)  4*5;
```

(`%o1`) 20

```
(%i2) 1 + 2 + 3 + 4 + 5 +
      6 + 7 + 8;
```

(`%o2`) 36

```
(%i3) 2*x- 5 + 3.5*x;
```

(`%o3`) $5.5\,x - 5$

```
(%i4) -2^4;
```

(`%o4`) -16

```
(%i5) (-2)^4;
```

(`%o5`) 16

```
(%i6) 5!;
```

(`%o6`) 120

Maxima does simple arithmetic both numerically, as in 4*5 returning 20, and symbolically, as in adding $2 \cdot x$ and $3.5 \cdot x$ to get $5.5 \cdot x$. In addition to the four primary arithmetic operations $+$, $-$, $*$, and $/$, *Maxima* can exponentiate (^) and take factorials (!). Other operations can be performed with function calls.[2] If our command is one that *Maxima* does not recognize (e.g., a misspelling of a command name or a capital letter that should have been lowercase, or a function not yet defined), *Maxima* usually returns the input as is entered, with no explanation.

The order of precedence for *Maxima*'s arithmetic operations is the same as that of most common programming languages. Parentheses can be used to arrange the desired order of operations, as in input `%i5` above. Often we need to refer to

[2]Centered dots indicate multiplication. Later in the book, variable names will contain a series of letters, like mpl to denote marginal product of labor. With our notation $m \cdot p \cdot l$ would denote a product of three variables, m, p, and l, while mpl is a variable.

the result of the most recently completed operation. *Maxima* allows us to refer to the result of the last evaluated expression with a percentage sign (`%`). Expressions resulting in syntax errors do not count.

An example follows. The `%` reference in *Maxima* is to the most recently executed command, not necessarily the one that appears directly above the command being executed. Thus, the user must keep track of the command sequence when using the feature.

```
(%i7)  (x + 3)^2;
```

(%o7) $(x+3)^2$

```
(%i8)  % - 5;
```

(%o8) $(x+3)^2-5$

Suppose that we want to refer to the first of the preceding two expressions after the second has been executed. The following accomplishes this task:

```
(%i9)  %o7;
```

(%o9) $(x+3)^2$

The output created by executing the first command is assigned the internal name `%o7`, which allows this expression to be recalled at any place in the *Maxima* session.

1.2.2 The wxMaxima *Screen*

The *wxMaxima* screen, shown below, consists of five parts. First, the title bar identifies the file being used, if a session has been saved as a named file. Next, a series of menus provides a relatively easy way of implementing most of *Maxima*'s important commands. The notes at the end of this chapter contain a link to a set of tutorials that provide a good overview of how to use the interface, while demonstrating much about *Maxima*'s capabilities. The third row in the *wxMaxima* screen contains a set of commonly used icons. The first opens a *wxMaxima* session that has been previously created and saved. (The tutorials at the *wxMaxima* web site consist of *wxMaxima* sessions that combine *Maxima* commands and notes.) The second icon saves the current session. The third icon prints the current session, showing both input and output. The fourth icon allows for configuration of *wxMaxima*. The next two icons copy or cut a selection. The seventh icon pastes material from the clipboard. The eighth icon interrupts the execution of a command, which is useful if an infinite loop has been entered inadvertently. The next three icons control animation (not addressed in this text). The last one opens [2].

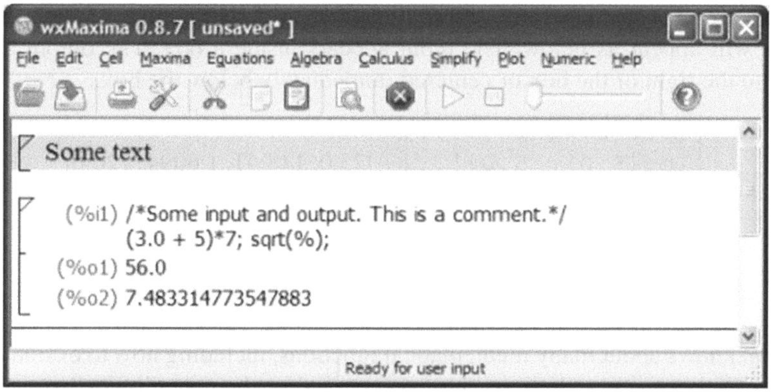

The fourth part of the screen is a window that contains input and output. The screenshot above shows two commands in an input group and the results of executing those commands. Below the cell that contains those commands and results is a horizontal line. This is the *wxMaxima* cursor. Clicking below or above a cell places the cursor there. Once the cursor appears, one can type input. Once the input group is complete, striking a combination of either the `Shift` key or the `Control` key and the `Enter` key submits the input to *Maxima* and the resulting output is returned. The result of this entry appears below the commands.

The use of input groups has a number of advantages, but it can cause some difficulty in matching input and output. Therefore, the set of commands entered into an input group should be chosen judiciously. Text cells inserted among input cells can help to keep track of what commands are being entered and to discuss the resulting output. To insert a text cell, click before or after an input cell and either use the menu entry `Edit / Cell / New Text Cell` or right-click on the mouse (`ctrl`-click in Macintosh) and select from the pop-up menu.[3]

An example of an input/output cell appears below. Four commands are entered. Note that only one command number is printed; the rest can be inferred.

```
(%i10)  5.0^2; exp(5.0); log(5.0); 5.0!;
```

(%o10) 25.0
(%o11) 148.41
(%o12) 1.6094
(%o13) 120.0

We could enter a list of commands, enclosed in square brackets and separated by commas. The entire list is then executed as a single command with the output also appearing as a list, with the same number.

```
(%i14) [5.0^2, exp(5.0),  log(5.0), 5.0!];
```

(%o14) [25.0, 148.41, 1.6094, 120.0]

---

[3]Some earlier versions use the `F6` key to enter a text cell.

Hereafter, the text does not report command or output numbers. The input and output will appear as below. The input appears inside a box. The output appears either to the right of the box or centered immediately below the box.

```
[5.0^2, exp(5.0),
log(5.0), 5.0!];
```

$[25.0, 148.41, 1.6094, 120.0]$

1.2.3 Mathematical Functions

Maxima knows about many mathematical functions, including how to evaluate them for specific argument values and how to manipulate them symbolically—to differentiate and integrate, apply identities, etc. Some of these functions are illustrated below. In addition, the inverse trigonometric and hyperbolic functions are available with the corresponding names appended by the character a: `asin`, `asech`, etc.

First consider these four built-in functions: finding an absolute value, extracting a square root, raising e (≈2.718) to a power, and extracting the natural logarithm of a number. For all cases, the number used is −44. In this case, *Maxima* should refuse to carry out the last operation, extracting the logarithm of a negative number.[4] *Maxima* is instructed to report the result of exponentiation as a floating-point number. Consider the cases below. In the first two listed items, *Maxima* reports exact answers. In the third, it reports a floating-point representation, as instructed via the `float` command. It cannot evaluate log(−44), so it repeats the input.

```
a: -44$ [abs(a),sqrt(a),
float(exp(a)),log(a)];
```

$[44, 2\,\sqrt{11}\,i, 7.78113\,10^{-20}, \log(-44)]$

Maxima offers the expected complement of trigonometric functions, such as the following. For $\theta = \pi/3$, *Maxima* reports $sin(\theta)$, $cos(\theta)$, and $tan(\theta)$. The `kill` command clears *Maxima*'s memory, and the `fpprintprec` command sets floating-point print precision.[5]

```
%theta : %pi/3$    [x1,x2,x3]:
  [sin(%theta),cos(%theta),tan(%theta)];
  kill(all)$ fpprintprec:5$
```

$[\frac{\sqrt{3}}{2}, \frac{1}{2}, \sqrt{3}]$

The list below reports the arcsin, arcos, and arctan for the results above. Of course, the arcsin of the sin of a value is the original value. Likewise for the other functions. Therefore each value must equal the defined value of θ, $\pi/3$.

---

[4] *Maxima* will return a complex number if instructed to extract the logarithm of a negative floating-point number.
Here is the output for $log(-44.0)$: 3.1416*%i+3.7842.

[5] Reducing floating-point print precision does not reduce the precision with which *Maxima* stores the variable's value in its memory. The printing precision level can be returned to the default value (16) with `fpprintprec:0`.

```
[asin(x1),   acos(x2),   atan(x3) ];
```
$[\frac{\pi}{3}, \frac{\pi}{3}, \frac{\pi}{3}]$

Functions can be embedded within each other. The first entry below reflects the fact that the natural logarithm of e^n is n. The % reference cannot be used inside a list, because *Maxima* treats the execution of all of the list's commands as a single operation. Therefore, we name the first expression and pass that name along with the third expression.

```
[temp: log( exp(2) ); sin( %pi/2);
temp + sin(%pi/2)];
```
$[2, 1, 3]$

As noted previously, *Maxima* reports exact values unless instructed otherwise. Consider, for example, the arc tangent of $\pi/2$.[6]

```
[atan(%pi/2),   float(atan(%pi/2))];
```
$[\operatorname{atan}\left(\frac{\pi}{2}\right), 1.0039]$

Maxima contains many other mathematical functions, as well as commands that allow us to manipulate expressions. The *wxMaxima* menus provide a well-organized listing of commands, and working through these menus is a valuable exercise. For a complete listing of commands, see [2].

1.2.4 Referring to and Evaluating Expressions

Often we need to bind an expression to a symbolic reference to that expression. *Maxima* offers three distinct ways of establishing such a bond. The first, used earlier and considered in more detail immediately below, is to permanently bind the expression to a name. Doing so creates a name for the expression that may be called at any later point in the *Maxima* session. A second possibility is that the bond is temporary, used only to define the environment within which some operation is to be carried out. Finally, a functional expression, complete with specified arguments, may be defined. The establishment of a permanent bond, or assignment, is accomplished with the *Maxima* operator `(:)`. The variable on the left side of the colon is bound the value to which the expression on the right side evaluates. Wherever that variable appears on the right side of a subsequent expression, it is replaced by its assigned value.

The examples below show that the name can be bound either to a value or to a variable. The `kill(all)` command removes all bonds. The `values` command results in a list of all names that are bound to expressions. Only x and u remain; any bonds that were created earlier have been removed.

[6] This aspect of the output can lead to confusion, because *Maxima* responds in the same way if the input is gibberish. For example, arctan(%pi/2) would yield the output $arctan(\pi/2)$, but arctan is not a meaningful command, so *Maxima* just reports the input without trying to evaluate it.

```
kill(all)$
[x : 3*5 - 1/2,
  x^2, 1.0*x^2];
[u: 4 + w, u^2];
values;
```

$[\frac{29}{2}, \frac{841}{4}, 210.25]$ $[w+4, (w+4)^2]$ $[x, u]$

The `kill` command can be used more selectively. The command below removes the binding of x but keeps that of u.

```
kill(x)$  y:1+w^2;     values;
```

$w^2 + 1$ $[u, y]$

When moving from one problem to another, it is best to clear previously assigned variables. Suppose that we define $w = 3 + 5 \cdot x$ but forget that x has been bound to the value 29/2. The variable w evaluates to 151/2, rather than being an expression stated in terms of the value of x.

```
[x:29/2,
  w:3+5*x,
  kill(x),
  w:3+5*x];
```

$[\frac{29}{2}, \frac{151}{2}, done, 5\ x+3]$

The `kill` command should be used with care, because the results of any work before its execution are no longer accessible. An accidental `kill(all)` command need not be disastrous, however, when one is using *wxMaxima*. The input that generated the bonds remains as long as the session is open. Thus, inadvertent deletion of output from earlier operations can be undone by executing the relevant input groups once more, or by using the `control r` combination (or the menu entry `Cell, Evaluate All Cells`) to rerun all commands in the current session. Saving a session as a named file often is wise.

Temporary bindings, established with =, are often used when we wish to evaluate an expression under a specific set of conditions. In the example below, x is bound temporarily to the value 1 and then to the value $e \approx 2.718$, and finally to a value slightly larger than e. In each case, y is evaluated given the temporarily assigned x value, using the `ev` command.

```
kill(all)$ y:log(x)$
[ ev(y, x= 1),
  ev(y, x = %e),
  ev(y, x=2.8)   ];
```

[0, 1, 1.029619417181158]

Maxima can be used to specify functional relationships in a way that allows direct entry of the function's arguments. This text contains many examples, such as defining the quantity demanded of a good as a function of price, income, and a set of parameters. To compare the bound expressions discussed above and the functional form, consider the example below. In the first command, the name `y` is bound (or, equivalently, assigned) to variable x; in the second, a function of x is defined. Note the use of the quote–quote operator (no spaces between the two quote marks) to convert the bound expression to the functional form: *Maxima* is instructed to create a functional relationship and to evaluate the expression named y.

```
kill(all)$
y:5+x; f(x) := ''y;
```

$x + 5$ $\quad$ $\mathrm{f}(x) := x + 5$

Evaluation of y for a given value of x is carried out differently, depending on how the relationship between the two variables is specified. Consider first using the name `y`. Suppose that we wish to evaluate expression `y` for $x = 3$. Any of the three commands below yields the desired result.

These four entries differ only slightly. In the first, the temporary condition $x = 3$ is simply appended to the name of the expression to be evaluated. In the second, the name of the expression to be evaluated appears inside the parentheses along with the condition. For more complex expressions, more conditions may be inserted. The `subst` command tells *Maxima* to substitute 3 for x in expression `y`. Later, we encounter other ways to enter the `subst` command. Finally, the functional relationship allows a simple and direct entry of x's value into the expression.

```
y, x=3;    ev(y, x = 3);
subst(3,x,y);    f(3);
```

8 8 8 8

All four alternatives yield the same answer. Among the first three, use of `subst` (for substitute) is probably the best. The first is easy and can be used for quick solutions. It cannot, however, be used to create items in lists, and it cannot be used in programming. The `subst` command is a bit more flexible than the `ev` (for evaluate) command.

Use of binding versus functional expression is often a matter of preference. Usually, when the value of y involves an expression in terms of other variables, rather than just numbers, this text uses functional statements. As the final section of this chapter shows, the functional expression can be especially useful in plotting.

1.2.5 Sequences, Lists, and Sets

Expressions separated by commas make up a sequence. In general, the arguments of *Maxima* commands are sequences, although in many cases the sequence may have only one element. In addition, some *Maxima* commands return sequences. For example, in solving an equation that has multiple roots, *Maxima* returns a sequence of roots. (The next section deals more fully with roots.) We may then select a particular root to work with. Special commands create sequences, convert a mathematical expression into a sequence of its parts, and carry out other manipulations. Except for the extraction of a particular element, however, they are not used in this text. Explanations and examples of other sequence operations may be found in [2].

1.2.5.1 Lists

A *list* appears as a sequence enclosed by square brackets, as in the three examples that follow. The list need not be bound to a name, but such binding can be useful if terms are to be used later in the session.

```
kill(all)$ list1:[a,a,b,c,d,1,2,2,7,9,9,"x = 2"];
list2 : makelist(j^2, j, 1, 7);
list3 : realroots( (1 + 2*x)^3 = 12*(1 + x^5) );
```

$$[a, a, b, c, d, 1, 2, 2, 7, 9, 9, x = 2] \qquad [1, 4, 9, 16, 25, 36, 49]$$

$$[x = -\frac{34153983}{33554432}, x = \frac{23811145}{33554432}, x = \frac{38415571}{33554432}]$$

The first list is entered directly. The second list is created by the `makelist` command. *Maxima* is instructed to square each of the digits from one through seven and store the results in `list2`. *Maxima* creates the third list in response to the `realroots` command, which is instructed to find the real roots for the expression that appears in parentheses. Suppose that the second of the three terms in `list3` is of interest, and that we want the numerical value and not the expression. Executing the pair of commands below generates that value. The first command instructs *Maxima* to extract the right-hand side of the third item in `list3`. This value could be bound to a name for future reference.

```
rhs(list3[3]);   float(%);
```

$$\frac{38415571}{33554432} \qquad 1.144873231649399$$

1.2.5.2 Sets

A *set* appears as a sequence enclosed by braces. Two differences distinguish lists from sets. First, a list is strictly ordered, whereas the order of elements within a set is arbitrary. This distinction shows up in a number of places in *Maxima*. For example, as in `list3` above, *Maxima* returns a set of solutions for the unknown values. The ordering of the items in the solution set is internally determined by *Maxima*. The order matters, for it allows reference to specific items in the list. The second difference between a list and a set is that a set does not contain any duplicated elements, whereas two or more elements of a list may be identical. *Maxima* automatically eliminates duplicate elements in a set.

```
set1 : {1,1,1,2,4,5,a,b,b,c};
set2 : setify(list1);
```

$$1, 2, 4, 5, a, b, c \qquad 1, 2, 7, 9, x = 2, a, b, c, d$$

The first set contains only seven items, even though ten were entered into the first command. *Maxima* removed the redundant 1's and *b*'s. Likewise, the `setify` command, which converts `list1` into `set2`, includes only the distinct items and removes the redundant ones. This input group did not begin with a `kill(all)` command because the first command refers to `list1`, created in an earlier cell.

1.2.5.3 More on Lists

Lists can be used to bind variable names to values. The second command in the box below creates a list that binds four names to four values. The second line contains three commands, themselves placed in a list. The first command recalls the value of a, resulting in the value 1. The second command recalls the value of b and makes it an argument in the `float` command, resulting in a floating-point approximation to e^2. The third command instructs *Maxima* to recall d and evaluate it for $x = \pi/4$.

```
kill(all)$  [a,b,c,d]:[1,exp(2),"some text",
sin(x)]; [a, float(b),subst(x=%pi/4,d)];
```

$$[1, e^2, \texttt{some text}, \sin(x)] \quad [1, 7.3891, \frac{1}{\sqrt{2}}]$$

As noted above and illustrated once more in the first example below, expressions may be used to create a list of values. In the example below, the value of x ranges from 0 to 2π in increments of $\pi/4$. The next cell shows a list of the resulting values of $sin(x)$.

```
kill(all)$   makelist(sin(i*%pi/4), i, 0, 8 );
```

$$[0, \frac{1}{\sqrt{2}}, 1, \frac{1}{\sqrt{2}}, 0, -\frac{1}{\sqrt{2}}, -1, -\frac{1}{\sqrt{2}}, 0]$$

The next example begins by defining a function that describes a population that begins with a value of 2 and grows at a rate of 5 % per period. The `makelist` command creates a list of six values of t, beginning with $t = 0$ and ending with $t = 50$. The map command is then used to determine the value of $f(t)$ at each of these six values. An alternative to mapping in this case is to enter `tList` into the function `f(t)`. The results of the last two commands appear as the second output line.

```
kill(all)$    fpprintprec : 3 $
f(t)  := 2*exp(0.05*t); tList:makelist(10*t,t,0,5);
fList : map(f, tList); f(tList);
```

$$\mathrm{f}(t) := 2\exp(0.05\,t) \qquad [0, 10, 20, 30, 40, 50]$$
$$[2, 3.3, 5.44, 8.96, 14.8, 24.4] \qquad [2, 3.3, 5.44, 8.96, 14.8, 24.4]$$

1.3 Solving Algebraic Equations

Maxima can solve algebraic equations, both linear and, within limits, nonlinear. It can be relatively easy to deal manually with a small number of relatively simple equations. With more equations or more complicated ones, however, the potential for making mistakes argues against manual solutions. *Maxima* solves sets of equations with ease and accuracy through its `solve` command. Begin with a single-equation example, a rational function of x. The syntax of the `solve` command is direct: enter the expression(s) to be solved and then the variable(s) for which it is (they are) to be solved.

1.3.1 *Single Equation*

The cell below solves $f(x) = 0$ for x, assigning the name `solnf` to the solution. Note that the `solve` command's first argument is `f(x)`, and not `f(x) = 0`. The latter could have been entered and would yield the same result (exercise: confirm this assertion).

The expression on the right-hand side of the equation could have been entered directly as an argument of `solve` instead of using the $f(x)$ reference. Also, a name could have been bound to the expression, and that name could have been entered. The cell below contains an example of each of these alternative approaches. The results illustrate *Maxima*'s ability to return either symbolic solutions (the first two examples) or numeric solutions (the third example).

The first and third output lines consist of the expression to be solved and the solution. When the expression is entered directly (the second example), only the solution is returned. In the second and third examples, the solution consists of a list of two items.

```
[f(x):=(x-a)^2/(x+b), solnf:solve(f(x),x)];
soln_exp: solve(A*x^2 + B*x + C, x);
[z: 5*x^2 + - 500, solnz: solve(z, x)];
```

$$[\mathrm{f}(x) := \frac{(x-a)^2}{x+b}, [x = a]]$$

$$[x = -\frac{\sqrt{B^2-4\,A\,C}+B}{2\,A}, x = \frac{\sqrt{B^2-4\,A\,C}-B}{2\,A}]$$

$$[5\,x^2 - 500, [x = -10, x = 10]]$$

1.3.1.1 Confirming Solutions

The output below shows two equivalent ways of checking the solution for the first equation. If the critical value of x is to be retained for future use, it can be bound to a name, and that name can be entered into $f()$. Otherwise, a direct reference to the solution can be entered as the argument in $f()$. Using either approach the result is $f(a) = 0$. This result confirms that $x = a$ is the solution.

```
[xcritical:rhs(solnf[1]),
f(xcritical), f(rhs(solnf[1])) ];
```

$$[a, 0, 0]$$

The substitute command `subst` can be used. Note that the information to be substituted is entered in two alternative ways. The first command instructs *Maxima* thus: "Substitute the right-hand side of the first item in `soln_exp` for x in the expression that constitutes the third entry." The second instruction is briefer: The expression `x = 2` tells *Maxima* to evaluate the expression given this value of x.

In both cases, the result is an expression whose value is not immediately obvious. Simplification, using `ratsimp` (`rational simplification`), is required.

```
[sol1: subst(rhs(soln_exp[1]), x, A*x^2+B*x+C),
  ratsimp(sol1)];
[sol2:subst(soln_exp[2],A*x^2+B*x+C),radcan(sol2)];
```

$$[\frac{\left(\sqrt{B^2-4AC}+B\right)^2}{4A} - \frac{B\left(\sqrt{B^2-4AC}+B\right)}{2A} + C, 0]$$

$$[\frac{\left(\sqrt{B^2-4AC}-B\right)^2}{4A} + \frac{B\left(\sqrt{B^2-4AC}-B\right)}{2A} + C, 0]$$

We use the evaluation command `ev` to evaluate the third solution. We enter the conditions for evaluation in two different ways. We enter the first one by hand, inserting the contents of `solnz[1]` into the `ev` command. In the second case, we enter the reference to an item in the list named `solnz[2]`. In both cases, $z = 0$.

```
[ev(z,x=-10),   ev(z, solnz[2])];
```

$[0, 0]$

1.3.1.2 Evaluating Expressions

To determine a numeric value for $f(x)$, given values of a, b, and x, rephrase the function as below. *Maxima* does not distinguish between the variable x and the parameters a and b. We may specify a and b and then determine the equation for the resulting function of x (as the second column of the table below). Also, we may specify a value for x along with those of a and b and determine the resulting numerical value of x (third column below). Unless the arguments are expressed as floating-point values, *Maxima* reports the exact value of $f()$. The fourth column reports the floating-point equivalent of the exact result, $\frac{49}{2}$.

We use the `matrix` command to create a table that shows the output. The `matrix` command is not integral to this material and may be ignored.

```
kill(all)$  matrix(["The function","w/Parameters",
 "Given x","Floating point"], [f(x,a,b):=
  (x-a)^2/(x+b),f(x,5,4),f(-2,5,4), f(-2.0,5,4)]);
```

$$\begin{bmatrix} \text{The function} & \text{w/Parameters} & \text{Given x} & \text{Floating point} \\ f(x,a,b) := \frac{(x-a)^2}{x+b} & \frac{(x-5)^2}{x+4} & \frac{49}{2} & 24.5 \end{bmatrix}$$

1.3.2 Systems of Equations

The `solve` command can extend to systems of equations like the one below. We solve these three equations simultaneously and assign the result to the list named `sol`. The `solve` command has much the same structure as when a single expression is being analyzed, except that the expressions (or functions) must be placed in a list and the variable names must form a second list.

```
kill(all)$ expr1:2*a*x+4*y-10*z$
expr2:2*x-y+2*z=25$    expr3:2*x+3*y-z=0$
sol:solve([expr1,expr2,expr3], [x,y,z]);
```

$$[[x = \frac{325}{5a+28}, y = \frac{25a-250}{5a+28}, z = \frac{75a-100}{5a+28}]]$$

Once we have named the list, reference to individual items is direct. Be aware that this is a "nested" list, meaning that the three-item solutions list is embedded in another list (in this case, it is the only item).

We insert the list of solutions, which is set of equality statements, into the three original expressions. We use `ratsimp` to determine the numerical value of the expressions. Compare the ways the original expressions are stated to the ways the results of evaluating them are stated. In the first case, the $= 0$ is implicit. In the third, $= 0$ is explicit. In the second, the left-hand side of the expression equals 25, not 0.

Checking the solution is quite easy, as the display below shows. Evaluate the three equations given the values in `soln`. We use `transpose(matrix())` to show these expressions in a table. As before, this use of these commands is to facilitate display and can be ignored.

```
subst([sol[1]],[expr1,expr2,expr3])$
transpose(matrix(%));   ratsimp(%);
```

$$\begin{bmatrix} \frac{650a}{5a+28} - \frac{10(75a-100)}{5a+28} + \frac{4(25a-250)}{5a+28} \\ \frac{650}{5a+28} + \frac{2(75a-100)}{5a+28} - \frac{25a-250}{5a+28} = 25 \\ \frac{650}{5a+28} - \frac{75a-100}{5a+28} + \frac{3(25a-250)}{5a+28} = 0 \end{bmatrix} \qquad \begin{bmatrix} 0 \\ 25 = 25 \\ 0 = 0 \end{bmatrix}$$

We may solve the system of equations for a given parameter value for a as follows.

```
subst(5.0,a,sol);
```

$$[[x = 6.1321, y = -2.3585, z = 5.1887]]$$

1.3.3 Numerical Methods

The `solve` command can deal with many equations, but it can fail for either of two reasons. Some equations that have analytical solutions are not included in *Maxima*'s library. Furthermore, many equations have no solution. *Maxima* offers a number of *numerical methods* that can be used to deal with expressions that are beyond `solve`'s abilities.

We introduce a pair of these techniques. The `find_root` option can find a root over a specified range. The *Maxima* manual, [2], states that if *Maxima* can evaluate a function (or an expression) over the range and if the function (or expression) is continuous over that range, then `find_root` is guaranteed to find the root (or one of the roots if more than one occurs over the range).

1.3.3.1 A Unique Solution

Consider this simple example that represents an inverse demand curve as exhibiting constant elasticity (the price elasticity is −2.5) and a supply curve with a constant slope. We see that applying `solve` simply returns the initial expression. We must turn to a numerical method to solve this expression.

```
kill(all)$    pd(x) := 12*x^(1/(-2.5)) ;
ps(x) := 5 + 2*x;    solve(pd(x)-ps(x), x);
```

$$\mathrm{pd}(x) := 12\, x^{\frac{1}{-2.5}} \qquad \mathrm{ps}(x) := 5 + 2\, x \qquad [0 = 2\, x^{\frac{7}{5}} + 5\, x^{\frac{2}{5}} - 12]$$

Before applying numerical methods, drawing a graph often helps. The graph suggests that the expression $pd(x) - ps(x)$ has a root at $x \approx 2.0$.

```
wxdraw2d(explicit(pd(x),x, 0.1, 10),
    explicit(ps(x),x, 0, 10))$
```

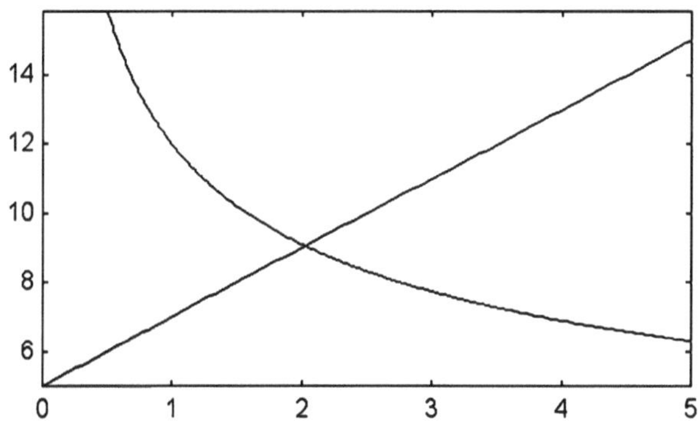

We first apply `find_root`. The third output item below confirms that the two curves intersect at $x \approx 2.0248$. The first two commands below fail to provide an answer, for different reasons. The first command instructs *Maxima* to evaluate a term that includes $\frac{12}{0^{0.4}}$, which it cannot do. Therefore, it returns the command as the first output line. The second command ignores the information in the graph and sets the search range such that the expression is positive at both ends of the range, so no root was found. This three-part response provides guidance in one's search for the range in which the root(s) can be found. Finally, the third use of `find_root` uses the information in the graph and the result is the value 2.02 ....

```
fpprintprec:16$  find_root(pd(x)-ps(x), 0, 4);
find_root(pd(x)-ps(x), 0.1, 1.5);
find_root(pd(x)-ps(x), 0.1, 4);
```

$$\mathrm{find\_root}\left(-2\, x + \frac{12}{x^{0.4}} - 5, 0.0, 4.0\right)$$

find_root: function has same sign at endpoints:

$$\mathrm{mequal}\left(\mathrm{f}(0.1), 24.94263717811496\right), \mathrm{mequal}\left(\mathrm{f}(1.5), 2.203396005006326\right)$$

– an error. To debug this try: debugmode(true);

$$2.024794552325446$$

An alternative to `find_root` is the Newton search procedure. To invoke this requires loading the program `newton1` (this program need not be loaded each time that `newton` is to be executed, just once per session). The `newton` alternative differs from `find_root` in that the user provides a point from which to begin seeking the root and a measure to indicate how finely the search is to be carried out. The two approaches do not yield exactly the same results. Compare output `%o7` with output `%o9`.

```
load(newton1)$  newton(pd(x)-ps(x),
x, 2, .00001);
```

2.024794550645138

1.3.3.2 Multiple Solutions

Not all equations have unique solutions. Consider the quadratic equation below, which has two roots.

```
P: -4*x^2 + 100*x - 200$
wxdraw2d(xaxis=true, explicit(P,x,0,25))$
```

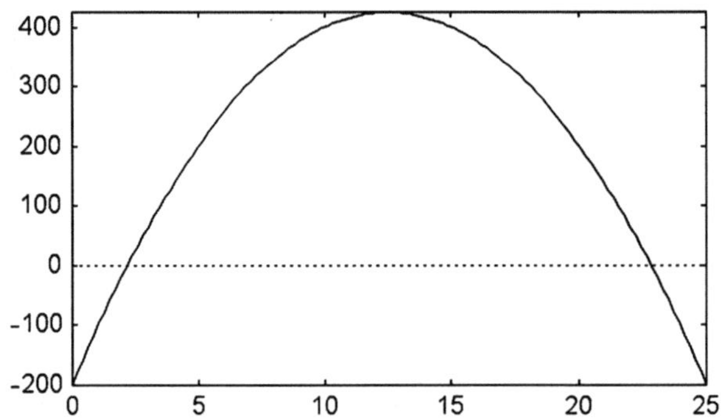

Either `find_root` or `newton` will return a different solution depending on the specification. With `find_root`, endpoints must be specified. First, search over the range $0.1 \leq x \leq 5$ and then over the range $3 \leq x \leq 30$.

```
[find_root(P,0.1,5), find_root(P, 3, 300)];
```

[2.192235935955849, 22.80776406404415]

The first `newton` command sets x below the lower value at which $f(x) = 0$, using the graph for guidance. The second command sets x above the upper value at which $f(x) = 0$.

```
[newton(P,x,0.5, .0001), newton(P,x,15, .0001)];
```

[2.192235913349343, 22.80776406404451]

1.3.3.3 Systems of Equations

An important attraction of the Newton method is that it extends to systems of equations. Consider the two equations below, from [2]. The graph suggests that these two functions have a real solution at $x \approx 0.5$ and $y \approx -1.75$, and another at $x \approx 4$ and $y \approx 2$. We limit the x range in $f(x, y)$, so $log(x)$ has real values.[7] Ignore the details of the graph construction for now.

```
kill(all)$    f(x,y) := x+3*log(x) - y^2;
  g(x, y) := 2*x^2-x*y-5*x+1;
wxdraw2d(xaxis=true,xlabel="x",key="f(x,y)",
implicit (f(x, y) , x, 0.01, 5, y, -5, 5 ),
 color = gray,line_width = 3, key="g(x,y)",
 implicit(g(x,y),x,0,5,y, -5,5))$
```

$$f(x, y) := x + 3\log(x) - y^2 \qquad g(x, y) := 2x^2 - x\,y + (-5)\,x + 1$$

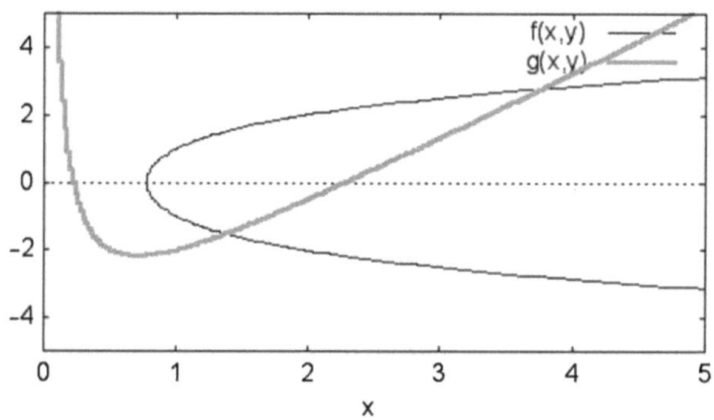

To analyze this system of equations load the package mnewton and execute it. The following must be provided: a list of expressions (or functions) to be evaluated, a list of variables, and a list of initial guesses. The solution that mnewton finds depends on the initial guesses, so it is important to use any available information when providing these values.

```
load("mnewton")$ mnewton([f(x,y),g(x,y)],[x,y],
   [3,2]); mnewton([f(x,y),g(x,y)],[x,y],[2,-1]);
```

$$[[x = 3.756834008012769, y = 2.779849592817897]]$$
$$[[x = 1.373478353409809, y = -1.524964836379522]]$$

[7]The option `draw_realpart=false` does not keep `implicit` from drawing the real part of a complex expression.

1.4 Calculus

Maxima can use symbolic methods to differentiate, integrate, find limits, and obtain series expansions of expressions. In addition, it can solve many kinds of linear differential equations and some nonlinear ones. Where analytical solutions are not feasible, *Maxima* can often numerically solve integrals and systems of ODEs.

This section introduces *Maxima*'s symbolic approach to differential and integral calculus. The next section considers ODEs.

1.4.1 Differentiation

Maxima can determine expressions for derivatives of various orders, and it can determine expressions for cross-derivatives.

1.4.1.1 Derivatives

Maxima's command for symbolic differentiation is `diff`. The first argument of the command identifies the expression that is to be differentiated. The expression may be entered directly, a name that has been bound (assigned) to the expression may be entered, or a functional call to the expression may be entered. An example of each appears below. In the first case the expression $y = log(x)/x$ is subject to differentiation.

The first command specifies that dy is required, so *Maxima* returns $dy = (dy/dx)dx$, where *Maxima*'s notation for dx is $del(x)$. *Maxima* interprets the second command as requesting the value of dy/dx, which it duly returns.

```
diff(log(x)/x);
diff(log(x)/x, x);
```

$\left(\frac{1}{x^2} - \frac{\log(x)}{x^2}\right) \operatorname{del}(x) \qquad \frac{1}{x^2} - \frac{\log(x)}{x^2}$

1.4.1.2 Multiple Variables

Consider an example with two variables, $z = x^a y^b$. The `diff` command, without reference to a variable, tells *Maxima* to fully differentiate this function with respect to x, y, a, and b. *Maxima* does not know, unless told, that a and b are constants. The second command instructs *Maxima* to produce the *partial* derivative of the expression with respect to x. In this example, the name z replaces the expression to which it is bound.

```
z : x^a*y^b$    diff(z);        diff(z, x);
```

$$b\,x^a\,y^{b-1}\operatorname{del}(y) + a\,x^{a-1}\,y^b\operatorname{del}(x) + x^a\,y^b\log(y)\operatorname{del}(b) + x^a\log(x)\,y^b\operatorname{del}(a)$$

$$a\,x^{a-1}\,y^b$$

We can inform *Maxima* that a and b are constants using the `declare` command. The two commands in the input group below follow the `declare` command with a request for the total differential of z, $dz = (\partial z/\partial x)dx + (\partial z/\partial y)dy$ *Maxima*'s notation does not distinguish between simple derivatives and partial derivatives.

```
declare([a,b],constant)$
diff(z);
```

$$b\,x^a\,y^{b-1}\,\mathrm{del}\,(y) + a\,x^{a-1}\,y^b\,\mathrm{del}\,(x)$$

Next, define a function $f(x, y) = x^4/y$. The example below shows the first and second derivatives with respect to x, denoted f_x and f_{xx}. It also shows the first derivative with respect to y and the cross-partial derivative, f_{yx}.

```
f(x,y) := x^4/y$
print(f, "=", f(x,y))$
```

$$f = \frac{x^4}{y}$$

The first derivative with respect to x and the second derivative with respect to x appear below. Note the use of brackets to generate subscripts.

```
print(f[x],"=",diff(f(x,y),x),
  ", and",f[xx],"=",
  diff(f(x,y),x,2))$
```

$$f_x = \frac{4\,x^3}{y}, \text{ and } f_{xx} = \frac{12\,x^2}{y}$$

The first derivative with respect to x and the cross-partial derivative, f_{yx} appear below. The command that yields the cross-partial derivative contains five arguments: the expression, the first variable with which the expression is to be differentiated, the level of differentiation, and the same two items for the other variable. For more variables, the extension is direct. (Exercise: Confirm that $f_{xy} = f_{yx}$.)

```
print(f[y],"=",diff(f(x,y),y),
  ", and  ",f[yx], "=",
  diff(f(x,y),y,1,x,1) )$
```

$$f_y = -\frac{x^4}{y^2}, \text{ and } f_{yx} = -\frac{4\,x^3}{y^2}$$

1.4.2 Integration

The syntax of indefinite integration resembles that of differentiation, with the `integrate` command replacing `diff`, except that multiple integration requires multiple calls to `integrate`. With the same function as above, we find the integral of the expression with respect to x and then the integral of the resulting integral with respect to y. In the latter instance, the integration over x is performed first since in nearly all cases Maxima evaluates the arguments to a command before invoking it. The output can also be generated in a single nested command. We continue to use the function $f(x, y) = x^4/y$.

```
integrate(f(x, y),x);    integrate(%,y);
integrate(integrate(f(x, y), x), y );
```

$$\frac{x^5}{5\,y} \qquad \frac{x^5\log(y)}{5} \qquad \frac{x^5\log(y)}{5}$$

1.4.2.1 Indefinite and Definite Integrals

The integral can be evaluated over a specified range. Two *definite integrals* appear below. Note the use of the names `Ix` and `Iy` to allow references within the lists. The second entry in each list below is the volume defined by the double integral over x and y.

```
[[Ix05:integrate(f(x,y),x,0,5),
   integrate(Ix05,y,1,2)],   [Iy02:
   integrate(f(x,y),y,1,2), integrate(Iy02,x,0,5)]];
```

$$[[\frac{625}{y}, \quad 625\log(2)], \quad [\log(2)\,x^4, \quad 625\log(2)]]$$

The limits of integration may include ∞ and $-\infty$. The following example uses the equation for the standard normal (z) distribution. The first definite integral determines the area under the function's curve over the range $z = 0$ to $z = 1.96$. The second integral shows that the area under the right half of the distribution is 0.5. The third integral shows that the area under the curve from ∞ to $-\infty$ is 1.0.

```
f(x)  :=  (1/sqrt((2*%pi)))*exp(-x^2/2);
float( integrate( f(x), x, 0,     1.96));
float( integrate( f(x), x, 0,     inf));
float( integrate( f(x), x, -inf, inf));
```

$$\mathrm{f}(x) := \frac{1}{\sqrt{2\,\pi}}\exp\left(\frac{-x^2}{2}\right) \qquad 0.47500210485178 \qquad 0.5 \qquad 1.0$$

Suppose that *Maxima* is instructed to integrate this expression: e^{-at^2}. It responds with a question: `Is a positive, negative, or zero?` This question reflects the fact that the result of the integral depends on the properties of the function's parameters. In economics, these parameters are nearly always real and are often positive. *Maxima* makes no such assumption, however. In the example used below, a negative value of a would render the integral divergent, so it could not be analyzed over the specified range. To avoid the interruption required for answering *Maxima*'s question, use the `assume` command.

```
assume( a > 0)$
integrate(exp(-a*t^2),t,-inf,inf);
```

$$\frac{\sqrt{\pi}}{\sqrt{a}}$$

Maxima's library of integrals is quite large.[8] Even, so, it cannot evaluate all integrals for which closed forms exist. Furthermore, not all functions can be

[8] It incorporates material from [1].

integrated analytically. Consider the expression e^{-x^3}. We seek the integral of this function between $x = 0$ and $x = 1$. The `integrate` command returns an expression that is far from informative. In many cases like this one, symbolic analysis does not yield a useful result (or yields no result at all). Quick examination of the graph indicates that the value is between 0.5 and 1.0. How? We can use numerical methods to approximate the area under the curve.

```
expr: exp(-x^3)$ integrate(
expr,x,0,1); wxdraw2d(
explicit(expr,x,0,1))$
```

$$\frac{\gamma\left(\frac{1}{3}\right) - \text{gamma\_incomplete}\left(\frac{1}{3},1\right)}{3}$$

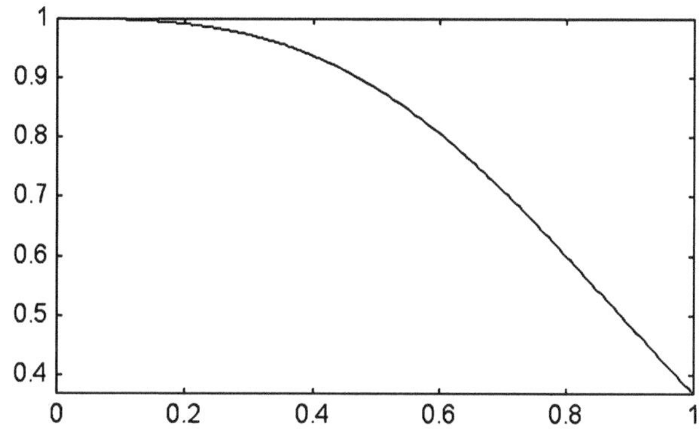

1.4.2.2 Numerical Integration

Maxima offers a range of options for numerical integration. The simplest one to use is the Romberg method. Also, a relatively complete set of specialized functions for numerical computation of one-dimensional definite integrals is contained in *QUADPACK*. The `romberg` command, applied to `expr`, shows that the area is approximately 0.808. For comparison, one of the *QUADPACK* routines, `quad_qags` is included. The syntax for `romberg` is identical to that of `integrate`, and the resulting output is similar: a single value. The `quad_qags` command is also the same as `integrate`, but its output is more complete: it shows the area under the curve, a measure of the imprecision of the estimate (quite low here), the number of iterations required to converge on the estimated value, and an error code (0 indicates no error).

```
romberg(expr,x,0,1); quad_qags(expr,x,0,1);
```

$$0.80751119008366$$
$$[0.80751118213967, 8.965175070537419\,10^{-15}, 21, 0]$$

The `romberg` command can be extended to integrals of more than one variable. We illustrate this computation using the function that we used with `integrate`, but `romberg` can be used for multivariate functions that `integrate` cannot evaluate. Like `integrate`, the `romberg` nesting can be extended if the function involves more than two variables.

```
f(x, y) := x^4/y$  romberg(
 romberg(f(x,y),y,1,4),x,0,10);
```

27725.88782757979

1.4.3 Limits

Maxima can compute limits for many functions. Consider this example: $f(x) = sin(x)/x$. In this example, this limit is the same when x approaches 0 from either above or below.

```
[f(x):=sin(x)/x,
limit(f(x),x,0)];
```

$[f(x) := \frac{\sin(x)}{x}, \quad 1]$

For many functions, however, the value of the limit depends on the direction from which the limiting value of x is approached. The following function, $g(x) = |x|/x$, approaches -1 as $x \to 0$ from below, and it approaches +1 as $x \to 0$ from above. Unless *Maxima* is informed regarding direction, it returns und, (i.e., undetermined).

```
f(x)  := abs(x)/x;              limit(f(x),x,0);
limit(f(x), x,0,minus);         limit(f(x),x,0,plus);
```

$$f(x) := \frac{|x|}{x} \qquad und \qquad -1 \qquad 1$$

If a function grows without bound as x approaches the specified value, *Maxima* indicates this with ∞ or $-\infty$. If, as in the first case below, *Maxima* cannot determine the sign, it returns `infinity`.

```
limit(1/x,x,0);
limit(1/x,x,0,minus);
limit(1/x,x,0,plus);
```

infinity $\quad -\infty \quad \infty$

1.4.4 Taylor and Maclaurin Series

Maxima's `taylor` command provides a Taylor series representation of a function. The command includes the expression being analyzed, the variable(s), the value(s) of the variable(s) about which expansion is taken, and the degree of the series. The output below shows the results of expansion around $x = 0$ (Maclaurin series) and $x = \pi$ of the expression $sin(x)/x$.

```
taylor(sin(x)/x,x,0,4);
T4: taylor(sin(x)/x, x,%pi,4);
```

$$1 - \frac{x^2}{6} + \frac{x^4}{120} + \dots$$

$$-\frac{x-\pi}{\pi} + \frac{(x-\pi)^2}{\pi^2} + \frac{\left(\pi^2-6\right)\,(x-\pi)^3}{6\,\pi^3} - \frac{\left(\pi^2-6\right)\,(x-\pi)^4}{6\,\pi^4} + \dots$$

Maxima uses ellipses (...) to indicate the presence of a remainder. This series must be truncated if the series is to be used in further analysis. The `taytorat` (`taylor to rational`) command, which truncates the series, appears below. Expanding the floating-point representation to shows clearly that the truncated series is a fourth-degree polynomial.

```
fpprintprec:5$ T4rat:taytorat(T4);
expand(float((T4rat)));
```

$$-\frac{\left(\pi^2-6\right)\,(x-\pi)^4 + \left(-\pi^3+6\,\pi\right)\,(x-\pi)^3 - 6\,\pi^2\,(x-\pi)^2 + 6\,\pi^3\,(x-\pi)}{6\,\pi^4}$$

$$-0.00662\,x^4 + 0.104\,x^3 - 0.487\,x^2 + 0.482\,x + 0.71$$

1.5 Differential Equations

Maxima can be used to solve differential equations. As with integration, it can analyze these functions either symbolically or numerically. The example below analyzes the behavior of a system that produces the differential equation specified below. The single quote before `diff` instructs *Maxima* not to attempt differentiation.

```
kill(all)$      expr:'diff(x(t),t)+%alpha*x(t)=0;
```

$$\frac{d}{d\,t}\,\mathrm{x}\,(t) + \alpha\,\mathrm{x}\,(t) = 0$$

The functional notation, $x(t)$, must be used. We solve this ODE twice. In the first case, the initial value of x, $x(0)$, is not specified. In the second case, the value is specified. The purpose of the second solution is to show the `atvalue` command that can be used to specify a value of $x(t)$ at some value of t, not necessarily at $t = 0$. The `atvalue` command includes the expression to be evaluated, the value of the independent variable for which it is being evaluated, and the value of the expression at that value. In this example, $x_0 = \pi$.

```
sol1:desolve(expr,x(t));
atvalue(x(t),t =0,%pi)$
sol2:desolve(expr,x(t));
```

$$\mathrm{x}\,(t) = \mathrm{x}\,(0)\,e^{-\alpha\,t} \qquad \mathrm{x}\,(t) = \pi\,e^{-\alpha\,t}$$

In both cases, the values of $x(0)$ and α uniquely define the time path of $x(t)$. Next we apply the solutions to the expression $dx/dt + \alpha * x = 0$ in order to check the solution. In both cases, values of 0 are returned, as must be the case if the solutions are correct.

```
diff(rhs(sol1),t ) + %alpha*rhs(sol1);
diff(rhs(sol2),t ) + %alpha*rhs(sol2);
```

0 0

1.5.1 Numerical Analysis

Many differential equations cannot be solved analytically. In such cases, numerical methods can provide insights into their behavior. For the simple case above, we apply the Runge–Kutta numerical method (*Maxima* command: `rk`), even though an exact solution is available, to show `rk`'s syntax. For this numerical analysis, let $x(0) = 1$ and $\alpha = 0.05$.

Naming the expression uses an assignment rather than the functional expression notation used by `desolve`. Within the `rk` command itself are these items: the name of the expression (or the expression itself), the dependent variable, the initial value of the variable, and a list of four items that define the domain of the computations. Here, the independent variable is `t`, which ranges from 0 to 4 in increments of 1.

The `rk` command produces a list of five pairs of output. Each pair identifies the value of `t` and the value x that is implied by that value of `t`. Exercise: Confirm that these five values are approximately those implied by `sol1` given $x(0) = 1$ and $\alpha = 0.05$.

```
dxdt:0.05*x$   rk(dxdt,x,1,[t,0,4,1]);
```

$$[[[0, 1], \quad [1, 1.0513], \quad [2, 1.1052], \quad [3, 1.1618], \quad [4, 1.2214]]$$

1.5.2 Second-Order ODEs

The example below shows how *Maxima* treats a second-order ODE. This example is the equation for a function for which the initial value is x_0 and the rate of change (first derivative) is v_0. The `atvalue` command informs *Maxima* that $dx(t)/dt = v_0$ when $t = 0$. A specific value of v_0 could be entered in place of the name v_0. Again, the expression is restated in terms of the solution, and it yields the requisite value of 0. In this case, the application of `ratsimp` is required to show that the expression does equal 0. Also, it is necessary to state the assumption that $\omega > 0$ (ω is the Greek letter omega), using the `assume` command, in order to avoid having *Maxima* interrupt the execution with a request for additional information.

```
kill(all)$                              assume(%omega > 0)$
expr : 'diff( x(t), t, 2) + %omega^2*x(t) = 0;
atvalue( diff(x(t), t, 1), t= 0, v0 )$
sol : desolve (expr, x(t));   ratsimp(
    diff( rhs(sol), t, 2) + %omega^2*rhs(sol));
```

$\frac{d^2}{d\,t^2}\,x\,(t) + \omega^2\,x\,(t) = 0 \qquad x\,(t) = \frac{\sin(\omega\,t)\,v0}{\omega} + x\,(0)\,\cos\,(\omega\,t) \qquad 0$

1.5.3 Extensions

This section shows that the analysis of ODEs can be extended in two directions, by allowing higher-order equations and by using numerical analysis. In addition to these extensions, *Maxima* allows for the analytical and numerical analysis of systems of ODEs. Also, it offers graphical methods that complement the algebraic and numerical methods illustrated above. A detailed treatment of these topics is in [3], Chap. 6.

1.6 Simplifying and Manipulating Expressions

Mathematical expressions seldom occur in the form that is most useful for the analyst. Sometimes, they are overly long or complicated. Other times, an expansion of a terse statement of the expression can yield insights. This section shows some *Maxima* commands that can be applied to expressions in ways that can make these expressions more informative.

1.6.1 Simplification

Maxima can generate some long and involved expressions. As seen in examples earlier in the chapter, the `ratsimp` (`rat`ional expression `simp`lify) command does make such expressions more manageable. According to [2], `ratsimp` simplifies "the expression ... and all of its sub-expressions, including the arguments to non-rational functions. The result is returned as the quotient of two polynomials in a recursive form; that is, the coefficients of the main variable are polynomials in the other variables." The `ratsimp` command is a member of a set of *Maxima*'s simplification functions, which includes `trigsimp` (`trig`onometric `simp`lification) and `radcan` (`rad`ical-to-`can`onical to simplify expressions that contain logarithms, exponentials, and radicals). The simplification commands, when given a variable name that has been bound to an expression, will return a modified expression if *Maxima* recognizes a way to simplify the expression. Unless the result of `ratsimp` is assigned to the variable name, however, the original expression is retained.[9]

```
matrix(["Expression",Simplification","Expression"],
  [a: (x^2 - 4)/(x + 2),  ratsimp(a),  a] );
```

$$\begin{bmatrix} \text{Expression} & \text{Simplification} & \text{Expression} \\ \frac{x^2-4}{x+2} & x-2 & \frac{x^2-4}{x+2} \end{bmatrix}$$

The value of a that is stored in *Maxima*'s memory is not changed by `ratsimp(a)`, as output `%o3` in the cell above shows. The `ratsimp` command puts an expression into the form of numerator over denominator and divides out the greatest common divisor of numerator and denominator, as the example below illustrates.

```
matrix(["Expression","Simplification"],
 [f(x):= 2+(x^2+3*x+2)/((x+2)*(x-1)),ratsimp(f(x))] );
```

$$\begin{bmatrix} \text{Expression} & \text{Simplification} \\ \mathrm{f}(x) := 2 + \frac{x^2+3x+2}{(x+2)\,(x-1)} & \frac{3x-1}{x-1} \end{bmatrix}$$

When `ratsimp`, `radcan`, or `trigsimp` cannot simplify an expression, *Maxima* returns the original expression.

1.6.2 Factoring and Expanding Expressions

As its name suggests, *Maxima*'s `factor` command factors polynomials and expressions containing polynomials. Its first argument is the expression. If complex factors are to be extracted use `gfactor`, which factors the expression over the Gaussian integers.

```
transpose(matrix(["Expression","w/factor",
"w/gfactor"],  [f(x):=(x^4 - 16),
factor(f(x)),gfactor(f(x))]));
```

$$\begin{bmatrix} \text{Expression} & \mathrm{f}(x) := x^4 - 16 \\ \text{w/factor} & (x-2)\,(x+2)\,\left(x^2+4\right) \\ \text{w/gfactor} & (x-2)\,(x+2)\,(x-2\,i)\,(x+2\,i) \end{bmatrix}$$

The command `expand` distributes operations, which can include multiplications and divisions, over additions and subtractions. Thus, it can reverse the effect of

[9]As before, the `matrix` command is used to generate a table. Removing this command and ending each command with a semicolon would generate three output lines.

`factor` for a polynomial. For polynomials, Development Team [2] recommends `ratexpand`, which uses a more efficient algorithm. The `expand` command works for a large number of functional types. When it cannot expand the expression, as in the first example below, it returns the initial expression.

```
expand(sin(%alpha +%beta));
expand(exp(y+log(x+1)));
ratsimp(%);
```

$$\sin(\beta + \alpha) \qquad x\,e^{y} + e^{y} \qquad (x+1)\,e^{y}$$

In the set of examples above, *Maxima* cannot expand the first expression, $sin(\alpha + \beta)$. It can expand the second expression, $e^{log(x+1)}$, yielding $xe^{y} + e^{y}$, or $(x + 1)e^{y}$. The manual, [2], shows ways to optimize the `expand` command for specific purposes.

An option that can help in expanding an expression is `expop` (*expo*nent, *p*ositive), which specifies the highest positive exponent that is automatically expanded. (A similar option, `expon`, specifies the highest negative exponent.) The following example shows a result of using `expand` without `expop` (for which the default value is 0) and with `expop` set at a value higher than two. The cell below contains four items, $(x + 1)^3$ expanded when `expop` equals its default value, a statement to set `expop` to 30, $(x + 1)^3$ expanded again, and a statement to reset `expop` to 0.[10]

```
[expand(x+1)^3,
expop:30,expand(x+1)^3,
expop:0];
```

$$[(x+1)^3, \quad 30, \quad x^3 + 3\,x^2 + 3\,x + 1, \quad 0]$$

1.7 Plotting

Maxima's graphic commands are among the most useful features for studying economic models. *Maxima* uses the open-source `gnuplot` software for most of its graphing. Two different interfaces to *gnuplot* are available. These are `plot` and `draw`. Chapter 8 of [2] discusses `plot`; Chap. 48 discusses `draw`. The default in *wxMaxima* is `plot`. Users wishing to take advantage of *wxMaxima*'s menu options may use the `plot` menu and its dialog boxes. This option, as the name suggests, uses `plot`, not `draw`. The development that follows, like the examples that have appeared earlier, uses the `draw` interface with *gnuplot*. Compared to `plot`, `draw` is more verbose, and it makes the connection between the options selected and the graph's appearance more apparent.

[10] Entering `expop = 3` or any higher value would give the same result.

1.7.1 Two-Dimensional Graphs

The basic command for plotting two-dimensional data is `wxdraw2d`. The `wx` part of `wxdraw2d` instructs *wxMaxima* to include the graph as part of the *wxMaxima* session's input and output. The `draw` part instructs *wxMaxima* to employ `draw` and not `plot`. Finally, `2d` informs *wxMaxima* to construct a two-dimensional graph. Three-dimensional graphs are introduced below. The command `draw` can be used without the `wx`. The result is a graph drawn in a separate window, one that can be manipulated and saved as a separate graphics image. This window must be closed before continuing with the *wxMaxima* session, but the contents can be saved as a graphics file.

The command can be used to graph a single function or expression, or a set of functions and/or expressions on a single graph. It can also be used to create parametric plots, such as the trajectory of a projectile near the earth's surface, and plots that consist entirely of points. The following example illustrates simple plotting using the Maxwell–Boltzmann probability distribution function.

```
f(x, a) := sqrt(2/%pi)*(x^2*exp(-x^2/(2*a^2))/a^3);
```

$$\mathrm{f}(x,a) := \sqrt{\frac{2}{\pi}}\,\frac{x^2\exp\left(\frac{-x^2}{2\,a^2}\right)}{a^3}$$

The graph below is produced with the minimal input into the `wxdraw2d` command. The command is `wxdraw2d( )$`. The dollar sign suppresses a bit of output that serves no purpose here.

```
wxdraw2d( explicit( f(x,2.5), x, 0, 20)   )$
```

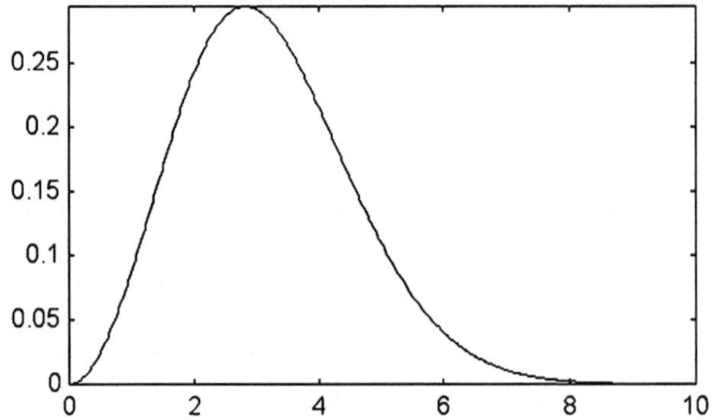

The rudiments of `wxdraw2d`'s structure appear in the illustration above. The command is `wxdraw2d()$`. The minimal input within the parentheses is a statement about the function type (an explicit function in this example), the function's arguments, the name of the variable over whose range the graph is to be plotted, and the range's endpoints. The instruction here instructs *Maxima*

to plot this explicit function when $a = 2.5$, over the range $x = 0$ to $t = 20$. Other expressions could be added within the parentheses, each adjoining pair separated by a comma. Using the same function, we plot three graphs below. Each one refers to one of the three values of a: 1, 2.5, and 5. The only necessary change is to add two more `explicit` entries. The `key` entries are not required, but do make interpretation easier. Likewise, the `color` and `line_width` entries.

```
wxdraw2d( key = "a = 1", explicit( f(x,1),x,0,20),
  key = "a = 2.5",  line_width = 2,
  explicit( f(x,2), x,0,20),key="a = 5",
  color = gray40, explicit( f(x,5), x, 0, 20)  )$
```

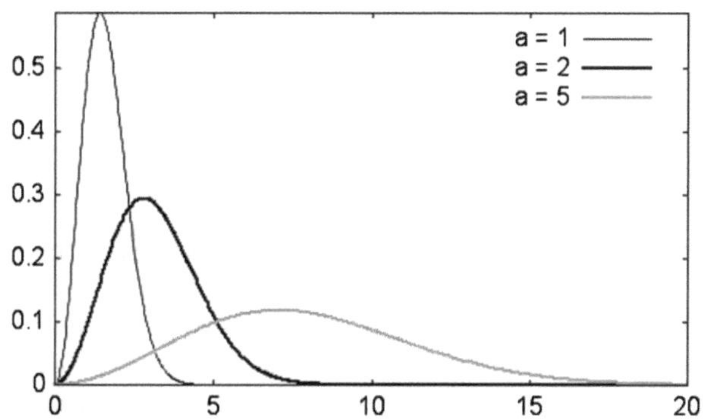

The function below, named `slopes`, takes on a value of 0 for x up to $x = a$. At $x = a$, the function jumps to x^2 and equals x^2 for values between a and $2a$. At $2a$, the function bends and takes on a constant slope. Note the use of `if then elseif then`. Any required number of `elseif` terms could be added.[11]

```
/*Define the function*/
   f(a,x):=if x<a then 0 elseif x<2*a then x^2
     else (2*a)^2 + (x - 2*a)$
/*Values for graphing */ [a1,a2,a3]:[2,5,8]$
   a_max:max(a1,a2,a3)$  y_max:f(a_max,2.5*a_max)$
/*Draw the graph*/    wxdraw2d( yrange=[0,1.25*y_max],
 xlabel="x",ylabel="y",key=string(f(a1,x)),
 explicit(f(a1,x),x,0,2.5*a_max),line_width=2,key=
 string(f(a2,x)),explicit(f(a2,x),x,0,2.5*a_max),
 color=gray50,key= string(f(a3, x)),explicit(f(
  a3,x),x,0,2.5*a_max),dimensions=[640,360])$
```

[11] *Maxima* offers fairly sophisticated programming capabilities.

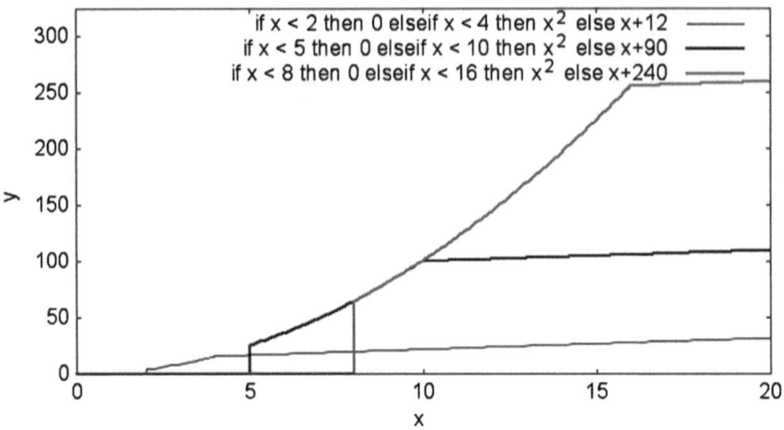

1.7.2 Graphing Points

The illustration above consists of three parts. The first part defines the function described above. The second part specifies three values of the parameter a, and it determines the maximum value of y, given that the maximum x is $2a$. These specifications are passed along with the `wxdraw2d` command, as specified in the next paragraph. The `wxdraw2d` command begins with a definition of the `yrange`. The upper end of that range is set to a value large enough to make room for the legend. The next entry, the `key` definition, differs from the one used above in that the text is not entered directly, but results from expressing the function as a `string` of text. The upper limits of the range of the independent variable in the three `explicit` statements now relate to a named value, a_max, rather than being entered as numbers.

In the resulting graph, the later-entered `explicit` function is placed atop the previously entered one(s) over ranges for which the functions overlap. The graphs above use the `explicit` command to define the function to be graphed. Three other kinds of functional relationships may be drawn in `wxdraw2d`: implicit functions, parametric functions, and relationships that appear as data points. The example below combines an `explicit` functional representation with two that involve points.

```
/*Data*/ xL:makelist(x*%pi/10,x,0,40)$
   yL: makelist(sin(x*%pi/10),x,0,40)$
   rL:makelist(random(1.0),k,0,40) $
/*Graph*/        wxdraw2d(user_preamble=
  "set key bottom center",key="sine,points",
  points(xL,yL),key="sine, line",
  explicit(sin(x),x,0,4*%pi),point_type=6,color=
  gray40,key="random points",
  points_joined=true,points(xL,rL) )$
```

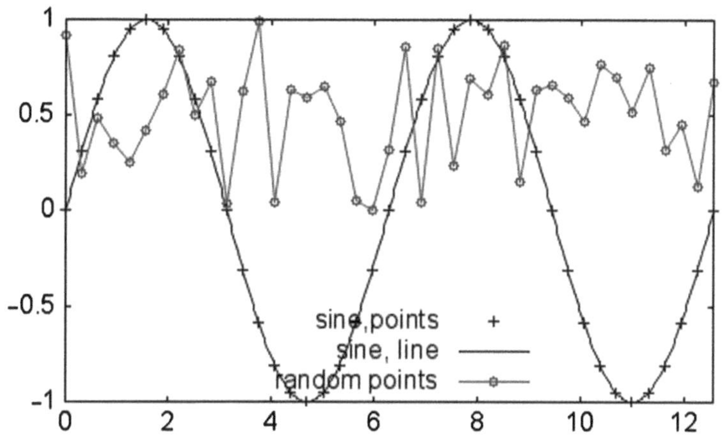

The input group above consists of two parts. The first constructs three lists, and the second plots the data in those lists along with an explicit function. The first list, xL, is a list of x values, ranging from 0 to 4π in increments of $\pi/10$. The second list, yL, consists of the sine of the 41 values in xL. The third list, rL, consists of a set of 41 random values, between 0 and 1, one for each of the 41 values in xL. Inside the wxdraw2d command, a user_preamble is used to place the key below the graph of the random numbers. Then the first pair of points, xL and yL are graphed. Next the explicit function $sin(x)$ is graphed for values of x from 0 through $4\pi4$. The third relationship to be graphed is the set of random numbers. For this series, the points_joined = true option is used.

1.7.3 Three-Dimensional Graphs

The example below shows a simple application of wxdraw3d. In it a Gaussian function intersects a plane.

```
g(x,y):=20*exp(-x^2-y^2)-10; p(x,y):=x+y;  wxdraw3d(
  xlabel="x",zlabel="z",ztics=10,key="Gauss",explicit(
  g(x,y),x,-3,3,y,-5,5),color=gray,key="Plane",
  surface_hide=true,explicit(p(x, y),x,-5,5,y,5,5))$
```

$$g(x,y) := 20\exp\left(-x^2 - y^2\right) - 10 \qquad p(x,y) := x + y$$

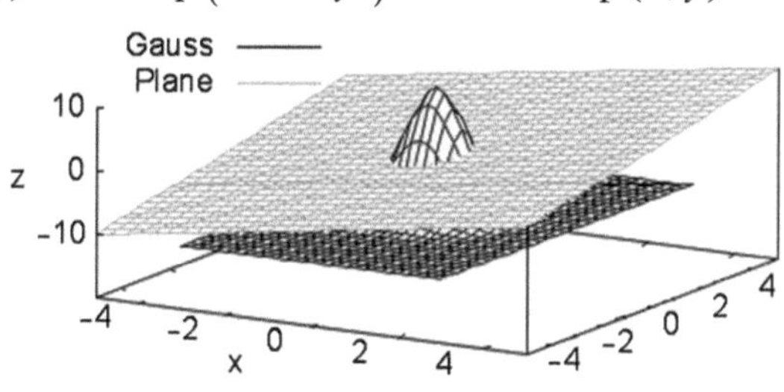

1.7.4 Graphs of Implicit and Parametric Functions

Some relationships cannot be represented as explicit expressions. In some cases, two or more variables together yield a specific result, but neither of the two can be expressed as a function of the other. A simple example is an expression for a circle: $x^2 + y^2 = a$ is an *implicit* function. Variable y cannot be expressed as a function of x, nor can x be expressed as a function of y. The graph below shows how the `implicit` command can be used to plot this implicit function. Note the two different ways that the expression is entered.

```
f(x,y, a):= x^2+y^2-a$
wxdraw2d( key = "Little Circle",xlabel="x",
ylabel="y",implicit(f(x,y,5),x,-5,5,y,-5,5),
line_width=2, key = "Big Circle",
  implicit(x^2 + y^2 = 12, x,-5,5,y,-5,5) )$
```

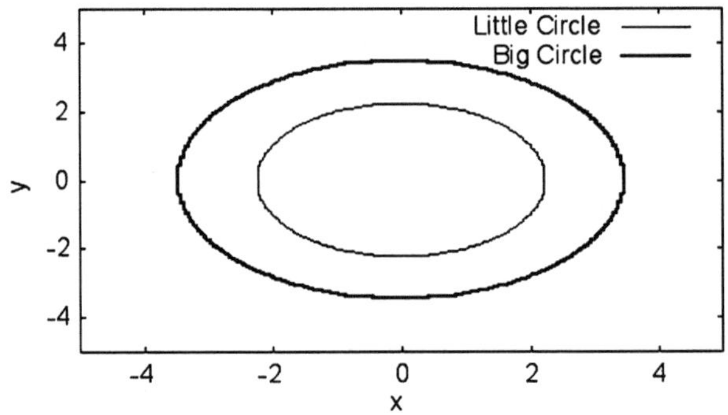

Neither circle above looks round. Their oval appearance reflects an aspect of the `draw`: it does not assume that x and y are in the same units (usually, they are not), and it selects an aspect ratio that is relatively pleasing to the eye. We can force the units to be the same by setting the aspect ratio. The `user_preamble` can be used to control *gnuplot* directly. In this case `set size ratio -1` forces x and y units to be equal. We add the `dimensions` specification to make the whole frame square. Exercise: Execute the commands in the cell below with the `dimensions` option removed.[12]

```
f(x,y,a):=x^2+y^2-a$    wxdraw2d(dimensions=[300,300],
key ="Little Circle", xlabel="x", user_preamble=
 "set size ratio -1",ylabel="y",implicit(f(x,y,5),
 x,-5,5,y,-5,5),line_width=2,key="Big Circle",
implicit(x^2 + y^2 = 12, x,-5,5,y,-5,5) )$
```

[12]You can remove this temporarily with comment brackets, thus: `/* dimensions = [300,300] */`

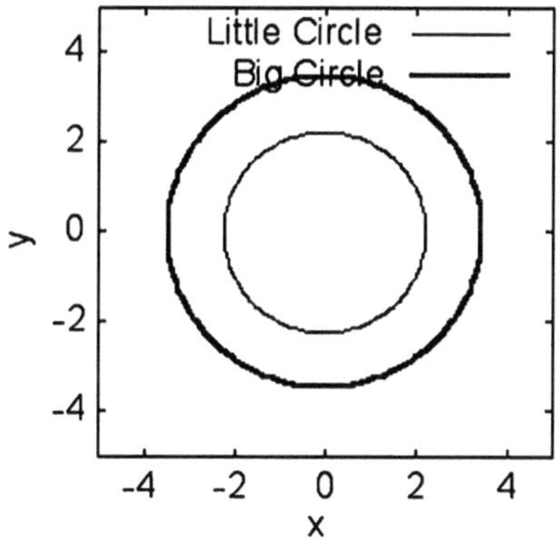

Sometimes x and y are related to each other, but not directly. Rather each of the two is a function of a third variable, which we call t. Suppose that $x = sin(t)$ and $y = cos(t)$, where t is an angle. We have a second way of drawing a circle.

```
wxdraw2d(user_preamble="set size ratio -1",
parametric(sin(t), cos(t), t, 0, 2*%pi),xlabel="x",
ylabel="y", dimensions=[300,300] )$
```

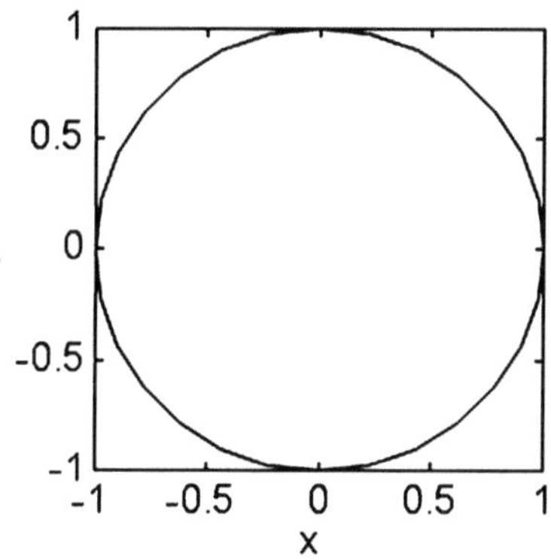

1.7.5 Scenarios

A variant of the `draw` command can be used to create graphs and assign names to them. These graphs can be displayed later in sets. Suppose that we have a function for which we would like to see the behavior of the following: the original function, the first derivative of that function, and the average value of the function.

Let $y = -0.5 \cdot x^3 + 25 \cdot x^2 + 100 \cdot x$. The display below shows the original function in the left panel and the per-unit functions in the right panel. The command `gr2d` produces the two named objects that appear in the graph, and `wxdraw` generates the display. We have embedded the commands to create the per-unit function inside the `gr2d` command, but we could have created named expressions before we created the graphic object and used the names.

Be aware that the units in the two panels differ. The panels could be displayed vertically (the default). Exercise: Remove `columns = 2` from the `wxdraw` command, and insert `dimensions = [480,480]`.

```
y: -0.5*x^3 + 25*x^2 + 100*x;
original: gr2d(xlabel="x",ylabel="y",
  explicit(y,x,0,50))$
perUnit: gr2d(xlabel="x",ylabel="y per unit of x",
  yrange = [-1000,1000], xaxis=true,
  key= average",explicit(y/x, x, 0,50),line_width=2,
  key="derivative",explicit(diff(y,x),x,0,50)  )$
wxdraw(original, perUnit,columns=2)$
```

$-0.5\,x^3 + 25\,x^2 + 100\,x$

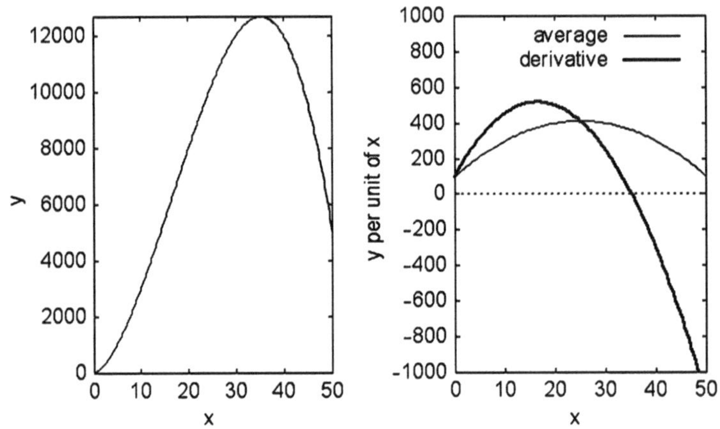

The `draw` command set can become quite large. Later in the text, we delete parts of the sets in order to reduce clutter. Ellipses (. . .) will indicate that one or more commands are omitted. The values of the omitted commands (usually `xrange`, `yrange`, `line_width`, `key`, or the use of `user_preamble` to place the key) can be determined from the graph that `draw` generates.

1.7.6 Default Settings

Each chapter in this text is generated from a single *wxMaxima* file. Each file begins with a set of commands that control the appearance of output. Most of these relate to `draw`. The first line loads `draw`. The next two lines set the default values for graphs. The fill color (for graphs that involve shading) is set to a medium gray. The default color for lines is `gray10`, which looks like black. Black lines can result in sets of points appearing as black blobs. The `draw_realpart` is set to false so that `draw` does not graph the real part of complex numbers. The font and font size are then set.

The commands in the last line do not relate to `draw`. We have used the last one, `kill(all)`, already. It appears at the beginning of each file in order to ensure that the session begins with a clean slate. The `ratprint:false` command keeps *Maxima* from printing output regarding intermediate steps when it processes rational numbers. The `fpprintprec:5` command sets the print precision for floating-point values. The default value is 16.

```
load(draw)$    set_draw_defaults(fill_color=
  gray40,color=gray10,draw_realpart=false,
  font="Verdana", font_size=10)$
 ratprint=false$ fpprintprec:5$ kill(all)$
```

References

An annotated list of useful sources appears in *Further Readings*.

1. Abramowitz M, Stegun IA (eds) (1964) Handbook of mathematical functions. U.S. Government Printing Office, Washington
2. Development Team (2013) Maxima reference manual. http://maxima.sourceforge.net/documentation.html
3. Woolett EL (2012) Maxima by example. http://maxima.sourceforge.net/documentation.html

Further Readings

Leydold J, Petry M (2011) Introduction to maxima for economics. http://statmath.wu.ac.at/~leydold/maxima/ provides a good overview of *Maxima* and introduces its use in economics

Maxima Development Team (2013) Maxima reference manual. http://maxima.sourceforge.net/documentation.html. The standard reference, provided in five languages, is also available inline via the *wxMaxima* help menu or the F1 key

Urroz GE (2009) Maxima book. http://www.neng.usu.edu/cee/faculty/gurro/Maxima.html. A good, though slightly dated, introduction to *wxMaxima*

Vodopivec A (2013) *wxMaxima*. http://andrejv.github.io/wxmaxima/index.html. This website, maintained by *wxMaxima*'s creator, contains workbooks that provide guidance, along with links to related material

Woolett EL (2012) Maxima by example. http://maxima.sourceforge.net/documentation.html. The material at this site introduces *Maxima* and applies it to topics that are direct to physicists, but develops techniques that economists often use

Chapter 2
Simple Economic Models

Abstract This chapter illustrates two models using *Maxima*. The first model is the production possibilities curve, which is a simple model of economy-wide production. The second model is that of competitive markets, where demand and supply determine the equilibrium price and quantity of a good. We also extend this model to examine disequilibrium and the effects of shifts in demand or supply, along with the relevance of elasticities, the impact of taxes, and the value that markets generate.

2.1 Production Possibilities

Perhaps the most abstract model of an economy is the production possibilities curve (PPC), which reduces the economy to two goods and places only quite general conditions on the functional relationship between the two goods. As abstract as it is, this model offers important lessons.

2.1.1 An Illustrative Example

We impose a specific functional form on the PPC, $y = a + b \cdot x^{2.5}$, where $a > 0$ and $b < 0$. We state this as an *explicit function*, so that y appears as a dependent variable. This representation is for convenience only. The levels of x and y are jointly determined, and their values must satisfy this equation: $y-(a+b \cdot x^{2.5}) = 0$.

The model appears below. Before graphing the function, we determine the range over which nonnegative values x correspond to nonnegative values of y, because only points in the first quadrant make economic sense. For this purpose, we use `find_root`. We enter the function and the range over which *Maxima* is to search. We use a quite high maximum value.[1]

[1]If this value is set too low, an error message will occur, showing the values of y at the endpoints. Executing the command again with an increased maximum value of x will lead to a solution.

M.R. Hammock and J.W. Mixon, *Microeconomic Theory and Computation: Applying the Maxima Open-Source Computer Algebra System*,
DOI 10.1007/978-1-4614-9417-1_2, © Springer Science+Business Media New York 2013

The output generated by this set of commands shows the original function and the value of x at which y is closest to zero.

```
[f(x,a,b):=a+b*x^2.5, [a1,b1]:[500,-0.01],
x1max : find_root(f(x, a1, b1), 0, 1000)];
```

$$[\mathrm{f}(x,a,b) := a + b\,x^{2.5}, \quad [500, -0.01], 75.786]$$

The wxdraw2d command below plots a single explicit function, the PPC. The x and y ranges are set at 1.1 times the maximum value of x (x1max from above) and of y (a1). Inside the explicit command, x ranges from 0 to x1max.

Examining the graph points out the first lesson from this simple model: some levels of output cannot be attained given the resource supplies and technology that define this PPC. Even if all resources are used (no unemployment) and they are used efficiently, all points above the curve are ruled out.

```
wxdraw2d
  (xlabel="Good x",
  ylabel="Good y",
  yrange=[0,1.1*a1],
  xrange=[0,
    1.1*x1max],
  line_width=2,
  explicit(f(
   x,a1,b1),
   x,0,x1max))$
```

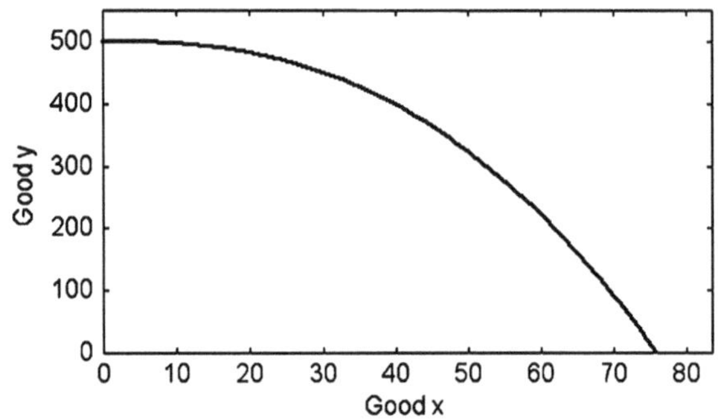

2.1.2 *Opportunity Costs*

The second implication of the graph is the existence of trade-offs. Once the PPC has been reached, more x can be had only at the cost of less y and vice versa. Opportunity costs exist.

The next cell evaluates the function f(x,a1,b1) for four values of x. The first input line defines the initial values x1 and x2, and the change in x from each of these starting points Δx.[2] Then values for y are returned. Thus, the function is evaluated at approximately these values of x : 15.16, 18.95, 37.89, and 41.68. The resulting values of y are approximately 491.1, 484.4, 411.6, and 387.8. As x increases, y decreases, reflecting the opportunity cost of increased x production.

[2] The values are defined in terms of x1max. This is not necessary. Specific numbers can be entered if you prefer, as long as the values are within the range determined above. The changes are discrete, so the Greek letter Delta, Δ is used.

```
[[x1, x2, Deltax] : [x1max/5, x1max/2, x1max/20],
[y1, y1d] : [f(x1, a1, b1), f(x1 + Deltax, a1, b1)],
[y2, y2d] : [f(x2, a1, b1), f(x2 + Deltax, a1, b1)]];
```

[[15.157, 37.893, 3.7893], [491.06, 484.38], [411.61, 387.83]]

The next graph adds these four points to the PPC. The graph shows that the decrease in *y* production increases as we move from $x =$ `x1` to $x =$ `x2`. That is, this PPC displays the *increasing* opportunity cost of *x* as *x* increases.[3]

```
wxdraw2d(xlabel="Good x",ylabel="Good y",yrange=
  [0,1.1*a1],xrange=[0,1.1*x1max],line_width=2,
  explicit(f(x,a1,b1),x,0,x1max),point_size=2,
  point_type=asterisk,points([x1,x1+Deltax,x2,
  x2+Deltax], [y1,y1d,y2,y2d]) )$
```

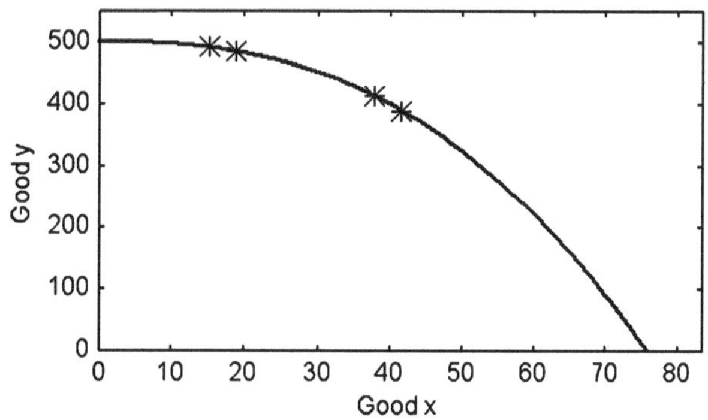

The cell below calculates the opportunity cost of *x* over the ranges considered above. Note the negative sign: Decreased *y* is a positive cost.

```
oc1: -(y1d-y1)/(Deltax)$   oc2: -(y2d-y2)/(Deltax)$
["OpportunityCosts, 2 Ranges:", oc1, oc2];
```

[Opportunity Costs, 2 Ranges:, 1.7631, 6.276]

The graph below shows opportunity cost as a continuous function of *x*. The opportunity cost at any point on the PPC is the negative of the PPC's slope at that point. The `diff` command takes the first derivative of the function, providing a functional specification of the slope.

Developing this function requires two steps, both of which appear on the first input line below. First apply `diff`. Then define the function as the result of the `diff` operation. The "quote - quote" before the % tells *Maxima* to evaluate the result of taking the derivative, and not just to keep it in memory as an expression. Again, note the negative sign.

[3]Economic theory does not dictate that opportunity cost must increase. The simple Ricardian model of comparative advantage, with a linear PPC, exhibits constant opportunity cost. PPCs can bow in, at least over some range, so that opportunity cost decreases.

The graph shows two points, each at the midpoint of the ranges defined above.[4] We extend the graph just beyond the largest value defined above, to $x2 + 2 \cdot \Delta x$.

The graph shows that the PPC above generates continuously increasing opportunity cost. It also shows that at the two points on the graph, the result of computing opportunity cost over a discrete range and the result of computing it at a point are quite close to each other, for this function.[5]

```
g(x, a1, b1)  := -''(diff(f(x, a1, b1), x));
wxdraw2d(ylabel="Good y per unit of Good x",title =
  "Opportunity Costs: Function and Points",xlabel=
   "Good x",explicit(g(x, a1,b1),x,0,x2+2*Deltax),
   point_size=2,points([(x1+Deltax/2),(x2+Deltax/2)],
     [oc1,oc2]))$
```

$$g(x, a1, b1) := -\left(-0.025\, x^{1.5}\right)$$

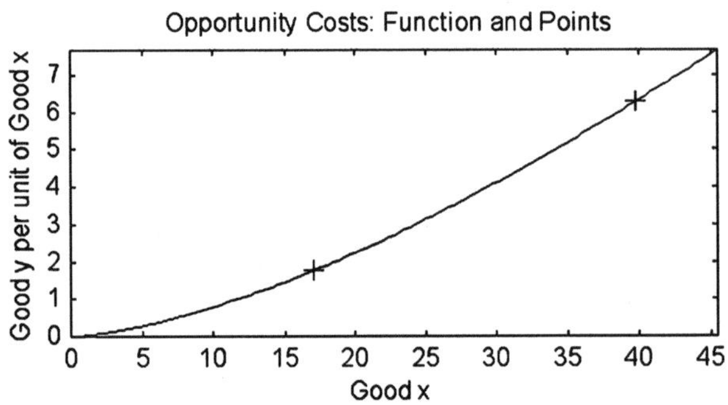

2.1.3 *Shifting PPC*

A change in either the resource availability or technology shifts the PPC. An increased resource base or an improved technology would shift the PPC outward (more y for each level of x produced). Whether the shift would be parallel or, more likely, take some other form depends on the details.

The illustration below shows a special case. Either technology is limited to x production or only resources used in x production increases. When all resources are allocated to y production, the quantity does not change.

Before drawing the graph, use `find_root` to set the range of x (to limit the output to nonnegative y values). The output includes both the initial x limit and its

[4] $(x1 + x1 + \Delta x)/2 = x1 + \Delta x/2$

[5] Evaluating the oc function at $x = x1 + \Delta x/2$ and at $x = x2 + \Delta x/2$ yields values that differ from the values computed over the ranges only at the third decimal point. These values are 1.76 and 6.274, respectively.

new counterpart. For graphing, we define ymax as the maximum of two values, a1 and a2. In the case below, a does not change so both values are equal. We define xmax as the larger of x1max and x2max.

```
[a2,b2]:[500,-0.0075]$ x2max:find_root(f(x,a2,b2),
 0,1000)$ [x1max,x2max]; ymax:max(a1,a2)$
 xmax:max(x1max,x2max)$
wxdraw2d(title="PPC shift: lower b value",xlabel=
 "Good x",ylabel = "Good y", yrange=[0,1.1*ymax],
 xrange=[0,1.1*xmax],key="Initial PPC",line_width=2,
  explicit(f(x,a1,b1),x,0,xmax),color=gray,key=
  "New PPC",explicit(f(x,a2,b2),x,0,xmax))$
```

[75.786, 85.028]

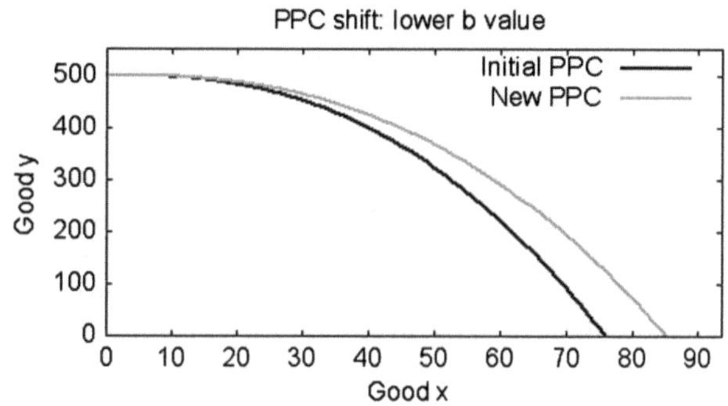

In the example depicted, the opportunity cost of x (the number of units of y given up per increased unit of x) falls, shifting the PPC out to the gray line, which has a smaller slope at each x value than the initial black line.[6]

Chapter 12 revisits the PPC, deriving a PPC from microeconomic foundations. In doing so, it investigates the nature of efficiency, identifying the conditions that are required for an economy to reach its PPC. It also shows how the price system can guide an economy toward the PPC.

2.2 Competitive Markets

The cell below contains two representations of the demand curve for a product and two corresponding representations of the supply curve. In each case the quantity equation occurs first, and the *inverse*, or price equation occurs next.

[6]This implies, however, that the opportunity cost of y increases: At each level of x, increasing y by an additional unit requires that more units of x be forgone than before the shift.

2.2.1 Demand Curves and Supply Curves

The demand curve $DQ = a \cdot p^E$ states the quantity demanded at each of a range of prices. The inverse demand curve $DP = (q/a)^{1/E}$ states the height of the supply curve, or the price that buyers are willing and able to pay, at each of a range of quantities. The supply curve can be represented in either of the two ways listed below, as well.

We use DQ to indicate "the demand curve, quantity" and DP to indicate "the demand curve, price." Likewise for the supply curve. Henceforth, variable names will often consist of a series of letters that indicate the variable's economic meaning. When two variables are to be multiplied, this text uses the $\cdot$ sign, a counterpart to the $\star$ in the *wxMaxima* commands.

The output below restates the input. Often, we suppress such output by ending commands with dollar signs rather than semicolons. We specify the parameter values below, but the following must be true: E must be negative (demand curves slope downward), a must be positive (only positive prices and quantities are allowed), b can be either positive or negative, and c is nonnegative. This last constraint is not mandated by economic theory. Also, if $b < 0$, c must be positive, so that some part of the supply curve lies in the first quadrant.

The first pair of commands creates the demand curve which defines the quantity demanded at each price, and its equivalent inverse demand curve which defines the price that buyers are willing and able to pay for each quantity. The second pair does for the supply curve and its inverse what the second set does for the demand curve and its inverse.[7]

```
[DQ(p,a,E):=a*p^E,        DP(q,a,E):=(q/a)^(1/E)];
[SQ(p,b,c):=b+c*p,        SP(q,b,c):=(q-b)/c];
```

$$[\mathrm{DQ}(p,a,E) := a\, p^E, \mathrm{DP}(q,a,E) := \left(\tfrac{q}{a}\right)^{\frac{1}{E}}]$$

$$[\mathrm{SQ}(p,b,c) := b + c\, p, \mathrm{SP}(q,b,c) := \tfrac{q-b}{c}]$$

The first command in the next cell evaluates the inverse demand curve given parameter values. It instructs *Maxima* to produce the equation for a specific demand curve. The second command maintains these values of a and E and adds the value of q. It instructs *Maxima* to find a point on the demand curve ($q = 40$, $p \approx 3.8981$). Similar exercises would produce counterparts for supply.

```
[DP(q, 1200,-2.5),DP(40,1200,-2.5)];
```

$$[\tfrac{17.048}{q^{0.4}},\ 3.8981]$$

[7] We could use *Maxima* to determine the equations for the inverse demand and supply curves. The accompanying exercise set takes this approach.

2.2.2 Equilibrium

We begin by solving the system of equations for the quantity and requiring *Maxima* to provide numerical value(s). The solution involves an expression that combines a polynomial (supply) and a transcendental function (demand).

For a relatively complex equation like this one, *Maxima*'s `solve` function might not suffice, so, we use `find_root`.[8] Iteration begins at $q = 0.0001$, avoiding division by zero, and ends at 500.[9] The resulting value is assigned the name `qE0`.

The equilibrium quantity is $qE0 = 35.765$. We determine the resulting price level twice, once by putting $qE0$ into the inverse demand curve and then by putting $qE0$ into the inverse supply curve. The comforting result is that both numbers are equal; the demand price equals the supply price.

```
[a0,E0,b0,c0] : [1200,-2.5,-5,10]$     [qE0:
  find_root(DP(q,a0,E0)-SP(q,b0,c0),q,.001,500),
["Prices",DP(qE0,1200,-2.5),  SP(qE0,-5,10)]];
```

[35.765, [Prices, 4.0765, 4.0765]]

Solving the system for quantity is arbitrary. We could have determined the price at which the demand price equals the supply price.

```
[pE0:find_root(DQ(q,a0,E0)-SQ(q,b0,c0),q,0.001,10),
["Quantities", DQ(pE0,1200,-2.5), SQ(pE0,-5,10)]  ];
```

[4.0765, [Quantities, 35.765, 35.765]]

Next, we create a list of eight prices, each stated as a multiple of the equilibrium price, and the associated quantities demanded, quantities supplied, and excess demand values. To do this we create two new equations, an equation for a specific demand curve and its counterpart for supply.

Maxima's `makelist` command is used to create a list of prices, `pList`. The prices range from $\frac{pE0}{5}$ to $\frac{8 \cdot pE0}{5}$. With this list in place, we use `maplist` to map the demand curve and supply curve equations onto the list. The results are `qdList0` and `qsList0`. For each, we map the relevant function onto the list. Once the lists are created, we usethe `append` command to add names. The excess demand list is created by subtraction.

---

[8]Actually, `solve` does work in this case, but it yields four solutions, three of which involve imaginary numbers or negative values. The fourth is the positive real solution that `find_root` provides directly.

[9]If the upper bound had been set too low, `find_root` would have returned an error message because the difference between the demand price and the supply price would have been positive at both ends of the range. If this happens, increase the upper end of the range and repeat the command. Or draw a simple graph to determine the approximate value.

```
[DQ0(p):= ''(DQ(p,a0,E0)),SQ0(p):=''(SQ(p,b0,c0))];
/*Create lists*/ pList0:makelist(pE0*i/5,i,1,8)$
  qdList0:maplist(DQ0,pList0) $
  qsList0 : maplist(SQ0, pList0) $
/*Append names */  pList0:append(["Price"],pList0)$
  qdList0:append( ["DQ"], qdList0)$   qsList0:
    append(["SQ"],qsList0)$ edList0:qdList0-qsList0$
/*Create table */
  transpose(matrix(pList0,qdList0,qsList0,edList0));
```

$$[\mathrm{DQ0}(p) := \frac{1200}{p^{2.5}}, \quad \mathrm{SQ0}(p) := 10\,p - 5]$$

Price	DQ	SQ	DQ-SQ
0.815	1999.3	3.153	1996.2
1.6306	353.44	11.306	342.13
2.4459	128.26	19.459	108.8
3.2612	62.479	27.612	34.867
4.0765	35.765	35.765	$-2.13163\,10^{-14}$
4.8918	22.673	43.918	−21.245
5.7071	15.422	52.071	−36.649
6.5224	11.045	60.224	−49.179

The table shows that for low prices, a quite large excess demand results. This reflects the fact that the demand curve selected is quite price elastic over its range. At $Price = 4.0765$, the excess demand is zero.[10]

Finally we draw the supply and demand curves and add the line that is defined above. Each of the two relationships involves an `explicit` command within `wxdraw2d`. The y range is limited to positive values and is truncated at twice the equilibrium price in order to emphasize the relevant part of the demand curve. Inside the `explicit` commands for the demand and supply curves, we limit the range to $1.5 \cdot qE0$.

To draw a line that highlights the equilibrium price and quantity, we use an `if ... then ... else` command. This draws a horizontal line for $q < qE0$ and has the line drop to zero at $q = qE0$. This function is drawn from $q = 0$ to $q = 1.05 \cdot qE0$. Notice that each individual curve must be drawn over its own specified range and that these ranges can differ among curves.

[10]Rounding error resulting from the use of a numerical method to find the equilibrium price results in the reported value, which is nearly zero.

```
eqline:if q < qE0 then pE0 else 0$ wxdraw2d( yrange=
 [0, 2.0*pE0], line_type=dots,ylabel="$ per unit",
 xlabel="Quantity: units per time period",explicit(
  eqline,q,0,1.05*qE0),line_width=2,line_type=solid,
  key="Demand",explicit(DP(q,a0,E0),q,0,1.5*qE0),
  key="Supply",color=gray,
  explicit(SP(q,b0,c0),q,0,1.5*qE0))$
```

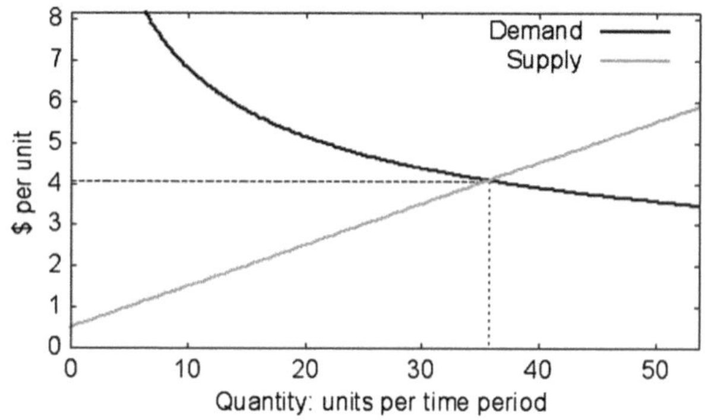

2.2.3 *Disequilibrium*

Confirm that for $q > qE0$ the supply curve lies above the demand curve. That is, confirm that the price that producers would have to be paid in order to produce this quantity exceeds the price that buyers are willing and able to pay for that quantity.

```
[1.1*qE0,  [DP(1.1*qE0,1200,-3),SP(1.1*qE0,20,5)]];
```

[39.342, [3.1245, 3.8683]]

Confirm that for $q < qE0$, the demand curve lies above the supply curve.

```
[0.9*qE0,  [DP(0.9*qE0,1200,-3),SP(0.9*qE0,20,5)]];
```

[32.189, [3.3406, 2.4377]]

2.2.4 *Price Ceilings*

Markets tend to drive the price toward the equilibrium price. Market tendencies can be overcome, however, by government policies that mandate a price other than the one that clears the market.

The cell below specifies that the price may not rise above $0.8 \cdot$ `pE0`. (The value of `multiple` can be changed.) The second input line determines the quantity demanded and the quantity supplied at the mandated price and assigns names to these values. For graphing, the maximum of these two values is assigned the name `maxq`.

```
multiple:0.8$ [pNE:multiple*pE0,
 [qNED,qNES]:
  [DQ(pNE,a0,E0),SQ(pNE,b0,c0)]];
maxq : max(qNED, qNES)$
```

[3.2612, [62.479, 27.612]]

The graph below shows the demand curve, the supply curve, and the two points derived above. The horizontal distance between the two quantities is the per-period shortage. The x and y ranges are set as multiples of the equilibrium p and x values, respectively, to allow for changing the value of `multiple`.[11]

```
wxdraw2d(yrange=[0,2.5*pE0],xrange=[0,1.25*maxq],
 xlabel="Quantity in units per time period",ylabel=
  "$ per unit",line_width=2,key="Demand",explicit(
   DP(q,a0,E0),q,.01,1.25*maxq),key="Supply",color=
   gray,explicit(SP(q,b0,c0),q,0,1.25*maxq),
   point_size=2,points_joined=true,color=black,
   line_type=dots,key="Excess Demand",
   points([qNED, qNES],[pNE,pNE]))$
```

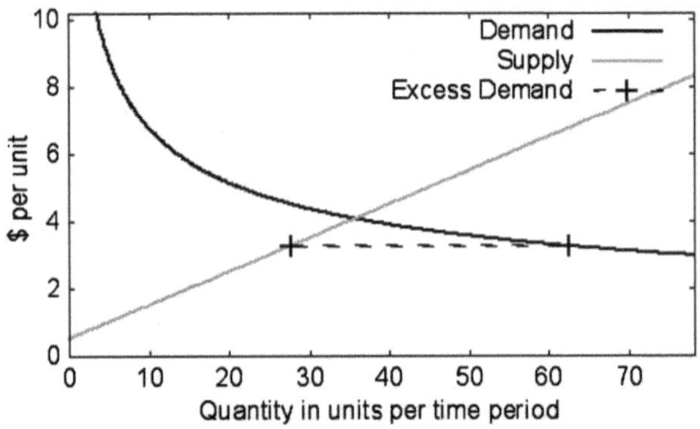

Sometimes black markets result from price ceilings. How a black market works (which price and quantity pair tends to result) depends on the specifics of the case. Suppose that black market sellers can buy the 27.612 units and auction them to the

[11]If `multiple` > 1, then a price floor is established. An exercise using these demand and supply curves but analyzing a price floor appears in the online exercises.

highest bidders. Then the solution is simple: the black market price is $4.521 per unit. We determine this by placing the quantity qNES into the demand curve.[12]

```
pBM : DP(qNES, a0, E0);
```

4.521

2.2.5 Shifts

Supply and Demand shifts are easily modeled numerically and graphically using the techniques developed in the previous section. The first input group shows the coefficients of the initial demand and supply curves, used above, and the coefficients that result in shifted demand and supply curves.

```
[a0, E0, b0, c0]  :  [1200, -2.5, -5, 10]$
[a0, E1, b1, c1]  :  [1200, -1.5, -5, 15]$
```

2.2.5.1 Demand Shift

Shift the demand curve. Either a or E could be changed. We leave a unchanged and change E from $E = -2.5$ to $E = -1.5$. That is, the demand curve becomes less responsive to price (elasticity falls).

The resulting equilibrium values appear below: $qE1 = 83.597$ and $pE1 = 5.9065$. Both quantity and price increase relative to their original values, $qE0 = 48.939$ and $pE0 = 3.5959$, which is to be expected as the demand curve shifts to the right.

It might not be apparent that decreasing price responsiveness (making the demand curve less elastic) should shift the curve to the right. In fact, it does so only for prices above $p = 1$. At $p = 1$, the quantity equals a, whatever the elasticity value. Thus, the two curves intersect at $p = 1$. (Exercise: Confirm this assertion graphically.) The relevant price range is higher for the new demand curve, and the quantity demanded falls less along the new demand curve than along the initial one.

```
qE1:find_root(DP(q,a0,E1)-SP(q,b0,c0),q,.001,500)$
pE1 : DP(qE1,a0,E1)$
matrix(["qE0","qE1","pE0","pE1"],[qE0,qE1,pE0,pE1]);
```

$$\begin{bmatrix} \text{qE0} & \text{qE1} & \text{pE0} & \text{pE1} \\ 35.765 & 64.914 & 4.0765 & 6.9914 \end{bmatrix}$$

[12]A more likely outcome is that some units will sell in the black market for a price higher than this mandated price, some will sell for a price between the free-market equilibrium price and the mandated price, and still others will sell at the mandated price.

The graph shows the shift in the demand curve and its effects. The new demand curve lies above (to the right of) the initial one, so both price and quantity tend to increase. The graph is the same as in the previous section with a few minor changes. First, variables *maxq* and *maxp* are defined for graphing purposes. One more `explicit` command is added.

```
maxq : max(qE0, qE1)$ maxp : max(pE0, pE1)$
wxdraw2d( user_preamble="set key bottom center",
 yrange=[0,1.5*maxp],ylabel="$ per unit",
 xlabel="Quantity: units per time period",key=
 "Initial Demand",explicit(DP(q,a0,E0),q,0,
  1.5*maxq),line_width=2,key="New Demand",
  explicit(DP(q,a0,E1),q,0,1.5*maxq),color=gray,
  key="Supply",explicit(SP(q,b0,c0),q,0,1.5*maxq))$
```

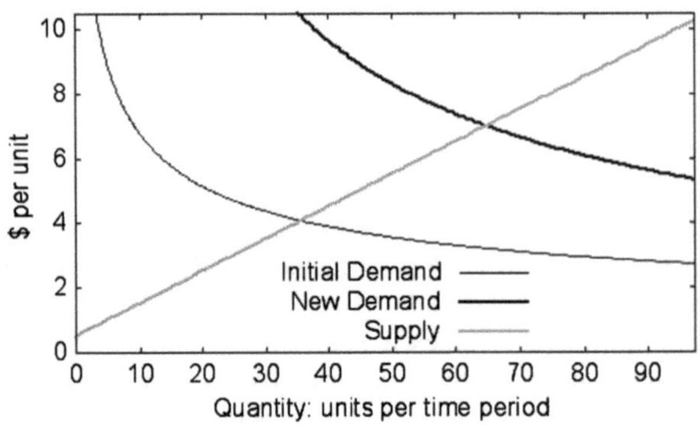

2.2.5.2 Supply Shift

The input cell below shows the effect of an increased supply. The table compares the initial quantity and price with the quantity and price after the shift in supply.

```
qE2:find_root(DP(q,a0,E0)-SP(q,b0,c1),q,.001,500)$
pE2 : DP(qE2,a0,E0)$
matrix(["qE0","qE2","pE0","pE2"],[qE0,qE2,pE0,pE2]);
```

$$\begin{bmatrix} \text{qE0} & \text{qE2} & \text{pE0} & \text{pE2} \\ 35.765 & 48.939 & 4.0765 & 3.5959 \end{bmatrix}$$

As an exercise, replicate the graph above, with a supply increase replacing a demand increase. Also, as further exercises, graphically analyze the effects of simultaneous demand and supply shifts. Confirm that when both curves shift, the direction of change is determinate for either price or quantity, and that the direction depends on the specific shift sizes for the other variable.

2.3 Elasticities

2.3.1 Price Elasticity

The responsiveness of the quantity demanded to price changes can be measured in terms of either slope or elasticity. The *slope* of the demand curve ($\frac{dq}{dp}$) is the number of units that quantity changes per one-unit change in price. The *price elasticity* is the percentage change in quantity per 1 % change in price.

Each measure has its merits. For a given product in a given setting, the slope is quite informative. For more general statements, elasticity becomes more useful, for two main reasons. First, it is free of units, so elasticities of demand for various products can be compared. Second, the elasticity is closely related to how total expenditures for a good change in response to a price change. This is particularly important for analyzing the behavior of price-searching firms (e.g., monopoly, monopolistic competition).

2.3.1.1 Constant-Elasticity Demand Curve

This section begins with the demand curve that has been used above. This demand curve exhibits constant elasticity. That is, the price elasticity of demand is the same at all points on the demand curve.

The demand curve appears twice in the input group below. It is evaluated for $a = 1200$ and $E = -2.5$ at two prices, $p = 9.9$ and $p = 10.1$. We use the midpoint formula $E_{arc} = \frac{q_2 - q_1}{q_{mean}} / \frac{p_2 - p_1}{p_{mean}}$.

The value of p_{mean} is 10 and the change in p is 0.2, so these values are entered directly. Their q counterparts are computed. The resulting "arc" elasticity is quite close to the "point" elasticity, -2.5. As an exercise, enter another pair of price values and confirm that the "arc" elasticity is nearly -2.5.

```
[[q1, q2]:[DQ(9.9,1200,-2.5),DQ(10.1,1200,-2.5)],
((q2 - q1)/(0.5*(q2+q1)))/(0.2/10)];
```

$$[[3.8913, \quad 3.7015], \quad -2.4996]$$

The graph below shows three constant-elasticity demand curves with different price elasticities. The first is elastic ($E = -2.5$), the second is unit elastic ($E = -1$), and the third is inelastic ($E = -0.25$). The three curves intersect when $p = 1$, with $q = a = 1,200$, as noted earlier.

```
wxdraw2d(yrange=[0,3],xlabel="Quantity",ylabel=
  "$ per unit",user_preamble="set key bottom left",
  key="Elastic",explicit(DP(q,1200,-2.5),q,1,1500),
  line_width=2,key="Unit Elasticity",explicit(
  DP(q,1200,-1),q,1,1500),color=gray,line_width=2,key
  ="Inelastic",explicit(DP(q,1200,-0.25),q,1,1500))$
```

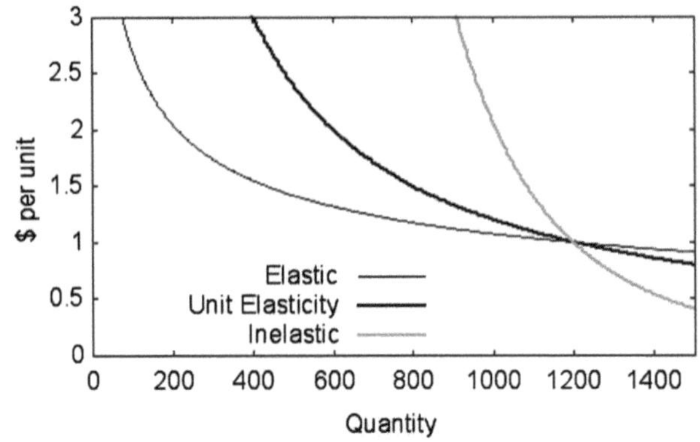

2.3.1.2 Elasticity, Price, and Expenditures

The price elasticity of demand determines the relationship between price changes and corresponding changes in total spending on the product or, equivalently, total revenue for the seller(s) of the product. When the demand curve is price elastic, price and total spending are inversely related; when the demand curve is price inelastic, price and total spending change in the same direction; and when the demand curve's elasticity is -1.0, total spending is the same at each price.

The input group below determines the marginal revenue (the change in total spending, or revenue, per one-unit change in quantity). The first statement informs *Maxima* that the price for which the product sells depends on the quantity. Then total revenue is defined.

Finally, marginal revenue is defined as the first derivative of total revenue.[13]

The result shows that $MR = p + \frac{dp}{dq} \cdot q$. Factoring p from this expression yields $MR = p \cdot (1 + (\frac{dp}{dq} \cdot \frac{q}{p})$, which is equivalent to $MR = p \cdot (1 + \frac{1}{E_p})$, where E_p is the price elasticity. Recall that $E_p < 0$, so demand is elastic when $E_p < -1$ and inelastic when $-1 < E_p \leq 0$.[14]

---

[13]We use binding rather than functional expressions because, in this setting, doing so is simpler. Binding the result of the `diff` command is somewhat easier than expressing the relationship as a function.

[14]Be aware that many textbooks, by tradition, use the absolute value of own-price elasticity of demand, so that it is reported as a positive number.

Given that $MR = p \cdot (1 + \frac{1}{E_p})$, when the demand curve is price-elastic, $1 + \frac{1}{E_p}$ must be positive, so quantity and total spending move in the same direction, implying that price and total spending are inversely related. Likewise, price-inelastic implies that price and total spending move in the same direction. Finally, if $E_p = -1$, $MR = 0$, so quantity and total spending are unrelated, as are price and total spending.

```
[depends(p,q),   TR:p*q,   MR:diff(TR, q)];
```

$$[[\mathrm{p}(q)], \quad p\,q, \quad \left(\frac{d}{d\,q}\,p\right)\,q+p]$$

The input/output group below shows that for the constant-elasticity curve defined above, the relationship defined above holds. To make the result of the `diff` command easily intelligible requires use of the `subst` and `factor` commands. The output shows that $MR = p + \frac{p}{E}$—so $MR = p \cdot (1 + \frac{1}{E}))$—or $MR = p \cdot \frac{E+1}{E}$.

```
[TR(q,a,E) :=q*(q/a)^(1/E), MR:diff(TR(q,a,E),q),
subst(p,(q/a)^(1/E), factor(MR))];
```

$$[\mathrm{TR}(q,a,E) := q\left(\frac{q}{a}\right)^{\frac{1}{E}}, \quad \frac{\left(\frac{q}{a}\right)^{\frac{1}{E}}}{E} + \left(\frac{q}{a}\right)^{\frac{1}{E}}, \quad \frac{p\,(E+1)}{E}]$$

The figure below shows two demand curves, one elastic and the other inelastic, and the associated marginal revenue curves. To facilitate graphing, we write the marginal revenue curve using the same functional notation that has been used for the demand curves. For all constant-elasticity demand curves, marginal revenue is a constant times price. For elastic curves, this constant is between 0 and 1. For inelastic curves, the constant is negative. In the examples below, $MR = 0.4 \cdot p$ for the elastic demand curve, and $MR = -\frac{1}{3} \cdot p$ for the inelastic demand curve.

```
MR(q,a,E):=''(DP(q,a,E)*(1+1/E));   wxdraw2d(yrange=
  [-2,5],xaxis=true, yrange=[-5,15],xlabel="Quantity",
  ylabel="$ per unit",key="Demand, Elastic",explicit(
  DP(q,1200,-2.5),q,1,1500 ),key="MR, Elastic Demand",
  color=gray,explicit(MR(q,1200,-2.5),q,1,1500),color=
  black,line_width=2,key="Demand, Inelastic",
  explicit(DP(q,1200,-0.5),q,1,1500 ),color=gray,
  key="MR, Inelastic Demand",
  explicit(MR(q,1200,-0.75),q,1,1500))$
```

$$\mathrm{MR}(q,a,E) := \left(\frac{q}{a}\right)^{\frac{1}{E}}\left(\frac{1}{E}+1\right)$$

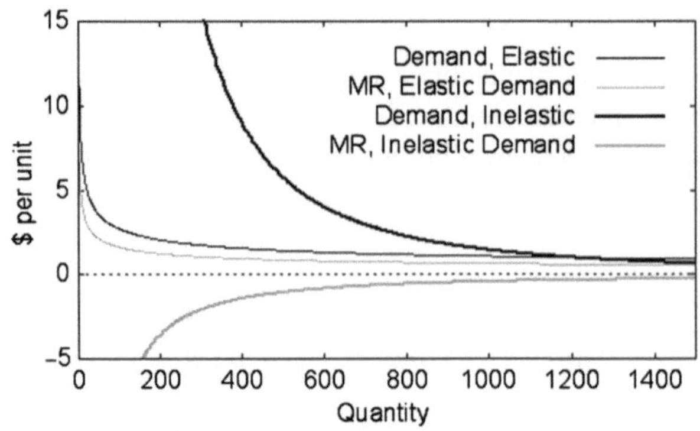

2.3.1.3 Linear Demand Curve

Linear demand curves have constant slopes and varying elasticities. The analysis below shows that $E_p = \frac{p}{b \cdot q}$. Thus, the elasticity is a negative value that approaches zero (the demand curve becomes less elastic) as q increases.

```
[p : a + b*q,TR : expand(p*q),
MR : diff(TR, q),E : (p/q)/(diff(p, q))];
```

$$[b\,q + a, b\,q^2 + a\,q, 2\,b\,q + a, \frac{b\,q+a}{b\,q}]$$

The demand curve has a point of unitary elasticity when $q = 0.5 \cdot q_{max}$, where q_{max} is the quantity demanded when $p = 0$. Thus, the linear demand curve is elastic for quantities less than one-half of the maximum quantity and inelastic for quantities greater than this value.

```
subst(q = 0.5*(-a/b),E);
```

$$-1.0$$

The example below illustrates these conclusions. This linear demand curve is named pL. Total revenue, marginal revenue, and elasticity are named TRL, MRL, and EL.

```
[pL : 400 - 2.5*q,  TRL : expand(pL*q),
MRL : diff(TRL, q),EL : -0.4*(pL/q)];
```

$$[400 - 2.5\,q,\ \ 400\,q - 2.5\,q^2, 400 - 5.0\,q,\ \ -\frac{0.4\,(400-2.5\,q)}{q}]$$

We graph the demand curve and the marginal revenue curves in one panel and the elasticity function separately. Elasticity is a pure, unit-free number, so it can be used to compare price elasticities across products. The graphs confirm that the elasticity equals -1 at the midpoint on the demand curve (where $q = 80$). Also, at that point $MRL = 0$.

```
D_MR:gr2d(xlabel="Quantity", ylabel = "$ per unit",
 xaxis=true,color=black,key="Demand",explicit(
 pL,q,0,160),line_width =2, key = "MR", explicit(
 MRL,q,0,160),ytics={-200,0,200},xtics ={0,80,160})$
Elasticity:gr2d(user_preamble="set key bottom right",
 xlabel="Quantity",ylabel="No units",yrange=[-5,0],
 ytics={["-5.0",-5],["-1.0",-1],["0.0",0]},
 color=black,explicit(-1,q,0,160),color=gray40,key=
 "Elasticity",explicit(EL,q,0,160),xtics={0,80,160})$
wxdraw(D_MR, Elasticity,dimensions=[360,360])$
```

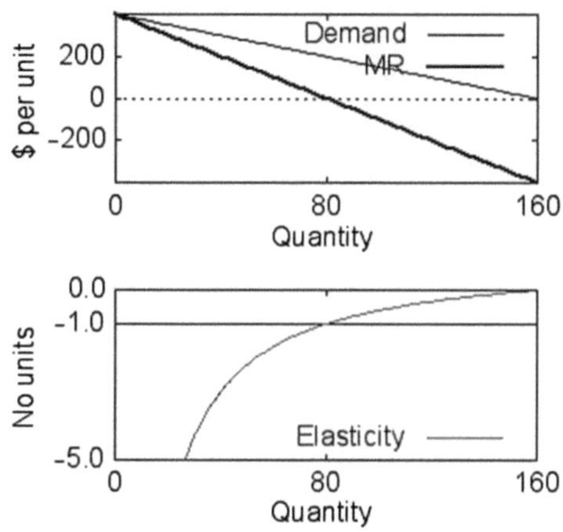

2.3.2 *Other Elasticities*

The own-price elasticity of demand is one of a number of elasticities that can provide useful information. This section considers some others:

- Income Elasticity of Demand: The percentage change in the quantity demanded for this product per 1 % change in income.
- Cross-Price Elasticity of Demand: The percentage change in the quantity demanded for this product per 1 % change in the price of some other related product.
- Price Elasticity of Supply: The percentage change in the quantity supplied per 1 % change in price.

In each case, "other things equal" assumption should be kept in mind.[15]

[15]Also keep in mind that elasticity is a quite general concept. Although economists define some other elasticities, such as output elasticity and cost elasticity, one should think of an elasticity as

We illustrate the elasticity concepts with constant-elasticity demand and supply functions. First, the demand curve (and its inverse) that appear below are direct extensions of the ones used earlier. Two arguments have been added: the price of a related good (pr) and income (y). Each of the exponents is an elasticity.

```
DQ(p,a,ED,pr,EDR,y,EDY):=a*p^ED*pr^EDR*y^EDY;
DP(q,a,ED,pr,EDR,y,EDY):=(q/(a*pr^EDR*y^EDY))^(1/ED);
```

$$\mathrm{DQ}\,(p, a, ED, pr, EDR, y, EDY) := a\, p^{ED}\, pr^{EDR}\, y^{EDY}$$

$$\mathrm{DP}\,(q, a, ED, pr, EDR, y, EDY) := \left(\frac{q}{a\, pr^{EDR}\, y^{EDY}}\right)^{\frac{1}{ED}}$$

The cell below specifies a set of parameter values. Also, it shows two sets of "other things," the price of a related good and income, that can shift the demand curve. Then a list of four expressions for the inverse demand curve is produced. The first demand curve is based on the initial values of pr, 2, and y, 1000. The second shows the result of increasing pr from 2 to 5, holding y constant. The third shows the effect of increasing y from 1000 to 1500, holding pr constant. The fourth shows the combined result of the two partially offsetting changes.

```
[pr0,y0]:[2,1000]$   [pr1,y1]:[5,1500]$
[a0,ED0,EDR0,EDY0]:[10,-0.8,-0.2,0.4]$
[DP(q,a0,ED0,pr0,EDR0,y0, EDY0),
 DP(q,a0,ED0,pr1,EDR0,y0,EDY0),
 DP(q,a0,ED0,pr0,EDR0,y1,EDY0),
 DP(q,a0,ED0,pr1,EDR0,y1,EDY0)];
```

$$\left[\frac{472.87}{q^{1.25}}, \quad \frac{376.06}{q^{1.25}}, \quad \frac{579.15}{q^{1.25}}, \quad \frac{460.58}{q^{1.25}}\right]$$

2.3.2.1 Income Elasticity

Most goods are normal goods, for which the demand changes in the same direction as income. These goods exhibit positive income elasticities of demand. Inferior goods have a negative income elasticity of demand.

Within normal goods, some goods have an income elasticity of demand that exceeds 1.0. These are sometimes called *luxury goods*. For a luxury good, spending on the good changes by a greater percentage than income, so the *share* of income spent on these goods varies directly with income.

For other goods, those for which the income elasticity of demand is less than 1.0 (inferior goods included), the share of income spent on the good changes inversely with respect to income level changes.

The example below illustrates a case in which the income elasticity of demand is less than one: $EDY0 = 0.4$. The income level increases by 50 %, so the rightward shift in the demand curve should be about 20 %. The value 162.27 in the second

---

a general way of specifying the response of a variable to a change, rather than a list of standard formulas.

demand curve equation below is 17.61 % larger than the value 137.97 in the first equation (recall that each of these is the number of units demanded when $p = 1$).

```
[DQ(p,a0,ED0,pr0,EDR0,y0,EDY0),
 DQ(p,a0,ED0,pr0,EDR0,y1,EDY0)];
```

$$[\frac{137.97}{p^{0.8}}, \frac{162.27}{p^{0.8}}]$$

Finally, we confirm that the increased income causes spending on this good to increase but that its share of total spending decreases. When $p = 3$ (any price could be used), buyers' spending equals 359.3 per period, 35.9 % of their income. When income increases to 1,500, spending increases to 440.06, only 29.3 % of the higher income level.[16]

```
EXP0:3*(DP(3,a0,ED0,pr0,EDR0,y0,EDY0))$
EXP1:3*(DP(3,a0,ED0,pr0,EDR0,y1,EDY0))$
matrix(["Income","Spending","Share"],
          [y0,EXP0,EXP0/y0],[y1,EXP1,EXP1/y1]);
```

Income	Spending	Share
1000	359.3	0.359
1500	440.06	0.293

2.3.2.2 Cross-Price Elasticity

The cross-price elasticity measures the effect of a change in the price of a related good. The cross-price elasticity $EDR0 = -0.2$ implies that the two goods are complements: an increased price of the related good leads to a decreased demand for the good in question. As shown below, the higher price of the related good shifts the demand curve to the left (i.e., downward).

```
wxdraw2d(yrange=[0,500],xlabel="Quantity",ylabel=
"$ per unit", key = "Initial Demand",
explicit(DP(q,a0,ED0,pr0,EDR0,y0,EDY0),q,.1,5),
line_width=2, key="Demand, higher pr",
explicit(DP(q,a0,ED0,pr1,EDR0,y0,EDY0),q,.1,5))$
```

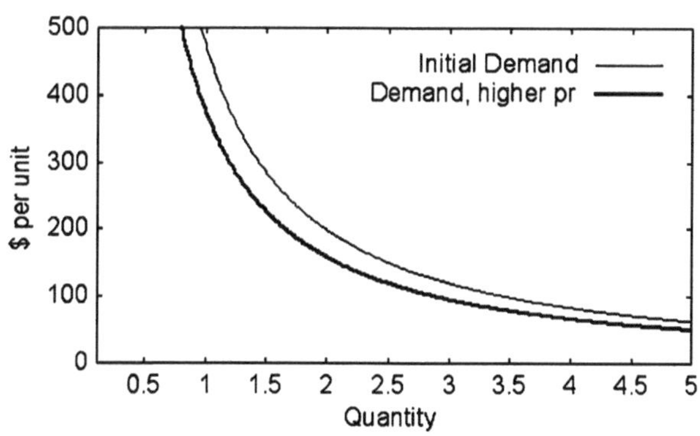

[16]For some specifications, the income shares might exceed 1.0. This result represents economic nonsense. If such a result occurs, reduce $a0$ and rerun the entire sheet, using `ctrl-r`.

2.3.2.3 Supply Elasticity

The concept of the price elasticity of supply differs little from that of its counterpart for demand. We consider it briefly in this section. The next section shows an important implication of these two elasticity measures taken as a pair, the incidence of excise taxes.

The input/output group below defines the equation, determines the equation of the supply curve for a set of parameters, and determines the quantity supplied given those parameters and a specific price.

```
[SQ(p,b,ES):= b*p^ES,    SP(q,b,ES):=(q/b)^(1/ES)];
[SQ(p, 40, 0.5),         SQ(1.5, 40, 0.5)];
```

$$[\mathrm{SQ}(p,b,ES) := b\,p^{ES}, \quad \mathrm{SP}(q,b,ES) := \left(\tfrac{q}{b}\right)^{\frac{1}{ES}}] \quad [40\,p^{0.5}, 48.99]$$

The graph below shows three supply curves. The curve exhibiting unitary elasticity is a ray through the origin.

```
wxdraw2d(user_preamble="set key top left", xlabel=
"Quantity",yrange=[0,2],ylabel="$ per unit",key=
"Unitary Elasticity",explicit(SP(q,40,1),q,0,75),
 key="Inelastic",line_width=2,explicit(SP(q,40,.5),
  q,0,75),key="Elastic",color=gray,
 explicit(SP(q, 40, 1.5), q, 0, 75) )$
```

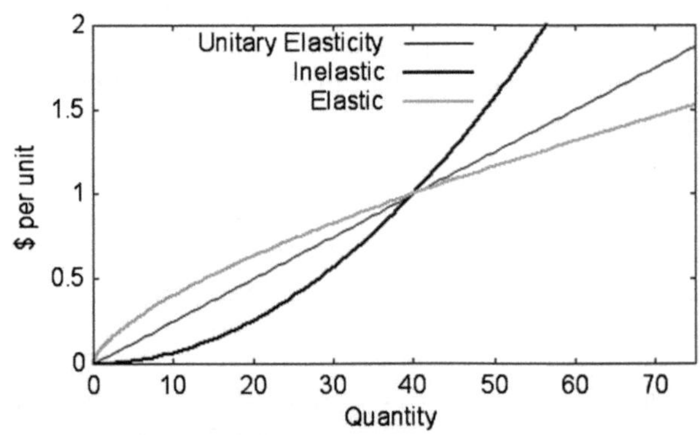

When the constant-elasticity supply curve exhibits unitary elasticity, its graph is a straight line. We can show that the following is also true: A linear supply curve that intersects the price (vertical) axis is elastic at all points and that a linear supply curve that intersects the quantity (horizontal) axis is inelastic at all points. This is true regardless of the linear supply curve's slope.

Let $q = a + b \cdot q$. Then $\frac{dq}{dp} = b$, and the supply elasticity equals $b \cdot \frac{p}{q}$. The cell below shows that, therefore, the elasticity equals $\frac{b \cdot p}{b \cdot p + a}$. If $a > 0$, the supply curve intersects the price axis at $p > 0$, and the elasticity is less than 1.0. If $a < 0$, then the supply curve intersects the horizontal axis at a positive value, the number of units that would be supplied even if $p = 0$, and the elasticity exceeds 1.0.

The third output entry shows that as price and quantity increase (rightward movement along the linear supply curve), the price elasticity approaches 1.0, irrespective of a's sign.

```
[q:a + b*p, b*p/q, limit(b*p/q,p,inf)];
```

$$\left[b\,p + a,\quad \frac{b\,p}{b\,p+a},\quad 1\right]$$

2.4 Taxes and Efficiency

2.4.1 *Tax Incidence*

An important implication of the demand/supply model is that a tax imposed on the sale or purchase of the product will be born in part by sellers and in part by buyers. Furthermore, the effects of the tax on buyers and sellers are independent of whether the tax is nominally placed on selling or on buying.

The expressions below are for the demand and supply curves. That is, all other variables are held constant, and their effects are embedded in the coefficients a and b. We consider an excise tax here, leaving the examination of an ad valorem tax as an exercise.

First, establish the equilibrium in the absence of taxes.

```
DQ(p,a,ED):=a*p^ED$ DP(q,a,ED):=(q/a)^(1/ED)$
  SQ(p,b,ES):=b*p^ES$    SP(q,b,ES):=(q/b)^(1/ES)$
 [a0, ED0,b0,ES0]:[1200,-4,40,0.75]$ qE:find_root(
   DP(q,a0,ED0)-SP(q,b0,ES0)=0,q,0.01,500)$
pE:find_root(DQ(p,a0,ED0)-SQ(p,b0,ES0)=0,
  p,0.01,500)$       [qE, pE];
```

$$[68.437,\quad 2.0463]$$

The results of an excise tax, $X0 = 0.50$ are as follows: the quantity traded falls from 68.437 units to 57.82 units, the price paid by buyers rises from 2.0463 to 2.1344, and the price received by sellers falls from 2.0463 to 1.6344. That is, the price paid rises by 0.0881, and the price received falls by 0.412. Thus, for this particular demand and supply pair, the tax falls disproportionately on sellers.

The equality in expression `qX0` shows the willingness-to-pay function on the left-hand side and the supply price on the right-hand side. The latter contains two parts: the original function, which reflects non-tax costs, and the per-unit excise tax.

```
X0:0.5$ qX0:find_root(DP(q,a0,ED0)=SP(q,b0,ES0)+X0,
 q,0.01,500)$ pDX0:DP(qX0,a0,ED0 )$ pSX0:SP(
  qX0,b0,ES0)$  [qX0,pDX0,pSX0]; [pDX0-pE,pSX0-pE];
```

$$[57.82,\quad 2.1344,\quad 1.6344]\quad [0.0881,\quad -0.412]$$

If the tax had been placed on the buyer of the good rather than the seller of the good, the price that buyers would be willing and able to pay to sellers at each quantity would be the original willingness to pay *less* the tax, $X0$. Thus the left-hand side of the expression above would become $DP(q, a0, ED0) - X0$ and the right-hand side would revert to $SP(q, b0, ES0)$. All solution values would be the same.

The graph shows the upward shift in the supply curve and the resulting equilibrium values. The supply curve shifts upward, parallel to the initial curve, by the amount of the excise tax, $X0$. If the tax had been placed on buyers, the demand curve would shift downward by $X0$, reflecting the lower net price that buyers are willing and able to pay to sellers.

```
pGrossNew : if q < qX0 then pDX0 else 0$
pNetNew: if q < qX0 then pSX0 else 0$
wxdraw2d(user_preamble="set key bottom right",
 line_type=dots,xlabel="Quantity",ylabel=
 "$ per unit",yrange=[0, 1.25*pDX0], explicit(
 pGrossNew,q,0,1.25*qE),explicit(pNetNew,q,
 0,1.25*qE),key="Demand",
 line_type=solid,explicit(DP(q,a0,ED0),
 q,0,1.25*qE),color="gray20",key=
 "Initial Supply",explicit(SP(q,b0,ES0),q,0,
 1.25*qE),line_width=2, key="Tax-laden Supply",
 explicit(X0+SP(q,b0,ES0),q,0,1.25*qE))$
```

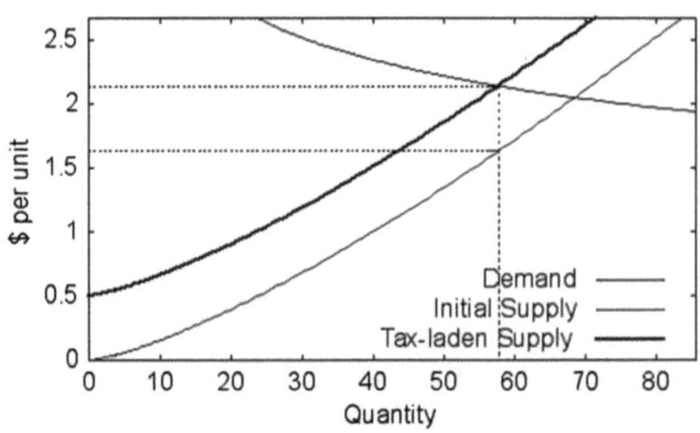

2.4.2 *Consumer Surplus*

Both buyers and sellers gain from exchange. The demand/supply model illustrates this gain. These gains are given the names Consumer Surplus and Producer Surplus. We begin our development of Consumer Surplus by relating the demand curve to marginal value.

```
DP(q,ED):= q^(1/ED); wxdraw2d(
 xlabel="Quantity", yrange=
 [0,10],ylabel="$ per unit",
 explicit(DP(q,-1.5),q,0,1.5))$
```

$$\mathrm{DP}(q, ED) := q^{\frac{1}{ED}}$$

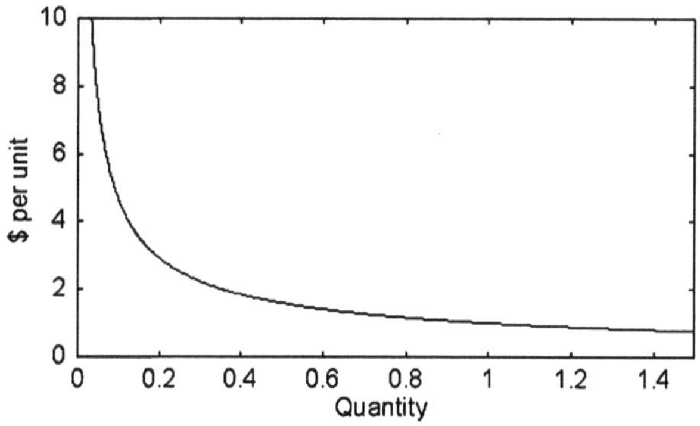

The `wxdraw2d` command shows a graph of the function over the range 0–1.5. (Remember: The units can be any size, perhaps in millions per year.) The graph's vertical axis is truncated so that the marginal value of, say, unit 0.0000001 does not appear (it is $46,416 in our normalized dollars).[17]

At each quantity, the height of the demand curve is the amount that some consumer is willing and able to pay for that unit. If we could sum all of these values, we would have a measure of value for a specified number of units of this good. Some practical difficulties appear, however.

- The marginal willingness to pay of the first few units might not be meaningful. Consider, for example, water. A single liter per week would not be enough to keep even one consumer alive, so any statement about how much consumers would pay for the first liter makes no sense. Fortunately, we seldom seek the total value of a good. Rather, we typically analyze a change that results in an increase or decrease in the quantity without reference to the first few units.
- A related problem is that some goods may not be available in continuously divisible units. We follow convention and assume goods are continuously divisible.[18]

---

[17]For this function, we can determine the area under the demand curve. We do not, because in general such an exercise lacks meaning. For some goods like water and food, the marginal value of the first few units is beyond measure.

[18]This convention is more reasonable that it might appear. Individuals who consume commodities like automobiles make two decisions, how many units to own and how often to buy new models. If a higher price of new good rises slightly and many individuals adjust be keeping their units a bit longer, then the market quantity demanded will decrease slightly, perhaps a few hundred units in a market in which millions of units are sold.

- We intentionally refer to the marginal willingness to pay as *a* measure of value, not *the* measure of value. This function can change if the income distribution changes, so using it as *the* measure would imply that the current income distribution is optimal.
- Taking willingness to pay as an appropriate measure of value requires that we accept a libertarian view of valuation: the consumer is judged to be the best judge of value. If paternalism is involved, the value that one might assign to a given quantity might be either higher or lower than the sum of all marginal willingness to pay.
- The buyers must receive all marginal benefits from consuming the product. That is, third parties, who do not purchase the good, must not gain or lose any benefit from the buyer's having purchased the good.[19] Similarly, the producers must pay all the marginal costs from producing the product.
- The units must be distributed to those consumers who place the highest marginal value on them. This will routinely happen with markets. When other distribution methods are used, the value will be smaller than the sum of the values of the first n units.
- This measure of value is only approximately the same as measures derived from consumer theory. In some important cases, the approximation may be quite close.

The caveats listed above constitute warning against mechanistic use of this measure of value. They do not imply that it cannot be useful.

2.4.2.1 Value of Additional Units

Consider an example in which the quantity is to be changed from one unit to some other number of units (0.8 units in the illustration). The value lost due to decreasing the quantity from 1 to 0.8 is the area under the demand curve between 1.0 and 0.8. The graph below shows three possible demand curves, all passing through the point (1, 1). The reason for having three curves is that one often does not know the demand curve. One does, however, know the current quantity of the good. We also suppose that the analyst knows the marginal value when $q = 1$. Further, we suppose that the analyst has an estimate of demand elasticity. We call that estimate $ED0$. We arbitrarily consider two other values of the elasticity, $ED1 = 0.75 \cdot ED0$ and $ED2 = 1.5 \cdot ED0$.[20]

---

[19]This caveat refers to a direct effect via the third party's utility function. Suppose that buyer A's purchase of the good precludes buyer B's purchase because A is willing and able to pay more than B. Then buyer B is in a sense worse off that if A had not been in the market, but this fact implies nothing about the validity of this measure of value.

[20]Choose $ED0$ such that no elasticity equals -1.0. The three elasticity values are part of the *Maxima* output.

```
EDO:-0.75$ [ED1,ED2]:[0.75*EDO,1.5*EDO]$ [q1,p1]:
 [0.8,DP(q1,EDO)]$ phigh:max(DP(q1,ED1),
 DP(q1,ED2))$ pmax: max(1, phigh)$
wxdraw2d(yrange=[0,1.2*pmax],xlabel="Quantity",
 ylabel="$ per unit",user_preamble=
 "set key bottom left",points_joined=true,
 line_type=dots,points([[1,0],[1,1]]),
 points([[q1,0],[q1,phigh]]),key="Median E",
 line_type=solid,
 explicit(DP(q,EDO),q,0,1.5*max(1,q1)),
 color=gray40,key="Less Elastic",line_width=2,
 explicit(DP(q,ED1),q,0,1.5*max(1,q1)),
 key="More Elastic",line_width=3,
 explicit(DP(q,ED2),q,0,1.5*max(1,q1)))$
```

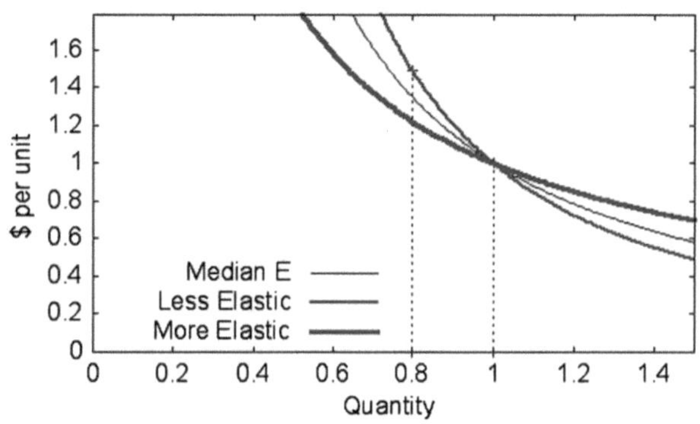

The graph above suggests an attractive way to bracket the actual value of the units in question: develop a menu of parameters (in this illustration, the price elasticity) and determine the implication of each set of parameters on the menu. To achieve this in our illustration, we integrate each of the three functions, beginning at $q = 1$ and ending at $q = 0.8$. Integration is accomplished with the `integrate` command. Inspection suggests that the area under the curve over this range (representing a 20–25 % change in quantity, depending on the direction of the move) is not very sensitive to the elasticity value. The "More Elastic" curve is twice as elastic as the "Less Elastic" curve. Even so, the difference in the two measures is 10.56 %.

```
[[change1,change2,change3]:float([integrate(
  DP(q,EDO),q,1,q1),integrate(DP(q,ED1),q,1,q1),
  integrate(DP(q,ED2),q,1,q1)]), change2/change3];
```

$$[[-0.232, \quad -0.244, \quad -0.22], \qquad 1.1056]$$

To illustrate the concept of Consumer Surplus, consider one of the three demand curves above and suppose that the price is $p = 1$. The graph below shows the area that corresponds to Consumer Surplus. The value of units 0 through 1, measured as willingness to pay, is the area under the demand curve over that range (this value may not be definable for reasons indicated above). The rectangle is the amount of that value that buyers pay for the product, and the remaining area is Consumer Surplus.[21]

```
wxdraw2d(yrange=[0, 5*pmax],xlabel="Quantity",ylabel=
"$ per unit",user_preamble="set key bottom left",
 line_type=dots,points_joined=true,color=gray10,
 points([[0,1],[1,1]]),points([[1,0],[1,1]]),
 label(["Consumer Surplus", 0.25, 2.5]),label(
  ["Amount paid", 0.2, 0.5]),line_type=solid,
 explicit(DP(q,ED0), q, 0, 1.5*max(1,q1))  )$
```

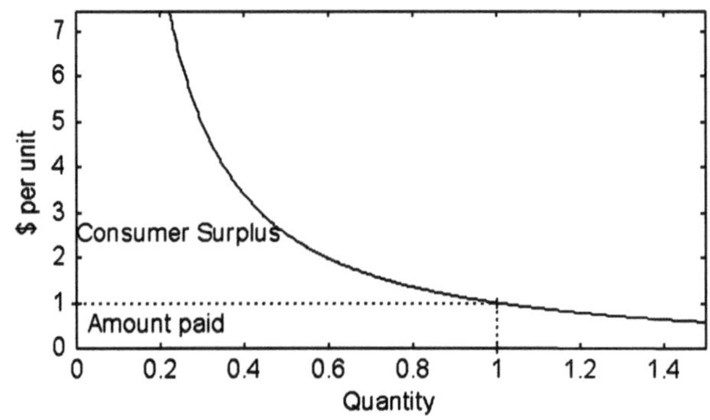

2.4.2.2 Prices and Consumer Surplus

In a market setting, the question that faces the analyst is often this: What is the change in Consumer Surplus due to a change in the good's price? The material below illustrates how this question can be addressed. We use the range of prices that is defined above: $p = 1$ and $p = p1$. At $p = 1, q = 1$ by construction. For $p = p1$, the demand elasticity determines the quantity demanded.

```
DQ(p,ED):=p^ED;
[q10,q11,q12]:
[DQ(p1,ED0),DQ(p1,ED1),
DQ(p1,ED2)];
```

$\mathrm{DQ}(p, ED) := p^{ED}$ [0.8, 0.846, 0.716]

[21]Note the use of labels for these areas.

Each label contains a list with three elements: the text that is to appear in the graph, the horizontal distance at which the label is to be centered, and the label's height. The alignment can be fine-tuned. See `label_alignment` in the *Maxima* user's guide.

The graph below is much like the one above. The `points` that are graphed have been changed, however, so that the trapezoidal area is defined by the two lines, a segment of the price axis, and a segment of the demand curve. As textbooks show, a price increase reduces Consumer Surplus in two ways: It increases the price that buyers must pay for the units that they still purchase at the higher price, and it reduces the quantity consumed so that buyers lose the Consumer Surplus on the marginal units. In the case illustrated here, the former effect is far larger than the latter one.

```
wxdraw2d( yrange = [0,1.2*pmax],xlabel="Quantity",
 ylabel="$ per unit",user_preamble=
 "set key bottom left",line_type=dots,points_joined=
 true,points([[0,1],[1,1]]),points([[0,p1],
 [max(q10,q11,q12),p1]]),line_type=solid,key=
 "Median E",explicit(DP(q,ED0),q,0,1.5*max(1,q10)),
 color=dark_grey,key="Less Elastic",line_width=2,
 explicit(DP(q,ED1),q,0,1.5*max(1,q11)),key=
 "More Elastic",color=gray40,
 explicit(DP(q,ED2),q,0,1.5*max(1,q12)))$
```

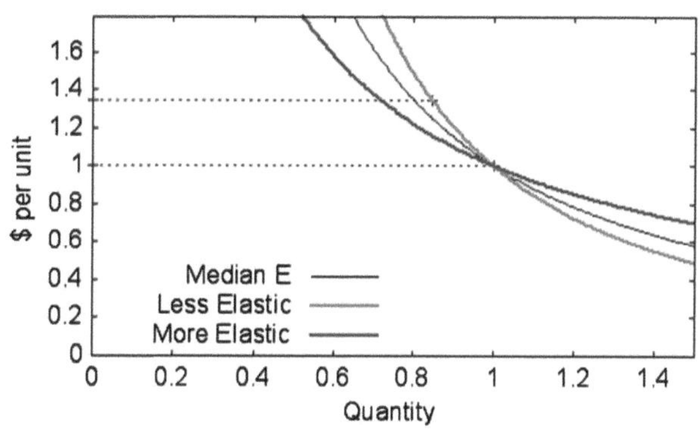

To determine the value of the change in Consumer Surplus for each of the three demand curves, we integrate the demand curve (not the inverse demand curve as above) over the range of prices. The three resulting values appear below. Again, the difference between the high-elasticity and the low-elasticity cases is small.[22]

```
[change4,change5,change6]:float([integrate(
  DQ(p,ED0),p,1,p1),integrate(DQ(p, ED1),p,1,p1),
 integrate(DQ(p,ED2),p,1, p1)]);  change5/change6;
```

$$[0.309, 0.318, 0.292]1.088$$

[22]This result is not general. Suppose the demand curves were shifted horizontally to the left (and truncated where they enter the second quadrant). Then the change in Consumer Surplus would be smaller than appears above, and the percentage differences among the three cases would be larger.

2.4.3 Producer Surplus

Producer Surplus is analogous to Consumer Surplus. It is the difference between the revenue received by sellers and the revenue required to induce them to produce a given quantity (which is also the opportunity cost of producing that quantity of the good). We consider the case of an industry of price-taking firms. For such firms the height of the supply curve defines the minimum price required to bring a specified quantity to market.

For purposes of illustration, we define this function in a fashion similar to that used for demand. The only difference is that price and quantity units have been defined so that (1, 1) is a point on the demand curve. For the same good, this implies that the supply curve must have a coefficient b so that it need not pass through (1, 1). We denote the price elasticity of supply ES.

```
[SQ(p,ES,b):=b*p^ES, SP(q,ES,b):=(1/b)*q^(1/ES)];
```

$$[\mathrm{SQ}(p, ES, b) := b\,p^{ES}, \mathrm{SP}(q, ES, b) := \tfrac{1}{b}\,q^{\frac{1}{ES}}]$$

The graph below shows Producer Surplus. The supply curve passes through the quantity $q = 1$ at a price determined by the income elasticity ($ES = 1$) and the coefficient ($b = 1.5$). These values are arbitrary. We use the resulting value, $p = 0.667$ repeatedly in drawing the graph, so we create a temporary binding of the name `ptemp` to this value. We use labels as above to indicate the areas that correspond to Producer Surplus and to the cost of producing the good in question. The height of the supply curve equals the marginal cost of the good, so Producer Cost is the area under the supply curve between $q = 0$ and $q = 1$. The graph does not show a change, but it is easy to see that a price change results in a changed Producer Surplus for the same reason that it results in a changed Consumer Surplus: producers receive a different payment per unit, and they sell a different number of units.

```
ptemp:SP(1,0.5,1.5)$ wxdraw2d(yrange=[0,1.5*ptemp],
 xlabel="Quantity", ylabel="$ per unit",
 user_preamble="set key top left",
 label(["Producer Surplus",.4,.8*ptemp]),
 label(["Producer Cost", 0.8, 0.2*ptemp]),
 label_alignment=left,label(
 ["Revenue = Producer Surplus+Producer Cost",
 0.05,1.4*ptemp]),points_joined=true,points(
 [[0,ptemp],[1,ptemp]]),points([[1,0],[1,ptemp]]),
 explicit(SP(q, 0.5,1.5),q,0,1.5*max(1,q1)))$
```

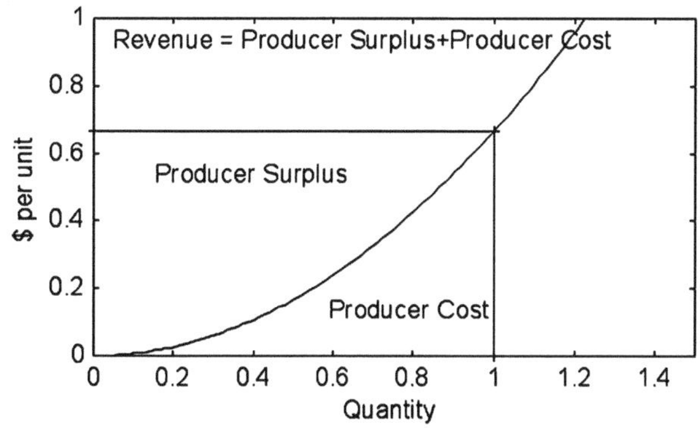

The next graph contains both demand and supply. It shows that the equilibrium quantity is just above 1.0 and that the equilibrium height of either the demand curve or the supply at the equilibrium quantity is just under 1.0.

```
wxdraw2d(yrange=[0,1.25*pmax],xlabel="Quantity",
 user_preamble="set key center right", ylabel=
"$ per unit",color=gray10,key="Demand",
 explicit(SP(q,0.5,1.5),q,0,2), line_width=2,
 key="Supply",explicit(DP(q, ED0), q, 0, 2) )$
```

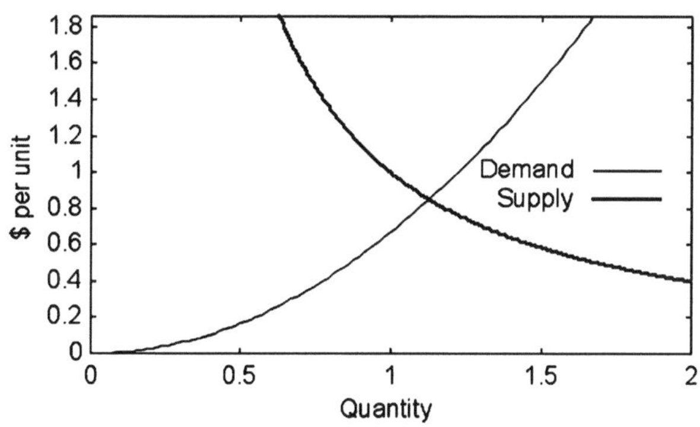

We confirm below what the graph shows: the equilibrium quantity is 1.1293 units and the equilibrium price is $0.85.

```
[q_e: find_root(SP(q,0.5,1.5)-DP(q,ED0),q,0.001,200),
[pd_e, ps_e] : [DP(q_e, ED0) ,SP(q_e, 0.5, 1.5)]];
```

[1.1293, [0.85, 0.85]]

2.4.4 Combined Surpluses and Efficiency

This section confirms that the sum of Consumer Surplus and Producer Surplus is maximized at the market equilibrium depicted above (where the demand and supply curves intersect). Denote the sum of the two surpluses as `surpluses`. To determine the functional form of the variable, we integrate the relevant functions over values of q.

```
assume(ES > 0, ED < -1)$
surpluses: integrate(
  DP(q, ED)-p+(p-SP(q,ES,b)),q);
```

$$\frac{q^{\frac{1}{ED}+1}}{\frac{1}{ED}+1} - \frac{q^{\frac{1}{ES}+1}}{b\left(\frac{1}{ES}+1\right)}$$

The derivative of `surpluses` with respect to q appears below. The first term is the height of the demand curve and the second term is the height of the supply curve.[23]

```
diff(surpluses, q);
```

$$q^{\frac{1}{ED}} - \frac{q^{\frac{1}{ES}}}{b}$$

The result above says that the demand price *less* the supply price must equal zero, or that the two must be equal. Thus, the condition is that the quantity must be the one at which the demand and supply curves intersect.

The sufficient condition for having determined a maximum value is that the second derivative be negative. The expression below shows the second derivative of `surpluses`. Both numerators are positive, as are b and ES. Because the own-price elasticity of demand $ED < 0$, this expression must be negative.

```
diff2 : diff(surpluses, q, 2);
```

$$\frac{q^{\frac{1}{ED}-1}}{ED} - \frac{q^{\frac{1}{ES}-1}}{b\,ES}$$

Finally we apply the values used above to confirm the results for a particular set of parameters. The embedded list in the output shows the equality of the demand price and the supply price. The other term shows the value of the second derivative, confirming that it is negative. The quantity that yields maximum surplus has been found.

```
[[q_e^(1/ED0), q_e^(1/0.5)/1.5],
subst([ED=ED0,ES=0.5,b=1.5,q=q_e],
    diff2)];
```

$$[[0.85, 0.85], -2.5097]$$

[23] This result can be predicted by examining the expression to be integrated, `surpluses`. Because p is subtracted from the height of the demand function and added in the definition of Producer Surplus, the function is DP(q, ED) - SP(q, ES, b). This function is integrated (the "antiderivative" is extracted) and then the derivate of the integral is extracted, returning the original function.

Further Readings

The table at in the preface lists chapters in microeconomics textbooks that relate to this chapter. The following can also be of value:

Baldani J, Bradfield J, Turner RW (1996) Mathematical economics. Dryden, Fort Worth, Chaps. 1–2.

Wainwright K, Chiang AC (2004) Fundamental methods of mathematical economics, 4th edn. McGraw-Hill, New York, Chaps. 1–2.

Chapter 3
Demand Theory: Preferences

Abstract The economic analysis of consumer behavior begins with a model of rational behavior. The consumer is assumed to have a well-defined *utility function* that, coupled with a specification of the constraints facing the consumer, implies a specific action. This chapter develops and illustrates the nature of a utility function. The next two chapters add specifications of the constraints that face the consumer and derive the model's implications.

3.1 A Utility Function

This chapter is the first of three that illustrate the simple economic theory of consumer choice. The illustration assumes the existence of an established *utility function* that expresses the consumer's preferences.[1]

This chapter examines the properties of this illustrative function. Chapter 4 analyzes the nature of *optimizing behavior*, the achievement of the highest possible utility level given available options. It introduces the budget constraint and shows the conditions that must be met if the consumer is to achieve the highest utility level possible when subject to that constraint.

Chapter 5 looks at how price and income changes affect the consumer's choices. Regarding a price change, we examine the two effects of the change, the substitution effect and the income effect. We derive the consumer's demand curve from the utility function. Then we return to the issue of consumer surplus that Chap. 2 introduced.

---

[1] "Established" here means that the utility function does not change due to any thought experiments that we undertake—specifically, income level changes and price level changes. It does not mean that the function cannot change as a result of outside forces like education or persuasion.

M.R. Hammock and J.W. Mixon, *Microeconomic Theory and Computation: Applying the Maxima Open-Source Computer Algebra System*,
DOI 10.1007/978-1-4614-9417-1_3, © Springer Science+Business Media New York 2013

The function that we use for most of this set of material is the *Constant Elasticity of Substitution* or CES function. Its first use in the economics literature was in the analysis of production (see [1]), as was the case with its cousin the Cobb Douglas function (see [2]), but it can illustrate aspects of consumer theory.

Before we proceed, one point requires emphasis: The function yields numeric values for utility (u), but these values have no economic meaning. The function could be modified in any way that would be "order preserving" and would still represent the same consumer.[2]

Using a function like this one to analyze consumer behavior embeds the assumption that the consumer can compare and rank any two pairs of x, y values, along with the assumption that the consumer's preferences are convex. "Convex" means that a line segment connecting any two points on the three-dimensional utility surface that represents the function graphically lies inside (below) this surface.

More narrowly, this particular function embeds the assumption that the elasticity of substitution between the two goods is the same at all points on the function. For now, we leave the term *elasticity of substitution* undefined, except to say that higher values of this term apply to closer substitutes.[3]

```
u(x,y,a,b):=(a*x^-b+(1-a)*y^-b)^-(1/b);
```

$$\mathrm{u}\,(x, y, a, b) := \left(a\, x^{-b} + (1-a)\; y^{-b}\right)^{-\frac{1}{b}}$$

3.1.1 Graphical Representation

The graph below shows a three-dimensional representation of this function for a set of parameters, a0 $= \frac{1}{4}$ and b0 $= \frac{4}{5}$. (Values may be entered in decimal form.) The implied elasticity of substitution is s0 $= \frac{5}{9}$.

```
a0:1/4$ b0:4/5$ [xmax,ymax]:[50,50]$
umax:u(xmax,ymax,a0,b0)$  wxdraw3d(view = [50,330],
 color=grey40, xlabel="x",ylabel="y",zlabel= "u",
 ztics={0,umax},xtics={0,xmax},ytics={0,ymax},
 explicit(u(x,y,a0,b0),x,0.01,xmax,y,0.01,ymax ))$
```

[2]Let the arbitrarily selected pair (x1,y1) yield a utility level $u1$, and let the arbitrarily selected pair (x2,y2) yield a utility level u2 for this function. An *order preserving* transformation of the function would yield different utility values for these two pairs, but the order of the two would remain the same as the order of u1 and u2.

[3]The elasticity of substitution is $s = \frac{1}{1+b}$. We demonstrate this relationship later when the term is formally defined.

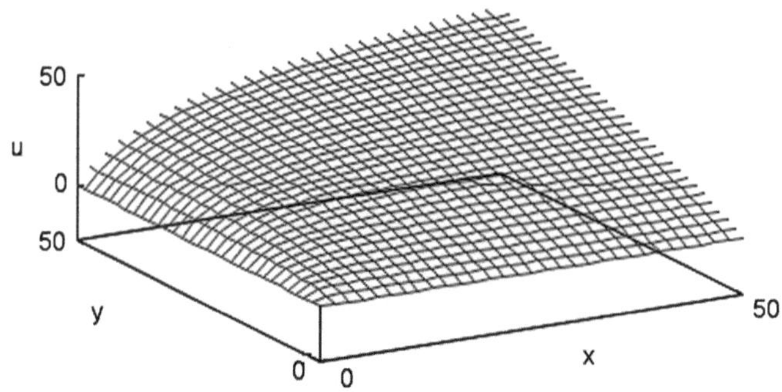

The graph reveals that utility increases whenever the quantity consumed of either of the two goods (or both) increases over the range displayed (this is true for all positive values of x and y). That is, both x and y are "goods" at all values. This utility function does not allow the possibility that, for some large quantities, the consumer would be better off with less of either good. A function that incorporates this possibility appears later in this chapter.

Each of the lines that form the grid below shows the utility level associated with various levels of one of the goods, given that the level of the other good is held constant. Reading in the x direction for example, each successive line shows the utility associated with each level of x consumption, holding constant the consumption of y. The graph shows two aspects of these lines: their height increases as y increases, and their slope increases as y increases.[4]

The two graphs below slice the graph above along the x and y axes. The result is four "utility of x" curves and four "utility of y" curves. This presentation emphasizes the two relationships noted immediately above. We generate two scenarios, UX and UY using the gr2d command. Then wxdraw produces the graphs.

```
UX: gr2d( user_preamble = "set key top left",xlabel=
"Good x", ylabel = utility",xtics=20, ytics=10,key=
"y=25",explicit(u(x,25,a0,b0),x,0.001,xmax),
 line_width=2,key="y = 15",explicit(u(x,15,a0,b0),
 x,0.001,xmax),line_width=1,color=gray40,
 key="y = 10",explicit(u(x,10,a0,b0),x,0.001,xmax ),
 line_width=2,color=grey40,key = "y = 5",
  explicit(u(x, 5, a0, b0),x,0.001,xmax ) )$
```

The UY is essentially identical to UX, so only the first and last lines of the input appear below.

[4]Note the axis rotation. We executed this input group with draw3d replacing wxdraw3d. This command produces a graph outside *wxMaxima* that can be rotated. We made note of the rotation and then applied the selected rotation to wxdraw3d. The range over which the explicit functions are graphed begins at $x = 0.01$ and $y = 0.01$, rather than zero, to avoid division by zero.

```
UY: gr2d( user_preamble = "set key top left",xlabel
    ... Some commands are deleted ...
explicit(u(5,y, a0, b0), y, 0.001, ymax))$
```

The two cross-sections of the utility function appear below. The different shapes reflect the asymmetry given that $a = \frac{1}{4}$.

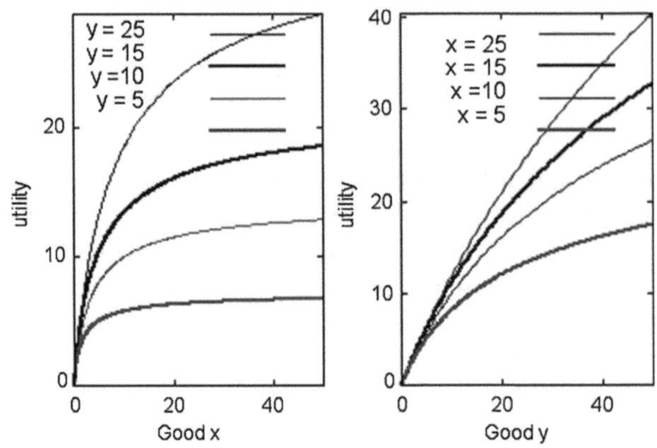

3.1.2 *Marginal Utility*

To consider the slopes (the rates at which utility changes per unit change in either x or y, holding the other constant) more closely, we determine the expressions for the *marginal utility* of x and of y. Mathematically, these are the first partial derivatives of the utility functions with respect to these variables.[5]

```
matrix( ["mux","muy"],
     [diff(u(x,y,a,b),x),diff(u(x,y,a,b),y)]);
```

$$\begin{bmatrix} \text{mux} & \text{muy} \\ a\,x^{-b-1}\left(\frac{1-a}{y^b}+\frac{a}{x^b}\right)^{-\frac{1}{b}-1} & (1-a)\;y^{-b-1}\left(\frac{1-a}{y^b}+\frac{a}{x^b}\right)^{-\frac{1}{b}-1} \end{bmatrix}$$

The term inside the inner parentheses in the expressions for the derivatives is the same as the term inside the expression for the utility function. The entire expression for the derivative can be rephrased as

$$\frac{\partial u}{\partial x} = a \cdot x^{-(1+b)} \cdot [\ldots]^{\frac{1+b}{-b}}$$

[5]Recall that *Maxima* uses the command `diff` for both derivatives and partial derivatives.

This expression can be restated as follows:

$$\frac{\partial u}{\partial x} = a \cdot \left(\frac{u}{x}\right)^{(1+b)}.$$

In a similar fashion, we can determine that

$$\frac{\partial u}{\partial y} = (1-a) \cdot \left(\frac{u}{y}\right)^{(1+b)}.$$

We write these two expressions as functions of the ratios $\frac{u}{x}$ and $\frac{u}{y}$, respectively.

```
mux(x,y,a,b):=a*((u(x,y,a,b))/x)^(1+b);
muy(x,y,a,b):=(1-a)*(u(x,y,a,b)/y)^(1+b);
```

$$\mathrm{mux}\,(x, y, a, b) := a \left(\frac{\mathrm{u}(x,y,a,b)}{x}\right)^{1+b}$$

$$\mathrm{muy}\,(x, y, a, b) := (1-a) \left(\frac{\mathrm{u}(x,y,a,b)}{y}\right)^{1+b}$$

The input group below produces a list that contains two items. The first is the *mux* function for a specified value of y, given the parameter values established earlier. The second item is the value of the function for $x = 15.0$. Note the use of decimals to force *Maxima* to return a decimal representation of the value. Placing the second expression inside the input list in a `float( )` command would have yielded the same result as writing the arguments with decimal numbers. The first item in the list shows that the *mux* is inversely related to the level of x.

```
matrix( ["mux expression", "mux value"],
  [mux(x,15,a0,b0), float(mux(10,15,a0,b0))] );
```

$$\begin{bmatrix} \text{mux expression} & \text{mux value} \\ \dfrac{1}{4\left(\frac{1}{4x^{\frac{4}{5}}}+\frac{3}{4\,15^{\frac{4}{5}}}\right)^{\frac{9}{4}} x^{\frac{9}{5}}} & 0.422 \end{bmatrix}$$

The graph below shows two *mux* curves. Both are downward sloping functions of x and the *mux* curve with $y = 25$ is higher than its counterpart with $y = 5$. Both results are as expected, given previous discussion.

```
wxdraw2d(yrange=[0,2],xlabel="x",ylabel="mux",key=
 "mux when y = 5", explicit(mux(x,5,a0,b0),x,
 0.001,50),line_width=2,key="mux when y = 25",
  explicit(mux(x,25,a0,b0),x,0.001,50))$
```

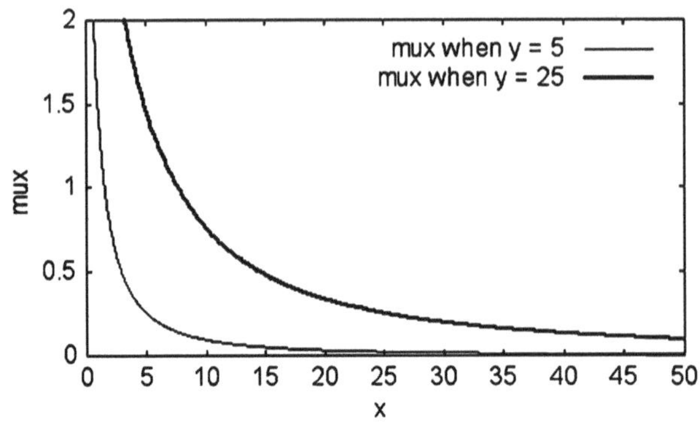

3.1.3 Cobb–Douglas and CES Functions

Maxima can be used to show that the Cobb–Douglas function $u = x^a \cdot y^{1-a}$ is a limiting case of the CES function: as $b \to 0$ (or, equivalently, as the elasticity of substitution approaches 1), the CES function approaches the Cobb–Douglas function. Taking the limit of the CES function as b approaches zero yields the first output in the cell below. The `radcan` (`rad`ical `can`onical) command converts this expression to the more standard form, making it clearly recognizable as the Cobb–Douglas function.

```
[limit(u(x,y,a,b),b,0),
 radcan(limit(u(x,y,a,b),
 b,0))];
```

$$[e^{a\log(x)-(a-1)\log(y)}, \quad x^a\, y^{1-a}]$$

3.2 Indifference Curves

This section develops the concept of the indifference curve. In our two-good model, an indifference curve consists of all combinations of the two goods that result in the same level of utility for the consumer. Thus, the consumer, offered any two combinations that lie on one of these iso-utility curves, would be indifferent between the two.

3.2.1 Graphical Representation

The graphs below show the same utility function representation as above. Here, the focus is on three contour lines (also called level curves or level sets) that collect points of equal height on the utility surface (equal levels of utility). The first graph shows these three contour lines on the utility surface.

The second graph shows the results of projecting these contour lines onto the base of the graph. Thus, the base in this graph contains a two-dimensional (x and y) representation of the three-dimensional utility surface.

```
G1:gr3d(color=gray50,view=[50,300],dimensions=[600,
 300],title="Contour Lines on Surface",xlabel="x",
 ylabel="y",zlabel="u",contour=surface,
 contour_levels={.4*umax,.6*umax,.8*umax},
 ztics={0,umax},xtics={0,xmax},ytics={0,ymax},
 explicit(u(x,y,a0,b0),x,.01,xmax,y,.01,ymax))$
G2:gr3d(color=gray50,view=[50,300],
 ... Some commands deleted ...
  explicit(u(x,y,a0,b0),x,0.001,xmax,y,0.001,ymax))$
wxdraw(G1,G2,columns=2)$
```

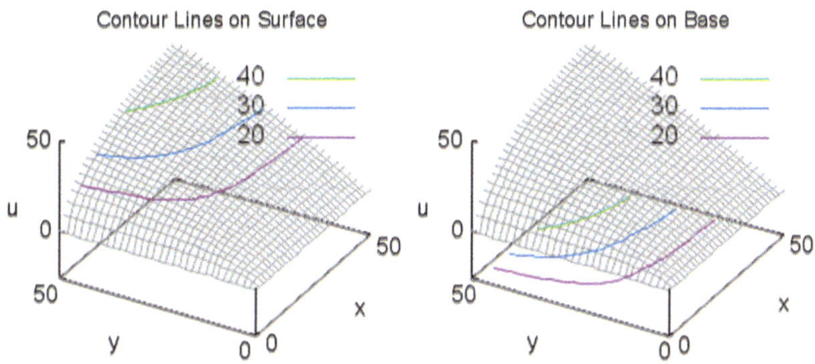

3.2.2 *Defining Indifference Curves*

A functional expression for indifference curves is useful. To derive such an expression, we use *Maxima's* `solve` command. We set the utility function equal to a value $u0$ and solve for y. Doing so requires answering a pair of questions posed by *Maxima*. The answer to the first question is that the term $1/b$ is not an integer. The answer to the second question is, "negative." This result can be derived from the utility function, given the condition that $y > 0$.[6] Its importance and implication appears in the solution below. The `declare` and `assume` commands pass information to *Maxima* that reduces the number of questions that must be

[6]The utility function is $u = (a \cdot x^{-b} + (1-a) \cdot y^{-b})^{-1/b}$. Therefore, $u^{-b} = a \cdot x^{-b} + (1-a) \cdot y^{-b}$, or $u^{-b} - a \cdot x^{-b} = (1-a) \cdot y^{-b}$. Thus, a positive value of y implies that the term on the left-hand side is positive. That is, $1/u^b > a/x^b$, and so $x^b > a \cdot u^b$. We use a placeholder name `u0` for utility because the name `u` has been assigned to the specified utility function.

answered before it generates a solution. Questions can be answered by simply typing the answer, followed by a semicolon, and hitting shift-enter.

```
assume(u0>0,a>0,a<1)$  declare(b,noninteger)$
soln: ratsimp(solve(u(x,y,a,b)=u0,y));
```

Is $\frac{1}{b}$ an integer? n; Is $x\ \left(a\, u0^b - x^b\right)$ positive, negative, or zero? n;

$$[y = (1-a)^{\frac{1}{b}}\, u0 \left(\frac{x^b}{x^b - a\, u0^b}\right)^{\frac{1}{b}}]$$

The next cell simplifies the right-hand side of the expression above and assigns the result to the function $ic(x, a, b, u0)$, using the "quote-quote" operator. We use this function to draw three representative indifference curves below. The dependent variable is y.

```
ic(x,a,b,u0):=
''(rhs(soln[1]));
```

$$\mathrm{ic}\,(x, a, b, u0) := (1-a)^{\frac{1}{b}}\, u0 \left(\frac{x^b}{x^b - a\, u0^b}\right)^{\frac{1}{b}}$$

Inspection of the ic function reveals that it yields an economically sensible value $(y > 0)$ only when the denominator is positive. We define an $xmin$ function to determine the value of x required for the denominator to be nonnegative. The function is $xmin \geq a^{1/b} \cdot u$. We solve for the equality and then impose inequality in an ad hoc fashion as required for graphing.

```
xmin(a,b,u):=a^(1/b)*u;  [u1,u2,u3]:[20,30,40]$
[xmin1,xmin2,xmin3]:float([xmin(a0,b0,u1),
xmin(a0, b0, u2), xmin(a0, b0, u3)])$
matrix([utility level",u1,u2,u3],
    ["minimum x value", xmin1,xmin2,xmin3]);
```

$$\mathrm{xmin}\,(a, b, u) := a^{\frac{1}{b}}\, u$$

$$\begin{bmatrix} \text{utility level} & 20 & 30 & 40 \\ \text{minimum x value} & 3.5355 & 5.3033 & 7.0711 \end{bmatrix}$$

The graph below shows three utility curves for a consumer. A variable ymax_ic is used to truncate the indifference curves at a y value large enough to show the curvature in all three indifference curves but not so large as to make the graph hard to read. All three curves bend rather sharply, becoming nearly vertical as x falls toward the minimum values defined above and nearly horizontal as x becomes roughly three times this value. This curvature reflects the fact that we have selected a b0 that results in a rather small elasticity of substitution $s = \frac{3}{4}$. We consider cases involving closer substitutes below.

```
ymax_ic:float(ic(1.1*xmin3,a0,b0,u3))$
wxdraw2d(title = "Three Indifference Curves",xlabel=
 "x",ylabel="y",xrange=[0,2*xmin3],yrange=
 [0,ymax_ic],key ="u = u1",explicit(ic(x,a0,b0,u1),
 x,xmin1+0.1,2*xmin3),
 ... Some commands deleted... )$
```

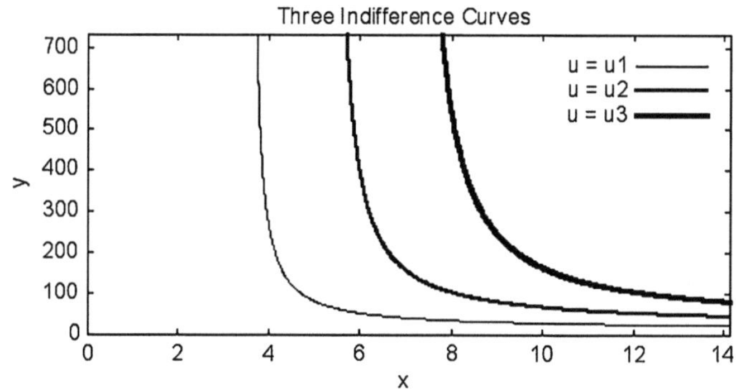

3.2.3 *Marginal Rates of Substitution*

The graph above shows that the slopes of the indifference curves change dramatically over the range observed. This curvature change says that as the amount of x available changes, so does the *change* in the amount of y required to keep the consumer at the same utility level. The change in y required to compensate for a one-unit change in x is called the *marginal rate of substitution* between x and y.

We examine this concept in some detail. First, consider discrete changes in the amount of x consumed. The input group below creates a set of lists. The first pair of lists, `xlist1` and `ylist1` creates the values of x and y at seven points on the indifference curve `u = u1` above.[7] These values are not printed. The next pair contains similar values, with x incremented by one unit. These values are printed. We then create a list of changes in y per one-unit change in x. Then we use *Maxima*'s `append` command to add titles and to position the items in a way that appears in the table. The `matrix` and `transpose` commands convert the lists into more easily read tables. The `append` command refers to lists, so even if a single item is added, it must be entered as a list.

The table shows that increasing x from 4.5355 units to 5.5355 units and reducing y by 56.002 units leaves the consumer at the same utility level (on the same indifference curve). In contrast, increasing x from 10.536 units to 11.536 units compensates for a reduction of only 1.6275 units of y. Thus, y is becoming more "dear" relative to x. We define "dear" more precisely in a moment.[8]

---

[7]The x list is entered directly into the function `ic` and so the result is a list.

[8]As an exercise, compute the last six items in the `y change` column based on the printed values of `ylist2` (the list immediately above the table). Your answers will differ from the tabled value due to rounding. Use *Maxima*, `y[i] - y[i-1]`.

```
xlist1:makelist(xmin1+i,i,1,7)$
ylist1:float(ic(xlist1,a0,b0,u1))$
xlist2:makelist(xmin1+i, i, 2, 8)$
ylist2:float(ic(xlist2, a0, b0, u1) )$
ychangelist:ylist2-ylist1$   ychangelist:
  append(["y change", "---"],ychangelist)$
xlist:cons(xmin1+1,xlist2)$
ylist:cons(ylist1[1],ylist2)$  transpose(matrix(
cons("x",xlist),cons("y",ylist),ychangelist));
```

x	y	y change
4.5355	118.51	− − −
5.5355	62.51	−56.002
6.5355	45.541	−16.97
7.5355	37.442	−8.0988
8.5355	32.714	−4.7279
9.5355	29.617	−3.0973
10.536	27.43	−2.1869
11.536	25.802	−1.6275

The table above refers to discrete changes. We can also determine the slope of the indifference curves at points. The input group below defines the slope as a function of x, the parameters a, b, and the specified utility level $u1$.[9]

Comparing the slopes at the eight values of x with the seven discrete change values above shows that the discrete change value listed next to an x value is between the slope at that value of x and the slope at $x - 1$.[10] Again, the lesson is the same: As x increases, the marginal rate of substitution of x for y decreases—one unit increases in x compensate for increasingly small decreases in y.

We name the negative of this slope the *marginal rate of substitution* or *mrs*. It is a measure of relative "dearness" as indicated by the consumer's willingness to accept more of one good in return for a reduction in the quantity of the other.

```
icslope(x,a,b,u):=''(
diff(ic(x,a,b,u),x))$
slopelist:float(icslope(
 xlist,a0,b0,u1))$ transpose(
  matrix(cons("x",xlist),
  cons("mrs",slopelist)));
```

x	mrs
4.5355	118.5
5.5355	26.176
6.5355	10.977
7.5355	5.9719
8.5355	3.7427
9.5355	2.5634
10.536	1.8659
11.536	1.4197

[9]To save space, the function is not printed.

[10]This reflects the fact that the value 56.002 above refers to a change over the range $x = 4.5355$ to $x = 5.5355$, and so forth.

The marginal rate of substitution can be related directly to the marginal utility functions defined above: the *mrs* equals the ratio $\frac{mux}{muy}$. Values of this ratio reappear as the final row of output below.

The value of mux at any (x, y) pair is the number of units (whatever the units are) that the consumer's utility level changes per one-unit change in x, with muy conveying the same information for y. If mux is, say, twice as large as muy, then two units of y are required to substitute for a unit of x. Apply this observation to the values below.

For this consumer, the bundle $x = 5.5355$ and $y = 62.51$ yields $mux = 2.5241$ and $muy = 0.0964$, so 26.184 units of y substitute for a single unit of x (this value differs from the tabled value, 26.176, due to rounding). When $x = 10.536$ and $y = 27.43$, the consumer enjoys the same utility level. Now, however, mux has fallen to 0.793 and muy has risen to 0.425, so the ratio of these two values implies that just under two units of y can substitute for a unit of x. Relative to their values at the initial point, x has become less dear and y more so.

```
muxlist:mux(xlist,ylist,a0,b0)$  muylist:
 muy(xlist, ylist, a0, b0)$ mrslist:muxlist/muylist$
names: matrix(["x", "y","mux","muy","mrs"])$
numbers: matrix(xlist,ylist,muxlist,muylist,mrslist)$
addrow(names, numbers);
```

x	y	mux	muy	mrs
4.5355	118.51	3.613	0.0305	118.5
5.5355	62.51	2.5241	0.0964	26.176
6.5355	45.541	1.8719	0.171	10.977
7.5355	37.442	1.4487	0.243	5.9719
8.5355	32.714	1.1577	0.309	3.7427
9.5355	29.617	0.948	0.37	2.5634
10.536	27.43	0.793	0.425	1.8659
11.536	25.802	0.673	0.474	1.4197

The cell below confirms that the elasticity of substitution is $s = \frac{1}{1+b}$. The elasticity of substitution is defined as

$$s = \frac{d(y/x)}{d(mrs)} \cdot \frac{mrs}{x/y},$$

the percentage change in the ratio of y to x that is required to effect a 1 % change in mrs.[11]

[11]Chapter 4 shows that the mrs equals the price ratio for the two goods and, therefore, does not depend on the consumer's choice. In this setting s is the percentage change in $\frac{y}{x}$ per 1 % change in the price ratio.

```
mrs:diff(ic(x,a,b,u),x)$
dyx:diff(ic(x,a,b,u)/x,x)$
dmrs:diff(mrs,x)$
 s: ratsimp((dyx/dmrs)*(mrs/
   (ic(x,a,b,u)/x)))$
matrix(["s =", s]);
```

$$\left[s = \frac{1}{b+1}\right]$$

3.2.4 Limiting Cases

The CES utility function can be used to approximate a wide range of utility functions. The graph below shows three important cases. In the first case $s \approx 1$, in which case the CES function is approximately the same as the Cobb–Douglas function $u = x^a \cdot y^{1-a}$, which is defined by a single parameter, a. In this case a move along an indifference curve that results in a 1 % change in the mrs also results in a 1 % change in $\frac{y}{x}$. An indifference curve for this function has the form

$$y = \left(\frac{u}{x^a}\right)^{1/(1-a)}.$$

In the second case s is quite small. The CES approaches a fixed-ratio function.

The graph below shows the three cases that are discussed above. The graph could also be produced, and with fewer lines of input, by using applying `implicit` to the following expressions: `u(x,y,a0,0.01)-u1`, `u(x,y,a0,-0.99)-u1`, and `u(x,y,a0,49)-u1`.

```
b4:99$  xmin4:float(xmin(a0,b4,u1))$  ymax_ic:float(
 ic(1,a0,-0.99,u1))$ wxdraw2d( ... Some commands
 omitted ... title="Indifference Curves for Three
 Consumers",key="s near 1",explicit(ic(x,a0,.01,u1),
 x,0.01,60), ... key="Small s",explicit(ic(x,a0,
 b4,u1),x,1.000000001*xmin4,60), ... key="Large s",
 explicit(ic(x,a0,-0.99,u1),x,0.01,60))$
```

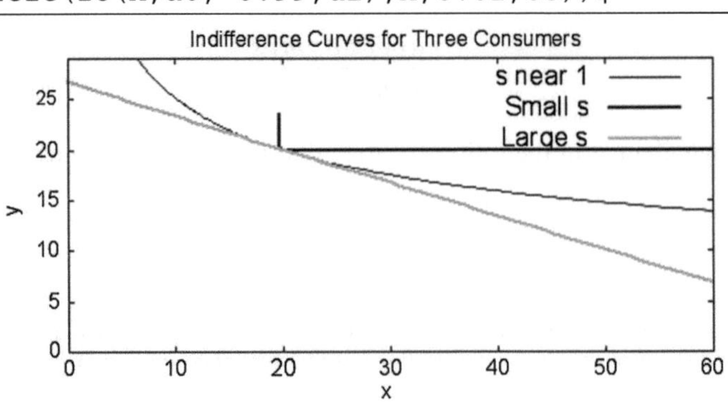

If b grows large enough, the elasticity of substitution approaches 0. The output below shows that as b approaches infinity, the minimum value of x approaches $x = u$. The isoquant would be L-shaped at $(x = u, y = u)$. Here the u values are in units of x and y, respectively, not in utility units. We require u units of utility, and each unit of utility requires one unit of x, so the amount of x required is

$$x = \frac{uo \text{ units of utility}}{1 \text{ unit of utility per unit of x}} = uo \text{ units of x.}$$

Likewise for y.

```
limit(xmin(a, b, u), b, inf);
```

u

To confirm how little substitution is possible given a sufficiently large b, we set $b = 149$ ($s = 1/150$). We determine the value of u for $x = y = 20$, which is $u = 20$. Then we reduce x by one half of a unit and increase y 1,000-fold. Then we reverse the position of x and y. In both cases, the 1,000-fold increase is insufficient to offset the (approximately 2.5 %) decrease. In both cases, the consumer is placed on a lower indifference curve.[12] The reduction in u is slightly less with the large increase in y. This reflects y's relatively large coefficient, $\frac{3}{4}$, *vs.* x's coefficient, $a = \frac{1}{4}$.

```
u1:u(20,20,a0,149)$ u2:float(u(19.5,20000,a0,249))$
u3:float(u(20000,19.5,a0,249))$ matrix(["u,corner",
 "u, much y", "u, much x"], [u1, u2, u3]);
```

$$\begin{bmatrix} \text{u,corner} & \text{u, much y} & \text{u, much x} \\ 20 & 19.61 & 19.522 \end{bmatrix}$$

3.3 "Bliss" and Negative Marginal Utility

As noted earlier, one characteristic of the CES utility function is that both goods (all goods, if more than two are included) have positive marginal utility levels no matter how much of the goods one consumes. This section considers an alternative illustration, one in which the marginal utility for x, y, or both can be negative. The function $w(x, y, a, bc, d, e)$ is defined here.

```
w(x,y,a,b,c,d,e):=a*x+b*x^2+c*y+d*y^2+e*sqrt(x*y^2);
```

$$\mathrm{w}(x, y, a, b, c, d, e) := a\,x + b\,x^2 + c\,y + d\,y^2 + e\,\sqrt{x\,y^2}$$

The next input group specifies a set of parameter values for this function and evaluates the function for two x, y pairs. For the first, the measure of utility, w is positive; for the second, it is negative.

[12]The value $b = 149$, implying $s = 0.004$, is used because larger values can result in an overflow error.

```
assume(x>0, y>0)$
[a1,b1,c1,d1,e1]:[10.0,-0.5,25.0,-0.8,5.0]$
w(x,y,a1,b1,c1,d1,e1);
w10:w(10.0,20.0,a1,b1,c1,d1,e1)
$w100: w(100, 20,a1,b1,c1,d1,e1)$
matrix(["w, x = 10", "w, x = 100"],[w10,w100]);
```

$$-0.8\,y^2+5.0\,\sqrt{x}\,y+25.0\,y-0.5\,x^2+10.0\,x$$

$$\begin{bmatrix} \text{w, x = 10} & \text{w, x = 100} \\ 546.23 & -2820.0 \end{bmatrix}$$

We should not be concerned about negative values of the utility function w. This does not mean that the consumer is suffering. We could add 5000 or any other number to the w function, making wxmaxw100 positive, without changing any of the analysis below.

The next step is to define the two marginal utility functions, named mwx and mwy.

```
mwx(x,y,a,b,c,d,e):= ''(diff(w(x,y,a,b,c,d,e),x));
mwy(x,y,a,b,c,d,e):= ''(diff(w(x,y,a,b,c,d,e),y));
```

$$\text{mwx}\,(x,y,a,b,c,d,e) := \frac{e\,y}{2\,\sqrt{x}}+2\,b\,x+a$$
$$\text{mwy}\,(x,y,a,b,c,d,e) := 2\,d\;y+e\,\sqrt{x}+c$$

We use the command mnewton to find an approximation to the solution of the two equations derived above. At this solution, the marginal utilities of both x and y are zero. The consumer has reached a "bliss point."

```
load("mnewton")$
soln:mnewton([mwx(x,y,a1,b1,c1,d1,e1)=0,
mwy(x,y,a1,b1,c1,d1,e1)=0],[x,y],[10,10]);
```

$$[[x=25.542,\quad y=31.418]]$$

Finally, determine the level of utility for the values of x and y that are determined above: when $mwx = mwy = 0$, $w = 718.92$. This value is computed for use in the next graph.

```
[[x0,y0]:[rhs(soln[1][1]), rhs(soln[1][2])],
   wmax:w(x0,y0,a1,b1,c1,d1,e1)];
```

$$[[25.542,\ \ 31.418],\quad 718.92]$$

Consider the function in three dimensions. The graph below shows the surface as a hill. The "far side" of the hill is not shown, to avoid clutter (the command `surface_hide = true` accomplishes this). The + sign shows the apex of the surface, where $w = 718.92$. The contour lines are selected so that two of them show indifference curves for the levels and the third shows an indifference curve that appears as a point (the "bliss point") on the graph.

```
wxdraw3d( zrange=[0,1.1*wmax],contour=both,view=
 [70,330],color=gray,dimensions=[600,500],xlabel=
 "Good x",ylabel="Good y",surface_hide=true,xtics=
 {0,2*x0},ytics={0,2*y0},contour_levels=
 {0.6*wmax,0.8*wmax,0.999*wmax},
 explicit(w(x,y,a1,b1,c1,d1,e1),x,0,2*x0,y,0,2*y0),
 point_size=2,color=black,points([[x0,y0,wmax]]))$
```

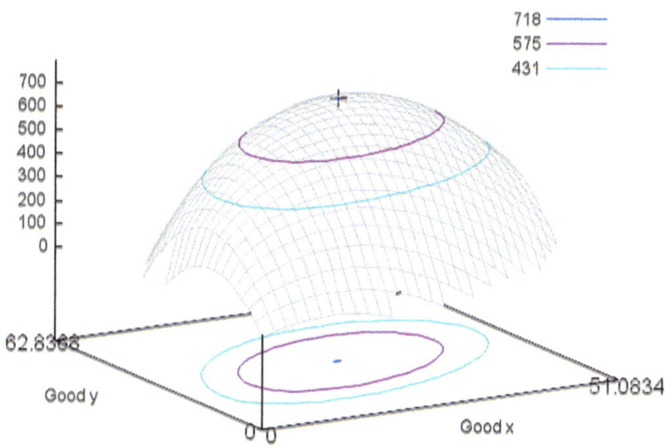

3.3.1 *Indifference Curves and the Bliss Point*

The code in the next input/output cell rephrases the information on the base of the graph above as a set of three indifference curves and a point (we add one indifference curve.) We divide the map of indifference curves into four sectors by graphing the x, y values at which $mwx = 0$ and at which $mwy = 0$. That is, we plot the lines that define where, at the margin, x and y change from goods to "bads."

All indifference curves tell essentially the same story. In the southwest sector, labeled `mux > 0 and muy > 0`, both x and y are goods. That is, in this range, taking units of one good from the consumer reduces the utility level unless more of the other good is given to the consumer. In the southeast and northwest quadrants, either x or y is, at the margin, a "bad." To the right of the curve `mux = 0`, for example, if the consumer is forced to consume more of x, the utility level can be maintained only by giving the consumer more of y, which is still a good.[13]

Once the consumer is moved beyond the "bliss point," the point at which $mux = muy = 0$ (also called a "satiation point"), into the area labeled `mux < 0 and`

[13]Be aware of the "at the margin" modifier here. The consumer would rather have 30 units of x and 10 units of y than no x and 20 units of y, so x does provide utility in total, but not at the margin.

`muy < 0`, both x and y become "bads" at the margin. The consumer would prefer $x0, y0$ to any point that represents more of either x or y, or both. If, at any point in this sector, we force the consumer to take more of either good, then keeping the consumer on the same indifference curve requires *taking away* some of the other good.

The graph below contains a relatively complex set of instructions. We review those that are relatively new. First, labels have been added. Each label contains a string of text, an x value, and a y value. These values are selected to place the text so that it is easily read. We use the `implicit` option for all of the graphs except to locate the "bliss point." *Maxima* calls on *gnuplot* to draw the graphs using a grid search procedure. The default search grid uses an `ip_grid = [50, 50]` to accomplish this plotting, but the resulting graph is not very smooth. We change the grid search parameter to `ip_grid = [100, 100]`.

```
wxdraw2d(user_preamble="set key bottom right",
 dimensions=[600,500],ip_grid=[100,100],yrange=
 [0,2*y0],xlabel="Good x",ylabel ="Good y",
 title= "Indifference Curves and Bliss Point",label(
 ["mwx>0 and mwy>0",x0/3,y0/2]),
 label(["mwx<0 and mwy>0",x0,y0/4]),
 label(["mwx<0mwy<0",2*x0,1.75*y0]),
 label(["mwx>0 and mwy<0",x0/2,1.75*y0]),
 key="Lowest w",line_width=2,color=black,
 implicit( w(x,y,a1,b1,c1,d1,e1)=.6*wmax,x,0,2.5*x0,y,
 0,2.5*y0),   ... Some commands deleted ...
 color=black,key="Bliss point",point_type=13,
 point_size=2,points([[x0,y0]])  )$
```

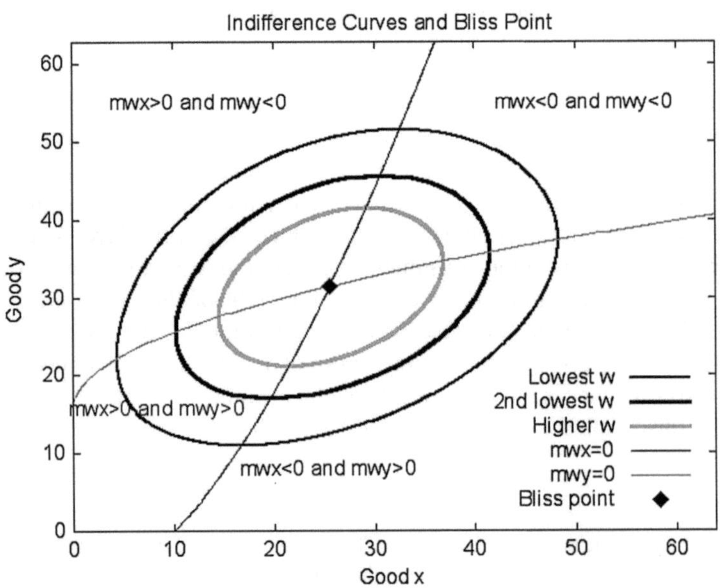

Given that a "bliss point" cannot be ruled out theoretically, we might question using the CES illustration, which allows for no such case. If we consider the consumer to be a free agent, however, we have warrant for such a functional form. No consumer would willingly move outside the southwest sector even if both goods were free. If both goods were free, the consumer would consume at the "bliss point," on the highest indifference curve (which degenerates to a point). If the consumer must pay for the goods and has a budget that does not include the "bliss point," then only the southwest sector matters. Henceforth, we analyze the case in which both x and y are goods (both marginal utility values are positive). That is, we need consider only cases like the southwest sector.

References

Each of the two references in this chapter is an important part of empirical research in economics. Also, the models developed in these articles are frequently used to illustrate economic principles, as they are used in this chapter.

1. Arrow KJ, Chenery HB, Minhas BS, Solow RM (1961) Capital-labor substitution and economic efficiency. Rev Econ Stat 43:225–250
2. Cobb CW, Douglas PH (1928) A theory of production. Am Econ Rev 18(Suppl.):139–165

Chapter 4
Demand Theory: Constraints and Optimization

Abstract This chapter introduces the budget constraint and combines the budget constraint with the preferences developed in the preceding chapter to provide a model of rational consumer choice.

4.1 Budgets and Budget Constraints

Suppose that a consumer purchases two products and has a fixed per-period money income. Also, assume that the quantities that the consumer selects do not affect the prices paid. Then the consumer's budget constraint is $px \cdot x + py \cdot y = m$, where x and y are in physical units per time period, px and py are in dollars per unit of x and y, respectively, and m is in dollars per time period.

```
budget(x,y,px,py,m):=px*x+py*y=m$
```

This budget line appears below. Note the use of `implicit` (and its syntax) rather than `explicit` as the type of function to graph. We restate the budget line as an explicit function below.

```
[px0,py0,m0]:[50,12,1200]$ wxdraw2d(xlabel="x",ylabel=
"y",key="Initial Budget Line",implicit(budget(
  x,y,px0,py0,m0),x,0,1.1*m0/px0,y,0,1.1*m0/py0))$
```

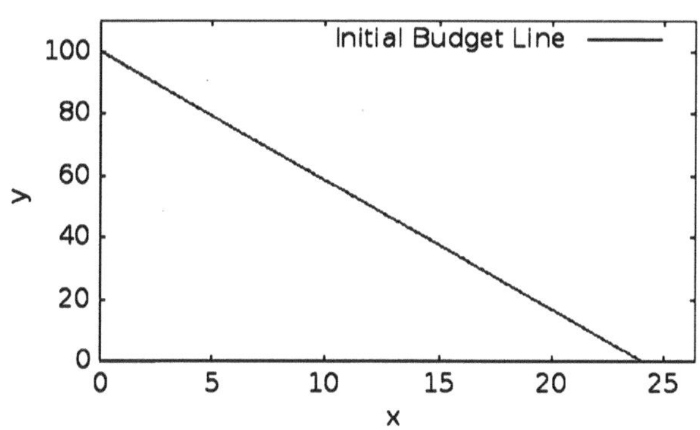

M.R. Hammock and J.W. Mixon, *Microeconomic Theory and Computation: Applying the Maxima Open-Source Computer Algebra System*,
DOI 10.1007/978-1-4614-9417-1_4,

Our analysis focuses on x, so we rephrase the budget line with y expressed as an explicit function of x. In the next cell, *Maxima*'s `solve` command returns the desired equation. This equation may be rewritten as $y = \frac{m}{py} - \frac{px}{py} \cdot x$.

4.1.1 Nominal vs. Relative Values

This statement of the expression reveals that real money income and relative prices matter, not nominal values. That is, revising m, px, and py in proportion leaves the budget line unaffected. The budget line is homogeneous, degree zero in px, py, and m.

For much of the analysis that follows, expressing the equation as a function with x, m, px, and py as its arguments is useful.

```
BL(x,m,px,py):=''(rhs(
solve(budget(x,y,px,py,m),
   y)[1]));
```

$$\mathrm{BL}(x, m, px, py) := -\frac{px\,x-m}{py}$$

The next cell confirms that the budget line is homogeneous, degree zero. Values of px, py, and m are changed by the proportion `k`. Exercise: Plot the budget line twice, first with the set of values for these three variables that is used above and then with a new set of values, each of which is the same multiple of the initial values.[1]

```
solve(budget(x,y,k*px,k*py,k*m),y);
```

$$[y = -\frac{px\,x-m}{py}]$$

4.1.2 Budget Lines with Rationing and with Subsidies

Considerations other than those included in the budget line above can affect the set of commodities from which the consumer selects. A consumer might receive quantity discounts for purchases above a certain level, might pay a tax on purchases above a specified level, might receive subsidies for units up to a specified level, or might be prohibited from purchasing more than a specified level. We consider the last two and leave the other two as exercises.

First, consider a quota. The budget line now consists of a line that is a segment of the initial budget line, with the consumer forbidden from selecting units beyond the specified quota, `x1` $= 15$ in this illustration.

---

[1]A minimal set of commands follows:

```
[px1,py1,m1]:[100,24,2400]$
first:gr2d(explicit(BL(x,m0,px0,py0),x,0,m0/px0))$
second:explicit(BL(x,m1,px1,py1),x,0,m1/px1))$
wxdraw(first,second,columns=2)$.
```

```
x1:15$  xmax:1.1*m0/px0$
wxdraw2d( ... Some commands omitted ...
explicit(BL(x,m0,px0,py0 ),x,0,m0/px0 ), ...
explicit(BL(x,m0,px0,py0),x,0,x1 ) )$
```

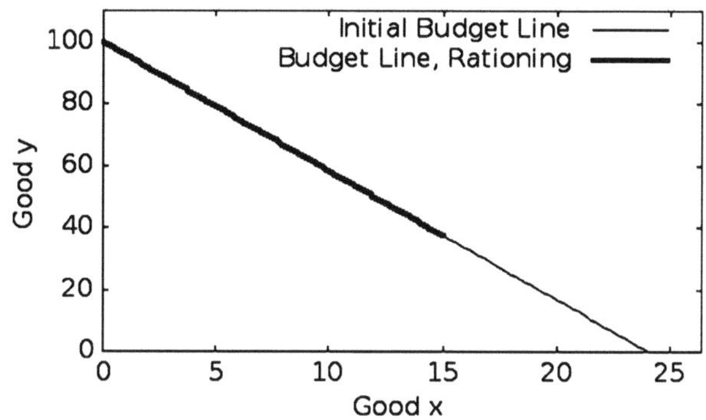

An ad valorem subsidy can be offered to all or part of the amount of a good that a consumer buys. Suppose that the first 15 units are subject to 40 % subsidy. The effect for the first 15 units is to reduce the budget line's slope from $\frac{px}{py}$ to $0.60 \cdot \frac{px}{py}$ After the 15th unit the relative price returns to $\frac{px}{py}$, and the new budget line is to the right of the initial one by the dollar amount of the subsidy, $px0 \cdot s \cdot x1$, divided by $px0$.

```
s:0.40$ xmax:(m0+(px0*s*x1))/px0$
wxdraw2d(... Some commands omitted ... explicit(
 BL(x,m0,px0,py0),x,0,xmax),line_width=3,key=concat(
 "New Budget Line Segment, x < ",string(x1)),
 explicit(BL(x,m0,px0*(1-s),py0),x,0,x1), ...
 explicit(BL(x,m0+s*px0*x1,px0,py0),x,x1,xmax))$
```

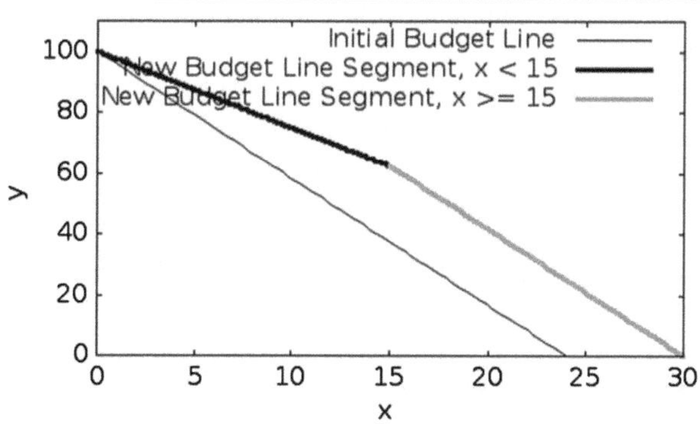

Often subsidies are provided in fixed quantities. The Supplemental Nutrition Assistance Program (a.k.a. food stamps) is an example in which the consumer can receive a fixed amount of food per period without reducing consumption of other goods. The budget line below reflects this sort of program. The consumer may select to consume all or part of the fixed amount. Beyond that amount, the consumer may purchase more at the market price. Thus, the new budget line consists of horizontal segment and a segment that is parallel to the initial budget line but to the right of that line by the number of units provided as a subsidy.

```
FS:5$ xmax:FS+m0/px0$  wxdraw2d( ... Some commands
 omitted ... explicit(BL(x,m0,px0,py0),
  x,0,m0/px0),...key="Fixed Subsidy Budget Line",
  points([[0,m0/py0],[FS,m0/py0]]),key="",
  points([[FS,m0/py0],[xmax,0]]))$
```

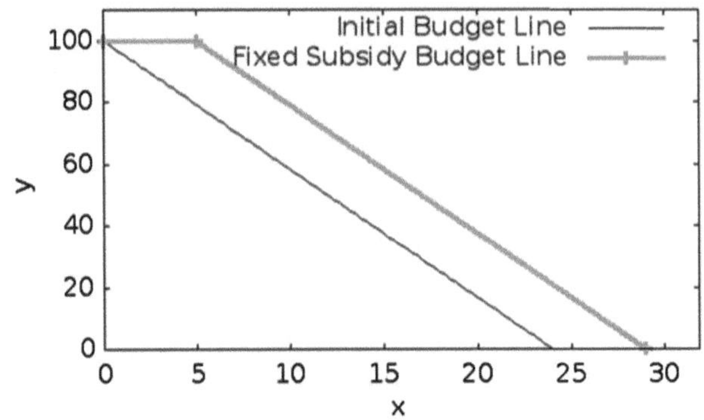

4.2 Indifference Curves and Choice

This section presents a model of consumer choice. Budget lines combine with the indifference curves to provide the basis for a model of rational behavior. Indifference curves and budget lines are not inherently related. The former reflect preferences and cannot be viewed directly. The latter relate to data beyond the consumer's control and can be viewed directly.

The act of rational choice connects the two. We begin with two input cells that bring forward previously developed material. The first shows important functional relationships: the utility function, the family of indifference curves implied by that function, and the family of budget lines.

```
u(x,y,a,b):=(a*x^-b+(1-a)*y^-b)^-(1/b)$
mux(x,y,a,b)  :=a*((u(x, y, a, b))/x)^(1+b)$
muy(x,y,a,b)  :=(1-a)*(u(x,y,a,b)/y)^(1+b)$
ic(x,a,b,u0)  :=u0*((1-a)*x^b/(x^b-a*u0^b))^(1/b)$
BL(x,m,px,py):=m/py-(px/py)*x$
```

The next input group recalls the parameter values used previously. It also shows the minimum value of x for any given utility level and three specific values of x's minimum value. The `matrix` command is used to generate the table and is not part of the analysis. The utility levels are 20, 30, and 40.

```
[a0,b0]:[1/3,4/5]$ [px0,py0,m0]:[50,12,1200]$
xmin(a,b,u0):=a^(1/b)*u0$ [u01,u02,u03]:[20,30,40]$
[xmin1,xmin2,xmin3]: float([xmin(a0,b0,u01),
xmin(a0,b0,u02),xmin(a0,b0,u03)])$
matrix(["xmin","xmin_u01","xmin_u02","xmin_u03"],
        [xmin(a,b,u0), xmin1,xmin2,xmin3]);
```

$$\begin{bmatrix} \text{xmin} & \text{xmin\_u01} & \text{xmin\_u02} & \text{xmin\_u03} \\ a^{\frac{1}{b}}\, u0 & 5.0656 & 7.5984 & 10.131 \end{bmatrix}$$

To develop a graphical view of the optimization process, we take the three utility levels specified above and the implied indifference curve functions as given and determine the minimum income level required to place the consumer on each of these three utility levels. To achieve this, we develop the necessary condition for achieving the highest utility level for a given income level or, conversely, of finding the lowest income consistent with a specified utility level.

4.2.1 *Optimization: Choosing the Best Combination*

The condition for optimization is this: The consumer must allocate money income in such a way that the marginal utility per dollar is the same for all goods. Suppose that this condition does not hold. Suppose, for example, that $mux = 15$, $px = 2$, $muy = 10$, and $py = 1$. (The marginal utility units cannot be specified, though they are real to the consumer. The price units are \$ per unit of the good.) In this case, $\frac{mux}{px} = 7.5$ per dollar and $\frac{muy}{py} = 10$ per dollar. The consumer would gain by increasing the quantity of y and decreasing the quantity of x. When this change occurs, mux increases and muy decreases.

The cell below determines the quantities of x and y that are required for the slopes of each of the three indifference curves to equal the price ratio. It also determines the implied money income levels required to achieve each of the specified utility levels.

```
assume(x>0)$    icslope(x,u0):=
''(radcan(diff(ic(x,a0,b0,u0),x)))$     [x1,x2,x3]:
 [find_root(icslope(x,u01)=-px0/py0,x,xmin1+0.1,50),
 find_root(icslope(x,u02)= -px0/py0,x,xmin2+0.1,50),
 find_root(icslope(x,u03)= -px0/py0,x,xmin3+0.1,50)]$
[y1,y2,y3]: float([ic(x1,a0,b0,u01),ic(x2,a0,b0,u02),
        ic(x3, a0,b0,u03)])$           [m1,m2,m3]:
   [x1*px0+y1*py0, x2*px0+y2*py0, x3*px0+y3*py0]$
matrix(["x","y","m"],[x1,y1,m1],[x2,y2,m2],
   [x3,y3,m3]);
```

$$\begin{bmatrix} x & y & m \\ 10.411 & 33.81 & 926.26 \\ 15.616 & 50.715 & 1389.4 \\ 20.821 & 67.62 & 1852.5 \end{bmatrix}$$

The ratio of marginal utilities, as we have seen, equals the marginal rate of substitution, or mrs, the indifference curve's slope. The first line of the input group above defines the mrs as a function of x and $u0$. The next four lines use `find_root` to determine the quantities of x at which $mrs = \frac{px}{py}$. This set of four lines is followed by a set of three lines that define the implied values of y. Finally, the last two lines define the money income levels required to afford the consumer the specified utility levels.

The cell below defines an explicit mrs function. It also confirms that the mrs is the same for each of the three input pairs defined above, and that this common value equals the price ratio.

```
mrs(x,y,a,b) := (a/(1-a)) * (y/x)^(1 + b);
numbers: float([mrs(x1,y1,a0,b0), mrs(x2,y2,a0,b0),
mrs(x3, y3, a0, b0), px0/py0])$   names:
  ["mrs1","mrs2","mrs3","px/py"]$
  matrix(names,numbers);
```

$$\mathrm{mrs}\,(x,y,a,b) := \frac{a}{1-a}\left(\frac{y}{x}\right)^{1+b} \qquad \begin{bmatrix} \mathrm{mrs1} & \mathrm{mrs2} & \mathrm{mrs3} & px/py \\ 4.1667 & 4.1667 & 4.1667 & 4.1667 \end{bmatrix}$$

4.2.1.1 The Income Consumption Curve

We can generalize these results, which show three values of y given the product prices and levels of x. The cell below sets the mrs function equal to the ratio of the two prices and solves the expression for y.[2]

```
declare(b,noninteger)$ assume(a>0,a<1,x>0,y>0,
px>0,py>0)$ soln:radcan(solve(mrs(x,y,a,b)=px/py,y));
```

$$[y = \frac{(1-a)^{\frac{1}{b+1}}\, px^{\frac{1}{b+1}}\, x}{a^{\frac{1}{b+1}}\, py^{\frac{1}{b+1}}}]$$

The cell below converts the expression above into the income consumption function.

$$yicc = \left(\frac{1-a}{a}\cdot\frac{px}{py}\right)^{\frac{1}{1+b}}\cdot x$$

[2]The `declare` and `assume` commands allow us to avoid a dialog with *Maxima*.

As the name indicates, the dependent variable defines value of y that lie on the income consumption curve.

```
yicc(x,px,py,a,b):=x*((px/py)*(1-a)/a)^(1/(1+b));
```

$$\text{yicc}\,(x, px, py, a, b) := x \left(\frac{\frac{px}{py}(1-a)}{a}\right)^{\frac{1}{1+b}}$$

The graph below combines the three indifference curves defined earlier with the budget lines based on the three values of m defined in the previous input group. The graph shows that the highest utility level attainable given an income level occurs at the mix of goods for which the pertinent indifference curve is tangent to the budget line. These points of tangency are (10.411, 33.810), (15.616, 50.715), and (20.821, 67.620).

```
wxdraw2d(title="Indifference Curves & Budget Lines",
xlabel="x",ylabel="y",xrange=[0,m3/px0],
 yrange=[0,m3/py0],key="IC and BL, u0 = u01",
 explicit(ic(x,a0,b0,u01),x,1,m3/px0),key = "",
 explicit(BL(x,m1,px0,py0), x, 0, m1/px0),
 ... Commands for u02 and u03 omitted ...
 line_type=dots,line_width=1,key="ICC",color=black,
 explicit(yicc(x,px0,py0,a0,b0),x,0,m3/px0) )$
```

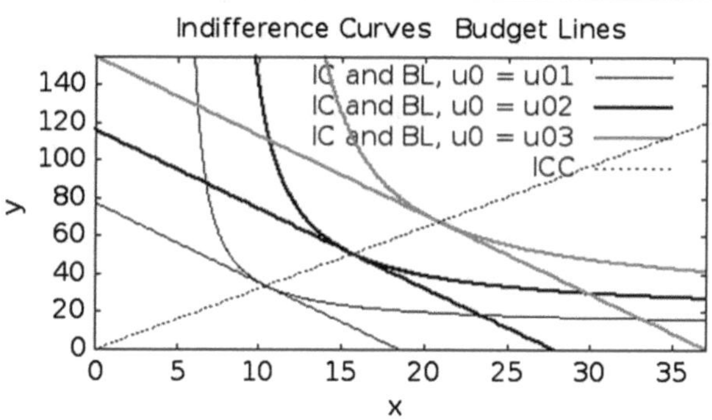

4.2.1.2 Elasticity of Substitution Revisited

The elasticity of substitution is the percentage change in $\frac{y}{x}$ that is associated with a 1 % change in the marginal rate of substitution (mrs). We now see that $mrs = px/py$, if the consumer is to maximize utility given a level of money income. Therefore, the elasticity of substitution is the percentage change in $\frac{y}{x}$ that results from a 1 % change in the relative price level.

For any function of the form $Y = A \cdot X^B$, B is the relevant elasticity. In the equality expression below, $Y = y/x$, $A = \frac{1-a}{a}$, and $B = \frac{1}{1+b}$. In our example

$b = \frac{4}{5}$, so the elasticity of substitution is $\frac{5}{9}$. Thus, a 9 % change in $\frac{px}{px}$ causes $\frac{y}{x}$ to change by 5 %. The output below confirms this, because

$$\frac{px^{5/9}}{py^{5/9}} = \left(\frac{px}{py}\right)^{5/9}.$$

```
[a0,b0]:[1/3,4/5]$  expr1:y/x=((px/py)*(1-a)/a)^
  (1/(1+b));   float((subst([a=a0,b=b0], expr1)));
```

$$\frac{y}{x} = \frac{(1-a)^{\frac{1}{b+1}}\, px^{\frac{1}{b+1}}}{a^{\frac{1}{b+1}}\, py^{\frac{1}{b+1}}} \qquad \frac{y}{x} = \frac{1.4697\, px^{\frac{5}{9}}}{py^{\frac{5}{9}}}$$

4.2.2 *Nonhomothetic Functions*

The graph above shows that the CES function generates an income consumption curve that is a ray through the origin. This curve reflects the characteristic that the marginal rate of substitution is a function of $\frac{y}{x}$ alone and not of their absolute sizes. Functions with this characteristic are *homothetic*. The indifference curves for homothetic functions are parallel displacements of each other.

The function below, due to [3], is not homothetic. The first function in the input group is the utility function. The second function is the implied expression for the indifference curves. The function for the indifference curves determines the value of y at any point on a specified indifference curve, so $y = e^{v0-x}$. The third command determines the slope of the indifference curve at any point; that is, it determines the marginal rate of substitution. For this function $mrs = -y$, for any value of x. Thus, mrs changes for any value of y/x, as x changes. This utility function is not homothetic.

```
v(x,y):=x+log(y)$  solve(x+log(y)=v0,y)$  icv(x,v0):=
''(rhs(%[1]))$  mrsv(x,v0):=''(diff(icv(x,v0),x))$
matrix(["Utility","Indifference Curve","mrs"],
   [v(x,y),  icv(x,v0), mrsv(x,v0)]);
```

$$\begin{bmatrix} \text{Utility} & \text{Indifference Curve} & \text{mrs} \\ \log(y) + x & e^{v0-x} & -e^{v0-x} \end{bmatrix}$$

The income consumption curve for this function is a horizontal line, the height of which is px/py. The next cell demonstrates this result.

```
soln:solve(mrsv(x, v)=-px/py, x)$
icc(px, py):=''(icv(rhs(soln[1]),v));
```

$$\text{icc}(px, py) := \frac{px}{py}$$

```
[v01,v02,v03]:[4,5,6]$      [xv1,xv2,xv3]  :
 [find_root(diff(icv(x,v01),x) =-px0/py0,x,0,50),
 find_root(diff(icv(x,v02),x) =-px0/py0,x,0,50),
 find_root(diff(icv(x,v03),x) =-px0/py0,x,0,50)]$
[yv1,yv2,yv3]:float([icv(xv1,v01), icv(xv2,v02),
  icv(xv3,v03)])$       [mv1,mv2,mv3]:
  [px0*xv1+py0*yv1,px0*xv2+py0*yv2,px0*xv3+py0*yv3]$
matrix(["x","y","m"], [xv1,yv1,mv1],[xv2,yv2,mv2],
  [xv3,yv3,mv3]   );
```

$$\begin{bmatrix} x & y & m \\ 2.5729 & 4.1667 & 178.64 \\ 3.5729 & 4.1667 & 228.64 \\ 4.5729 & 4.1667 & 278.64 \end{bmatrix}$$

The graph below illustrates the implication of this function's nonhomotheticity: The income consumption curve does not pass through the origin. For this function, any income change results in a change in the amount of x consumed but no change in the amount of y consumed.

```
wxdraw2d( xlabel = "Good x", ylabel = "Good y",
yrange = [0, mv3/py0], key = "v0 = v01",
explicit(icv(x, v01), x, 0, mv3/px0),  key = "",
explicit(mv1/py0 - (px0/py0)*x, x, 0, mv1/px0),
line_width = 2, key = "v0 = v02",
     ... Some commands omitted ...
key="ICC", color=black, line_width=1, line_type=dots,
explicit(icc(px0, py0), x, 0, mv3/px0)    )$
```

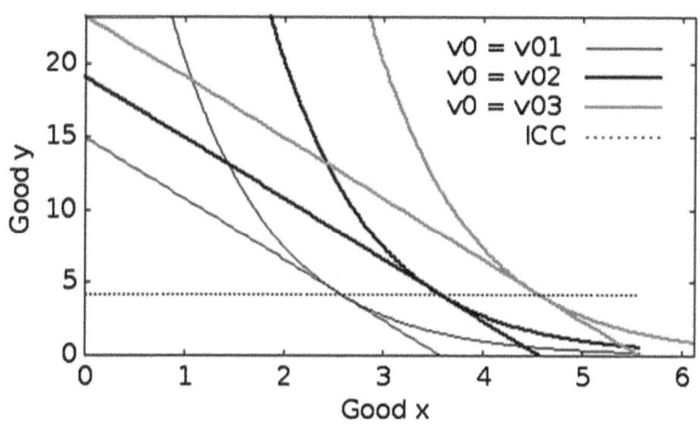

We can determine the income elasticity of demand for this consumer when a positive amount of good x is consumed. The horizontal income expansion path implies that all additional income is spent on this good: $\Delta x = \frac{\Delta m}{px}$, so $\frac{\Delta x}{\Delta m} = 1/px$. The elasticity equals $\frac{\Delta x}{\Delta m} \cdot \frac{m}{x}$ which equals $\frac{m}{px \cdot x}$. For small x values, this ratio is well above 1.0. Consider the three values above. The elasticity values appears below. The income elasticity decreases as m increases. We can generalize by noting that the elasticity in this case is the inverse of the share of income that the consumer spends on x. As m grows, this share grows, approaching one from below. The elasticity falls and approaches one from above.

```
EI:[mv1/(px0*xv1),mv2/(px0*xv2),mv3/(px0*xv3)]$
matrix(["Income","Income Elasticity"],
  [mv1,EI[1]],[mv2, EI[2]], [mv3, EI[3]]);
```

$$\begin{bmatrix} \text{Income} & \text{Income Elasticity} \\ 178.64 & 1.3887 \\ 228.64 & 1.2799 \\ 278.64 & 1.2187 \end{bmatrix}$$

4.2.3 *Corner Solutions*

The characterization of the ICC in the preceding section is incomplete. For $m/py < 4.1667$, the consumer cannot consume any combination of x and y on the ICC depicted in the graph. For lower income levels, this consumer buys only y, and the equality of the marginal rate of substitution and the price ratio no longer holds.[3] The consumer reaches a *corner solution*. For these lower income levels, the utility function is $v(0, y) = log(y)$.[4]

For these solutions, the highest attainable utility level is at the intersection of the income consumption curve and the vertical axis. We have established that this value is $y = \frac{px0}{py0}$, so utility levels must be at or below $log\left(\frac{px0}{py0}\right)$.

The input/output group below shows the utility levels, the relative price levels, the marginal rates of substitution, the quantities of y consumed, and the requisite money income levels.

For each of the three solutions, $mrs \leq \frac{px0}{py0}$. The expression of equality that must hold for an interior solution like those in the previous section is replaced by an inequality condition. Likewise, the analysis in previous sections[5] that is based on tangency conditions and the associated algebraic solutions must be replaced by programming solutions based on *Kuhn–Tucker conditions*, like the inequality stated above.

---

[3]The complete ICC consists of a segment of the vertical axis from 0 to 4.1667 and the horizontal line in the graph.

[4]We assert this for now. We demonstrate the result below.

[5]And in a section that follows, where Lagrangian multipliers are employed.

```
[v04,v05,v06] : float([log(px0/py0),
    0.8*log(px0/py0), 0.4*log(px0/py0)])$
p: float(px0/py0)$          [mrs4, mrs5, mrs6]:float(
[-mrsv(0, v04), -mrsv(0, v05), -mrsv(0, v06)])$
[y4, y5, y6] : float([exp(v04), exp(v05), exp(v06)])$
[mv4, mv5, mv6]:[py0*y4,py0*y5,py0*y6]$  matrix(
  ["Income","y","Utility","mrs","px/py"],[mv4,y4,v04,
    mrs4,p],[mv5,y5,v05,mrs5,p],[mv6,y6,v06,mrs6,p]);
```

$$\begin{bmatrix} \text{Income} & y & \text{Utility} & mrs & px/py \\ 50.0 & 4.1667 & 1.4271 & 4.1667 & 4.1667 \\ 37.585 & 3.1321 & 1.1417 & 3.1321 & 4.1667 \\ 21.237 & 1.7698 & 0.571 & 1.7698 & 4.1667 \end{bmatrix}$$

The graph below complements the preceding graph, which shows the horizontal segment of the income consumption curve. The graph below shows the combinations of the two goods for low income levels and demonstrates why only good y is consumed. For each of these three budget lines, the highest attainable utility level is reached at the vertical intercept, where $x = 0$. For any positive amount of x, the amount of y that must be given up exceeds the amount that the consumer can give up and remain on the same indifference curve. That is, the slope of the budget line exceeds the slope of the indifference curve for any positive value of x.

```
wxdraw2d(xlabel="Good x",ylabel="Good y",key=
 "v0 = v04", explicit(icv(x,v04),x,0, mv4/px0),
 key="mv = mv4", explicit(mv4/py0-(px0/py0)*x,
  x,0,mv4/px0),line_width=3,key="v0 = v05",explicit(
  icv(x,v05),x,0,mv4/px0),key="mv=mv5",
  explicit(mv5/py0-(px0/py0)*x,x,0,mv5/px0),
  color=gray,key="v0 = v06",explicit(icv(x, v06),x,0,
  mv4/px0),key="mv=mv6",explicit(mv6/py0-(px0/py0)*x,
  x,0,mv6/px0)    )$
```

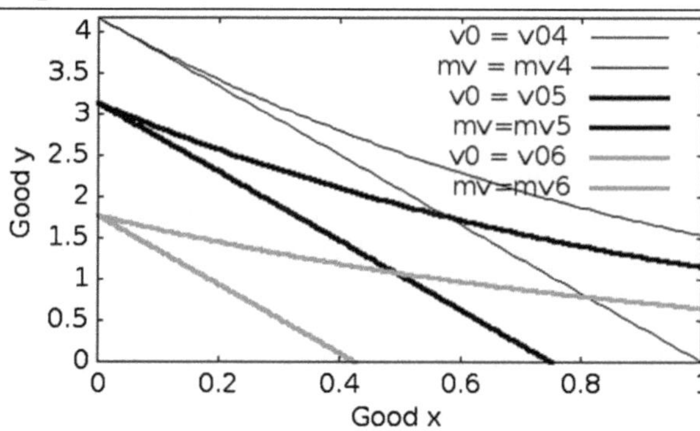

The CES utility can generate corner solutions. If the elasticity of substitution exceeds 1.0 ($b < 0$), then the indifference curves touch both axes, and corner solutions can occur. For a sufficiently low value of $\frac{px}{py}$ only x is consumed, and for a sufficiently high value only y is consumed. In the limiting case of perfect substitutes (the limit as $b \rightarrow -1$), when indifference curves are straight lines, all solutions are corner solutions.

The present illustration of a corner solution involves specialized consumption, consuming at the end of a budget line rather than at an interior point along that line. Corner solutions can also occur if the budget line exhibits discontinuities or kinks, like those above, that involve taxes or subsidies. In all such cases, the condition for optimal budget allocation involves a statement of inequality rather than one of equality.

4.2.4 An Inferior Good

In the CES case above, the consumption levels of both goods change in the same direction that income changes. Both x and y have positive income elasticities of demand and are called *normal goods*.

In the second case, considered immediately above, the income elasticity of demand for x is zero at low income levels and it exceeds 1.0 for higher income levels. Good y is a normal good at low income levels[6] and exhibits zero income elasticity at higher income levels. Thus, both x and y are normal goods over some income range and unaffected by income over some other range.

A third case is that of the *inferior* goods. The income elasticities of demand for such goods are negative. That is, income and the quantity consumed of inferior goods move in opposite directions. We illustrate this case with a utility function that results in one of the goods being inferior over some combination of income and prices. The utility function, due to [2], appears below.

```
u(x,y,a,b,c):=a*log(x)+b*log(y)+c*x*y;
```

$$\mathrm{u}(x, y, a, b, c) := a \log(x) + b \log(y) + c\,x\,y$$

We determine the quantity consumed of this good at two income levels, establishing that the good is inferior over the range considered. To do so requires that we find the equation for the marginal rate of substitution. The equation appears below. The mrs is the ratio of the two marginal utility functions, which are the first derivatives indicated below. We use the `ratsimp` command to simplify the expression. This is a second example of a non-homothetic utility function. (The two commands below could be nested into a single command.)

```
diff(u(x,y,a,b,c),x)/diff(u(x,y,a,b,c),y)$
mrs(x, y,a,b,c) := ''(ratsimp(%));
```

$$\mathrm{mrs}(x, y, a, b, c) := \frac{c\,x\,y^2 + a\,y}{c\,x^2\,y + b\,x}$$

[6] Over this range, its income elasticity of demand is 1.0. Why?

We cannot phrase y in terms of x using this expression, so we take another approach. We substitute the budget constraint, $y = m - p \cdot x$, directly into the function above. Doing so forces the solution to be consistent with that constraint. For simplicity, we define $py \equiv 1$, so that $p = \frac{px}{py}$.

```
mrs_aug(a,b,c,m,x,p):=''(mrs(x,m-p*x,a,b,c));
```

$$\text{mrs\_aug}\,(a, b, c, m, x, p) := \frac{c\,x\,(m-p\,x)^2+a\,(m-p\,x)}{c\,x^2\,(m-p\,x)+b\,x}$$

We now specify parameter values and solve the equation above for two income levels. *Maxima*'s `solve` does not yield meaningful solutions (the solutions involve imaginary numbers), so we use `find_root` to determine the approximate values of x for the two income levels.

The graph below shows the difference between mrs, the indifference curve slope, and $p0$, the budget line slope. Both of these slopes (and hence the difference between the two) are stated in terms of units of y given up per unit of x consumed. The values to be determined are the points of intersection with the x axis, where $mrs - p = 0$. The graph shows that only part of the function is relevant, that part for which $mrs \geq p0$. This information guides the use of `find_root`.

```
[p0, m0, m1]:[0.2,22.5,20]$ [a0,b0,c0]:[40,50,-0.1]$
 wxdraw2d( ... Some commands omitted ...   title=
  "Marginal Rates of Substitution",key=concat(
  "mrs-p0,m =",string(m0)),
  explicit(mrs_aug(a0,b0,c0,m0,
  x,p0)-p0,x,0.1,200), ... explicit(mrs_aug(
  a0,b0,c0,m1,x,p0)-p0,x,0.1,200))$
```

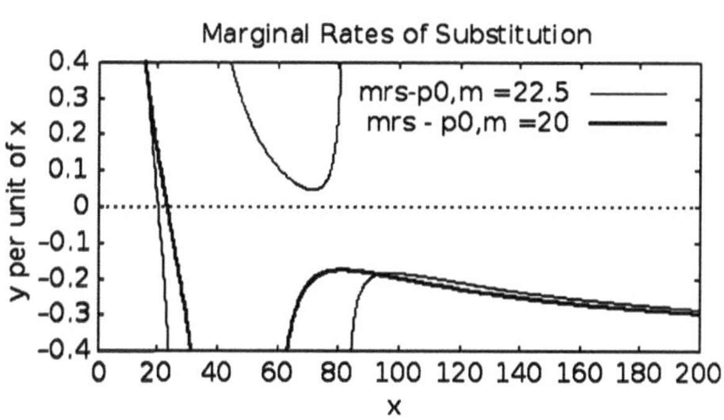

The graph above suggests that $mrs = p$ at values around $x = 20$. To find closer approximations, we execute `find_root` twice, once for each of the two income values, $m0$ and $m1$. When $x0$ or $x1$ has been determined, the corresponding y value is calculated from the budget line, and the resulting utility level is determined.

```
x0: find_root(mrs_aug(a0,b0,c0,m0,x,p0)-p0,x,1,30)$
    y0: m0-p0*x0$      u0:u(x0,y0,a0,b0,c0)$
x1: find_root(mrs_aug(a0,b0,c0,m1,x,p0)-p0,x,1,30)$
    y1: m1-p0*x1$      u1 : u(x1,y1,a0,b0,c0)$
matrix(["m", "x", "y", "u"], [p0*x0+y0, x0, y0, u0],
       [y1+p0*x1, x1, y1, u1]);
```

$$\begin{bmatrix} m & x & y & u \\ 22.5 & 20.145 & 18.471 & 228.72 \\ 20.0 & 23.226 & 15.355 & 226.72 \end{bmatrix}$$

The output above shows that when m decreases from 22.5 to 20.0, the quantity of x consumed *increases* from 20.145 to 23.225. The graph below depicts this result graphically.

```
wxdraw2d(user_preamble="set key bottom",xlabel="x",
 ylabel="y", line_width=2,point_type=asterisk,
 key = "Income consumption curve segment",
 points_joined=true,points([[x0,y0],[x1,y1]]),
 key="u = u0",line_width=1,
 implicit(u(x,y,a0,b0,c0)-u0,x,0.1,25,y,0.1,30),
 key="m0 - p0*x",explicit(m0-p0*x,x,0,50),
 color=gray, line_width=2,key="u =u1",
 implicit(u(x,y,a0,b0,c0)-u1,x,0.1,50,y,0.1,30),
 key="m1 - p0*x, m1 < m0",explicit(m1-p0*x,x,0,50))$
```

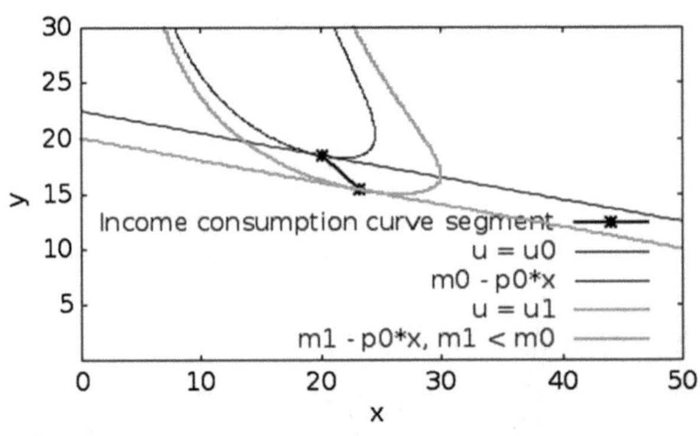

The two budget lines are tangent to the corresponding indifference curves at the points indicated by the bold asterisks. The bold line segment connecting these points approximates a segment of the income consumption curve. Unlike its counterparts in previous examples, this income consumption curve bends backward. As income falls, more x and less y is consumed. Thus x is an inferior good, and y is a normal good over this income range.[7]

[7] At least one good must be a normal good. Why?

4.3 Lagrangian Multipliers

Using graphs to illustrate and analyze the nature of optimization provides valuable insights, but it limits the number of variables that can be considered. An alternative is to use *Lagrangian multipliers* to examine optimization. We consider a fairly general application below and then we return to the two-good case.

To apply this method, *augment* the objective function (the function to be either maximized or minimized). Add a term that consists of the constraint, stated as an implicit function, multiplied by a term M, the Lagrange Multiplier. We name this augmented function L, the Lagrangian. L has the same value as u for any values of the arguments (x, y, z, and w) for which the constraint holds: For these values of the arguments, the term $M \cdot (\ldots)$ equals zero. Thus achieving a maximum value of L corresponds to achieving the extreme value of u given that the constraint is satisfied.

For the consumer to achieve the maximum value of L, the first partial derivatives of L with respect to all of its arguments must equal zero. This is the first-order or "necessary" condition. The next section considers the second-order or "sufficient" conditions. The consumer's utility level is a function of the consumption level of four goods: x, y, z, and w. The budget constraint is that $m = px \cdot x + py \cdot y + pz \cdot z + pw \cdot w$, where m is money income and the prices are px, py, pz, and pw.

The `depends` command asserts that u is a function of the four variables in the list. It does not specify the functional form.

```
depends(u, [x,y,z,w])$
L:u(x,y,z,w) + M*(m-px*x-py*y-pz*z-pw*w);
```

$$(-pz\,z - py\,y - px\,x - pw\,w + m)\;M + \mathrm{u}(x, y, z, w)$$

The five partial derivatives used to define the first-order condition for optimization appear below.

```
[Lx,Ly,Lz,Lw,LM]:[diff(L,x),diff(L,y),diff(L,z),
diff(L,w),diff(L,M)]$ transpose(matrix(["Lx",
"Ly =","Lz =","Lw =","Lm ="],[Lx,Ly,Lz,Lw,LM]));
```

$$\begin{bmatrix} \mathrm{Lx} = & \frac{d}{d\,x}\,\mathrm{u}(x, y, z, w) - px\,M \\ \mathrm{Ly} = & \frac{d}{d\,y}\,\mathrm{u}(x, y, z, w) - py\,M \\ \mathrm{Lz} = & \frac{d}{d\,z}\,\mathrm{u}(x, y, z, w) - pz\,M \\ \mathrm{Lw} = & \frac{d}{d\,w}\,\mathrm{u}(x, y, z, w) - pw\,M \\ \mathrm{Lm} = & -pz\,z - py\,y - px\,x - pw\,w + m \end{bmatrix}$$

For any of the first four equations above, setting the equation equal to zero implies that the Lagrangian multiplier M equals the relevant first partial derivative divided by the price of the relevant product. This, in turn, implies three things:

- The marginal utility per dollar spent on each good equals the Lagrangian multiplier M, so the values all equal each other: $\frac{du}{dx}/px) = \frac{\partial u}{\partial y}/py) = \frac{\partial u}{\partial z}/pz) = \frac{\partial u}{\partial w}/pw)$. This condition can be extended to cover any number of goods.

This result generalizes the result that was earlier derived using indifference curves and budget lines for two goods.[8]

- Consider any two goods, say x and w. Rearrange the terms in the equality stated just above, or divide the Lw expression by the Lx condition. Either way, the same result appears: $\frac{\partial u}{\partial w}/\frac{\partial u}{\partial x} = pw/px$. That is, the slope of the indifference curve between any two goods, holding constant the levels of the other goods, equals the ratio of the prices between these two goods. Again, this agrees with the result derived earlier, that optimization requires the tangency of an indifference curve and the budget line.
- We can interpret M as the marginal utility of income at this optimum set of quantities. (M's value might be different if either a different set of prices or a different income level leads to a different set of quantities.) To see that the multiplier M equals the marginal utility of income, suppose that the consumer's income changes by a small amount, dm. Also, suppose that the consumer changes consumption of x in response to the changed m. (Since marginal utility per dollar is the same for all goods, any good will do.) The change in x consumption is $dx = dm/px$. The implied utility level change is $du = mux \cdot dx$. Because $dx = dm/px$, $du = mux \cdot dm/px$. Therefore, $\frac{du}{dm} = \frac{mux}{px}$. Thus, the condition that $Lx = 0$ implies that the multiplier $M = \frac{du}{dm}$, the marginal utility of income.

The final first-order condition simply returns the constraint. Information from the other equations can be combined with this condition to determine the nature of the demand curve that optimization implies. Consider the optimization problem with the two-good CES utility function:

```
[u(x,y,a,b):=(a*x^-b+(1-a)*y^-b)^-(1/b),
 L :  u(x, y, a, b) + M*(m - px*x - py*y)];
Lx(x, y, a, b, px, py, m, M) := ''(diff(L, x));
 Ly(x, y, a, b, px, py, m, M)  := ''(diff(L, y));
 LM(x, y, a, b, px, py, m, M)  := ''(diff(L, M));
```

$$[\mathrm{u}\,(x,y,a,b) := \left(a\,x^{-b}+(1-a)\,y^{-b}\right)^{-\frac{1}{b}}, (-py\,y-px\,x+m)\,M+\frac{1}{\left(\frac{1-a}{y^b}+\frac{a}{x^b}\right)^{\frac{1}{b}}}]$$

$$\mathrm{Lx}\,(x,y,a,b,px,py,m,M) := a\,x^{-b-1}\left(\frac{1-a}{y^b}+\frac{a}{x^b}\right)^{-\frac{1}{b}-1}-px\,M$$

$$\mathrm{Ly}\,(x,y,a,b,px,py,m,M) := (1-a)\,y^{-b-1}\left(\frac{1-a}{y^b}+\frac{a}{x^b}\right)^{-\frac{1}{b}-1}-py\,M$$

$$\mathrm{LM}\,(x,y,a,b,px,py,m,M) := -py\,y-px\,x+m$$

[8] *Maxima* does not offer distinct notation for simple derivatives and partial derivatives. From an operational viewpoint the two do not differ. The analyst must be aware of the context.

Compare the first two equations in the second table above with the first set of equations in Sect. 3.1.2. The earlier expressions are the same as these except for the term $-px \cdot M$. From the general discussion above, we can conclude that application of the Lagrangian method to this CES illustration must lead to the results that have already been established.

4.3.1 Illustration of the Lagrangian Multiplier

We use this illustration to demonstrate a result not derived by using indifference curves and budget lines that the Lagrangian multiplier equals the marginal utility of income. As before, we use a numerical method to determine the optimum values of x and y and the associated value of the Lagrangian multiplier. Specifically, we use *Maxima*'s add-on module mnewton, which can solve systems of equations numerically. The mnewton command requires two lists, a list of expressions followed by a list of variables. We use the same parameter values that appear in previous sections.

```
[a0,b0]:[1/3,4/5]$[px0,py0,m0]:[50,12,1389.4]$
load("mnewton")$        soln0 : float(mnewton(
[Lx(x,y,a0,b0,px0,py0,m0,M),
Ly(x,y,a0,b0,px0,py0,m0,M),
 LM(x,y,a0,b0,px0,py0,m0,M)],
 [x,y,M],[20,20,0.01]))[1]$
```

The initial solution appears below. Based on the values above, $u = 30$.

```
[x0,y0,M0]:[rhs(soln0[1]),rhs(soln0[2]),
  rhs(soln0[3])]$matrix(["x0","y0","M0","u0"],
  [x0,y0,M0,u0:u(x0,y0,a0,b0)]);
```

$$\begin{bmatrix} \text{x0} & \text{y0} & \text{M0} & \text{u0} \\ 15.616 & 50.716 & 0.0216 & 30.0 \end{bmatrix}$$

Change m by one unit (the change could be different, and need not be positive). As a result, the consumer changes the consumption levels of both goods slightly, and the utility level rises to $u \approx 30.022$, so the change in utility is $\Delta u \approx 0.022$. This value equals the value of M that mnewton returns, at least to the available number of decimal places.

```
soln1:float(mnewton([ Lx(x,y,a0,b0,px0,py0,m0+1,M),
Ly(x,y,a0,b0,px0,py0,m0+1,M),LM(x,y,a0,b0,px0,py0,
  m0+1,M)],[x,y,M],[20,20,0.01]))[1]$   [x1,y1]:
 [rhs(soln1[1]), rhs(soln1[2])]$  u1:u(x1,y1,a0,b0)$
u_change: u(x1,y1,a0,b0) -u(x0, y0,a0,b0)$  matrix(
 ["x1","y1","u1","u__change"],[x1,y1,u1,u_change]);
```

$$\begin{bmatrix} \text{x1} & \text{y1} & \text{u1} & \text{u\_change} \\ 15.627 & 50.752 & 30.022 & 0.0216 \end{bmatrix}$$

More generally, a Lagrangian multiplier states the change in the value of the objective function (utility here) per one-unit change in the constraint (change in income here).

4.3.2 Multiple Constraints

Using the Lagrangian approach generalizes the analysis in that it can be applied to any number of goods. It generalizes the analysis in a second way as well: Additional constraints can be added. Suppose, for example that consuming good x requires a given amount of time per unit consumed.

Likewise, suppose that consuming y requires a (presumably different) amount of time per unit consumed. Then a second constraint could be added. The new Lagrangian equation would be

$$L = u(x, y, a, b + M_1 \cdot (m - px \cdot x - py \cdot y) + M_2 \cdot (T - tx \cdot x - ty \cdot y)$$

where T is the time available for consuming these two goods, tx is the time required to consume a unit of good x, and ty is the time required to consume a unit of good y.

```
L:u(x,y,a,b)+M1*(m-px*x-py*y)+M2*(T -tx*x-ty*y);
```

$$M2\ (T - ty\ y - tx\ x) + (-py\ y - px\ x + m)\ M1 + \frac{1}{\left(\frac{1-a}{y^b} + \frac{a}{x^b}\right)^{\frac{1}{b}}}$$

The Lagrangian equation now contains four variables, the quantities of the two goods and the two multipliers. The first-order conditions are defined by setting the following partial derivatives equal to zero.

```
transpose(matrix(["Lx =","Ly =","LM1 =", "LM2 ="],
 [diff(L,x),diff(L,y),diff(L,M1),diff(L,M2)]));
```

$$\begin{bmatrix} \text{Lx} = & -tx\ M2 - px\ M1 + a\ x^{-b-1} \left(\frac{1-a}{y^b} + \frac{a}{x^b}\right)^{-\frac{1}{b}-1} \\ \text{Ly} = & -ty\ M2 - py\ M1 + (1-a)\ y^{-b-1} \left(\frac{1-a}{y^b} + \frac{a}{x^b}\right)^{-\frac{1}{b}-1} \\ \text{LM1} = & -py\ y - px\ x + m \\ \text{LM2} = & T - ty\ y - tx\ x \end{bmatrix}$$

As the result above shows, the results of adding this constraint are two. First, each of the first two first-order conditions now contains an additional term. Second, a fourth first-order condition is added to the three that occur with a single constraint.

4.3.3 Expenditure Minimization

The constrained maximization problem stated thus far (maximize utility subject to a budget constraint) can be rephrased in terms of its *dual*: minimize the expenditures required to provide a specified utility level. The Lagrangian equation W below states this problem for the CES utility function. The Lagrangian multiplier is now specified as μ.

```
W:m-px*x-py*y+%mu*(u0-u(x,y,a,b));
```

$$\mu\left(30.0-\frac{1}{\left(\frac{1-a}{y^b}+\frac{a}{x^b}\right)^{\frac{1}{b}}}\right)-py\,y-px\,x+m$$

The partial derivatives required to determine the expenditures-minimizing combination of x and y appear below. Setting the first two equal to zero and solving for $\frac{y}{x}$ yields the now-familiar result

$$\frac{y}{x}=\left(\frac{px}{py}\cdot\frac{1-a}{a}\right)^{1/(1+b)}.$$

```
Wx:diff(W,x)$ Wy:diff(W,y)$ Wmu:diff(W,%mu)$
transpose(matrix(["Wx =","Wy =","Wmu ="],
[Wx,Wy,Wmu]));
```

$$\begin{bmatrix} Wx = & -\mu\,a\,x^{-b-1}\left(\frac{1-a}{y^b}+\frac{a}{x^b}\right)^{-\frac{1}{b}-1}-px \\ Wy = & -\mu\,(1-a)\,y^{-b-1}\left(\frac{1-a}{y^b}+\frac{a}{x^b}\right)^{-\frac{1}{b}-1}-py \\ Wmu = & 30.0-\frac{1}{\left(\frac{1-a}{y^b}+\frac{a}{x^b}\right)^{\frac{1}{b}}} \end{bmatrix}$$

The interpretation of the Lagrangian multiplier changes. The multiplier M was interpreted as the change in utility per one-unit change in income. Likewise, the multiplier from the dual problem, μ, is the change in expenditure implied by a one-unit change in the target utility level. Thus $\mu = 1/M$.

4.3.4 The Indirect Utility Function

The optimal combination of x and y for a CES utility function, given a pair of product prices and an income level, is

$$yxratio=\left(\frac{px}{py}\cdot\frac{1-a}{a}\right)^{\frac{1}{1+b}}.$$

The budget line is $m = px \cdot x + py \cdot y$. Substitute $yxratio \cdot x$ for y in the budget line and solve for the optimal value of x. The result is

$$xopt = \frac{m}{px + (py \cdot yxratio)}.$$

Next, substitute $\frac{y}{yxratio}$ for x and solve for y:

$$yopt = \left(\frac{m}{px/yxratio + py}\right)$$

These values are assigned the names `xopt` and `yopt` below. The output is suppressed. The *Maxima* expressions can be reduced to the equivalence of those above, but those expressions are quite messy.

```
yxratio:((px/py)*(1-a)/a)^(1/(1+b))$
xopt:m/(px+py*yxratio)$  yopt:xopt*yxratio$
```

Substitute these values into the consumer's utility function. The result is an expression for utility given the constraints and optimizing behavior. The resulting expression, the *indirect utility function* is expressed below.

This expression could be simplified, but its salient feature can be discerned in the current form. This feature is that utility is now stated in terms of the original utility function's parameters, along with the prices and income level that define the budget line. The values of x and y do not appear directly (hence, "indirect").

We confirm that, given the initial income level and prices, the attainable utility level is 30 units of utility. We also confirm that increasing m by \$1 increases utility by approximately 0.022 units, as we saw earlier.

```
v(px, py, m, a, b) :='' (u(xopt, yopt, a, b))$
v0: float(v(px0, py0, m0, a0, b0))$
v1:float(v(px0,py0,m0+1,a0,b0))$  chg:v1-v0$
transpose(matrix(["v =","Initial Utility =",
"New Utility =","change ="],
[u(xopt,yopt,a,b),v0, v1, chg]));
```

$$\begin{bmatrix} v = & \dfrac{1}{\left(\dfrac{1-a}{\left(\dfrac{m\left(\frac{(1-a)\,px}{a\,py}\right)^{\frac{1}{b+1}}}{\left(\frac{(1-a)\,px}{a\,py}\right)^{\frac{1}{b+1}}py+px}\right)^{b}}+\dfrac{a}{\left(\dfrac{m}{\left(\frac{(1-a)\,px}{a\,py}\right)^{\frac{1}{b+1}}py+px}\right)^{b}}\right)^{\frac{1}{b}}} \\ \text{Initial Utility} = & 30.0 \\ \text{New Utility} = & 30.022 \\ \text{change} = & 0.0216 \end{bmatrix}$$

Stating utility in terms of the parameters of the utility function and of the budget constraint allows direct investigation of the effect of changes in these parameters. Consider the case of a tax that raises a specified amount of revenue.

The tax could be imposed on the consumption of one of the two goods or it could be a *lump sum* tax, independent of post-tax consumption. We consider a 50% ad valorem tax on the consumption of good x. The tax could be a specific tax. Likewise, it could be imposed on y.

```
x0:float(subst([m=m0,px=px0,py=py0,a=a0,b=b0],
(xopt)))$  x1:float(subst([m=m0,px=px0*1.5,py=py0,
 a=a0,b=b0],(xopt)))$    taxRev : %*0.5*px0$
 x2:float(subst([m =m0-taxRev,px=px0,py=py0,
 a=a0,b=b0],xopt))$    matrix(["x /no tax",
 "x /ad valorem tax", "x /lump sum tax",
 "Tax revenue"],[x0,x1,x2,taxRev]);
```

$$\begin{bmatrix} \text{x /no tax} & \text{x /ad valorem tax} & \text{x /lump sum tax} & \text{Tax revenue} \\ 15.616 & 11.221 & 12.463 & 280.53 \end{bmatrix}$$

The cell above shows that at initial values of the system's parameters, $xopt = 15.616$. A 50 % tax on x would reduce x to 11.221 units and would generate \$280.53 in revenue per time period (25·11.221, except for rounding error). A per-period lump sum tax of this amount with no taxes on either x or y would result in $xopt = 12.463$ units.

We determine below that the consumer would prefer the *lump sum* tax. The *ad valorem* tax reduces the utility level from 30 to 23.675; the *lump sum* tax reduces utility to 23.943.

```
v0: float(v(px0*1.5, py0, m0, a0, b0))$
v1: float(v(px0, py0, m0-taxRev, a0, b0))$
matrix(["Utility/ad valorem tax",
"Utility/lump sum tax"], [v0, v1]);
```

$$\begin{bmatrix} \text{Utility/ad valorem tax} & \text{Utility/lump sum tax} \\ 23.675 & 23.943 \end{bmatrix}$$

4.4 Second-Order Conditions

The application of Lagrangian multipliers results in a set of expressions that provide the first-order (often called necessary) conditions that must be met for optimization to be realized. The application above takes as given that another set of conditions (second-order, or sufficient) are met. If these conditions are not met, then the values which satisfy the first-order conditions need not lead to a maximum value for the objective function.

This section develops the first-order and second-order condition and applies the analysis to the utility maximization model that we have already sketched. First, it reviews the first-order condition for a single-variable equation and examines the second-order condition. Then it looks at first-order and second-order conditions for establishing extreme values for an unconstrained function of more than one variable. Finally, it examines an optimization problem in which the consumer spends a fixed amount of income on three goods.

4.4.1 Functions of One Variable

The expression in the next cell is a quadratic function of x. Its first derivative, assigned the name `dfdx`, is linear. The value of x at which `dfdx` $= 0$, $x = -\frac{b}{2 \cdot c}$, is positive if b and c have opposite signs.

The second derivative of this function is $\frac{d^2 f}{dx^2} = 2 \cdot c$, so the first derivative is a monotonic function of x, always increasing if $c > 0$ and always decreasing if $c < 0$. As the graph below shows, the sign of c determines whether `xsoln` generates a maximum value for $f(x)$ or a minimum value.[9]

```
f(x,a,b,c):=a+b*x+c*x^2$     dfdx:diff(
  f(x,a,b,c),x)$  xsoln:rhs((solve(dfdx, x ))[1])$
d2fdx2:diff(f(x,a,b,c),x,2)$  matrix(["f(x)","df/dx",
"x,f=0","d2f/dx2"],[f(x,a,b,c),dfdx,xsoln,d2fdx2]);
```

$$\begin{bmatrix} f(x) & df/dx & x, f=0 & d2f/dx2 \\ c\,x^2+b\,x+a & 2\,c\,x+b & -\frac{b}{2\,c} & 2\,c \end{bmatrix}$$

The next cell specifies parameters for two parabolas. Both reach extreme values when $x > 0$. The value of x for which $\frac{df}{dx} = 0$ is determined for each, as is the resulting extreme value for each function.

```
[a0,b0,c0]:[500,5,-0.5]$   [a1,b1,c1]:[150,-1.5,1.0]$
xsoln0:subst([ a=a0,b=b0,c=c0],xsoln)$      xsoln1:
  subst([a=a1,b=b1,c=c1],xsoln)$  xmax:max(xsoln0,
  xsoln1)$  ymax:max(f(xsoln0,a0,b0,c0),
  f(xsoln1,a1,b1,c1))$
ymin:min(f(xsoln0,a0,b0,c0),f(xsoln1,a1,b1,c1))$
matrix(["a","b","c","x, dy/dx = 0",
"Extreme Value"],[a0,b0,c0,xsoln0,f(xsoln0,a0,b0,c0)],
[a1,b1,c1,xsoln1,f(xsoln1,a1,b1,c1)] );
```

[9] We consider only functions with a single local maximum or minimum value. Such functions have a global maximum or minimum value, which corresponds to the local value. Some functions can have multiple local maximum and minimum values and may or may not have a global maximum or minimum value. As an exercise, plot $g(x) = sin(x) \cdot x$ over a range of at least $-2 \cdot \pi$ to $2 \cdot \pi$. This function has many local maximum and minimum values but no global maximum or minimum value.

$$\begin{bmatrix} a & b & c & x, dy/dx = 0 & \text{Extreme Value} \\ 500 & 5 & -0.5 & 5.0 & 512.5 \\ 150 & -1.5 & 1.0 & 0.75 & 149.44 \end{bmatrix}$$

The graph confirms that the extreme (maximum or minimum) values of the function occur at the x values calculated above. It also shows that when $c < 0$ the extreme value is a maximum, and when $c > 0$ the extreme value is a minimum.

```
wxdraw2d( ... Some commands omitted ...
 explicit(if x < xsoln0 then f(xsoln0,a0,b0,c0)
  else 0,x,-10,xsoln0),line_type=dots, key=
  concat("Minimum:c =",string(c1)),
 explicit(if x < xsoln1 then f(xsoln1,a1,b1,c1)
   else 0,x,-10,xsoln1), line_type=solid,key=
  string(f(x,a0,b0, c0)), line_width=2,explicit(
  f(x,a0,b0,c0),x,-10,5*xmax),  ...
 explicit(f(x,a1,b1,c1),x,-10, 5*xmax))$
```

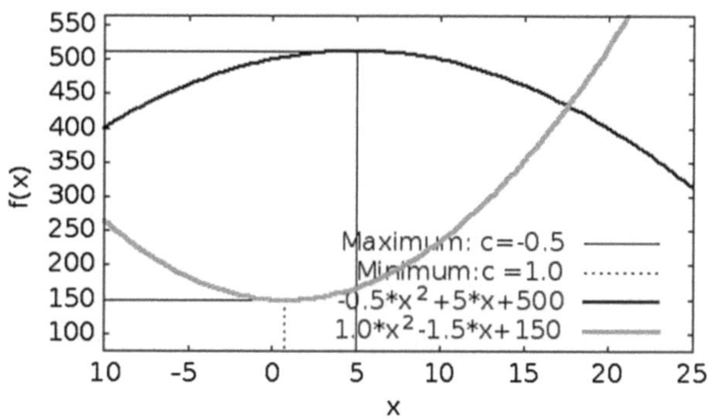

We may express the conditions illustrated by the graph more formally. If $y = f(x)$, then for a value of x to generate an extreme value of y, $\frac{dy}{dx} = 0$ at that value of x. The value of y must be unchanged for (at least) a small change in x's value. The change of y is $dy = f_x \cdot dx$, where $f_x \equiv \frac{dy}{dx}$.

For y to achieve a maximum value at a given x value, dy must be positive for smaller values of x, zero at the given value (this is the first-order condition), and negative for larger x values. Thus, $dy = f_x \cdot dx$ must be a decreasing function of x. That is

$$\frac{d(dy)}{dx} = \frac{df_x}{dx} \cdot dx$$

must be negative for positive values of dx. Therefore

$$\frac{d(dy)}{dx} = \frac{d^2y}{dx^2} \cdot dx < 0,$$

so $\frac{d^2y}{dx^2} < 0$. By similar reasoning, $\frac{d^2y}{dx^2} > 0$ implies that the extreme value is a minimum, as in the second illustration in the graph above.

4.4.2 *Functions of More than One Variable*

Consider a function of the three variables x, y, and z. For the function $u = u(x, y, z, w)$ to achieve an extreme value, it must be true that (small) changes in any of the variables must not affect u's value. That is,

$$du = (u_x \cdot dx) + (u_y \cdot dy) + (u_z \cdot dz) = 0,$$

which requires that the three first derivatives must all equal zero.[10]

Deriving the second-order condition for this function is a direct extension of the derivation for the single-variable case, but the resulting expression is much longer and more involved.

$$\begin{aligned} d^2u = {} & (u_{xx} \cdot dx + u_{xy} \cdot dy + u_{xz} \cdot dz) \cdot dx + \\ & (u_{yx} \cdot dx + u_{yy} \cdot dy + u_{yz} \cdot dz) \cdot dy + \\ & (u_{zx} \cdot dx + u_{zy} \cdot dy + u_{zz} \cdot dz) \cdot dz. \end{aligned}$$

Even applying Young's theorem to reduce the number of cross-partial derivatives leaves an equation that offers no obvious guidance.

We state without proof that a matrix that consists of the terms

$$u_{xx}, u_{xy}, u_{xz}, \ldots, u_{zy}, \text{ and } u_{zz}$$

like the one below can be evaluated to determine the nature of the function at the values of x, y, and z that the first-order condition yields. This matrix is an example of a *Hessian* matrix, a square matrix that consists of all the second-order derivatives of the function being examined.

```
H1: matrix([u[xx],u[xy],u[xz]],
     [u[yx],u[yy],u[yz]],
     [u[zx],u[zy],u[zz]]);
```

$$\begin{bmatrix} u_{xx} & u_{xy} & u_{xz} \\ u_{yx} & u_{yy} & u_{yz} \\ u_{zx} & u_{zy} & u_{zz} \end{bmatrix}$$

Before addressing the second-order conditions, we review two important aspects of a square matrix: the determinant and the principal minors.

[10]Notation: $\frac{\partial u}{\partial x} \equiv u_x$. Likewise for y and z.

4.4.2.1 The Determinant

The determinant of a square matrix is a value that results from a process of cross-multiplication of the matrix's terms. For a 2×2 matrix, the value results from simple cross-multiplication, as shown below. Also shown is the determinant of the Hessian matrix created above.

```
[matrix([a,b],[c,d]),determinant(matrix([a,b],
  [c,d]))];
determinant(H1);
```

$$\left[\begin{bmatrix} a & b \\ c & d \end{bmatrix}, a\,d - b\,c\right]$$

$$u_{xx}\,\left(u_{yy}\,u_{zz} - u_{yz}\,u_{zy}\right) - u_{xy}\,\left(u_{yx}\,u_{zz} - u_{yz}\,u_{zx}\right) + u_{xz}\,\left(u_{yx}\,u_{zy} - u_{yy}\,u_{zx}\right)$$

4.4.2.2 Principal Minors

To determine whether second-order conditions are met requires examining a Hessian matrix's *principal minors*. Principal minors are sub-determinants of the initial matrix (the matrix itself is one of its own principal minors). A matrix with n rows (and n columns) has n principal minors. The principal minors for the 2×2 matrix above appear below. $|a| = a$, and

$$\begin{vmatrix} a & b \\ c & d \end{vmatrix} = (a \cdot d) - (b \cdot c).$$

The Hessian matrix H1 has three principal minors, $|u_{xx}|$,

$$\begin{vmatrix} u_{xx} & u_{xy} \\ u_{yx} & u_{yy} \end{vmatrix},$$

and

$$\begin{vmatrix} u_{xx} & u_{xy} & u_{xz} \\ u_{yx} & u_{yy} & u_{yz} \\ u_{zx} & u_{zy} & u_{zz} \end{vmatrix}.$$

The second-order condition for an $n \times n$ matrix depends on whether a maximum value or a minimum value of the function is indicated. The following table states both the first-order and second-order conditions for such a matrix. The table assumes that $u = u(x_1, x_2, x_3, \ldots, x_n)$. The principal minors of the Hessian are the terms $|H_1|, |H_2| \ldots$.

Tests for extreme values		
Condition	Maximum value	Minimum value
First-order	$u_{11} = u_{22} = \ldots = u_{nn} = 0$	$u_{11} = u_{22} = \ldots = u_{nn} = 0$
Second-order	$\lvert H_1\rvert < 0, \lvert H_2\rvert > 0, \lvert H_3\rvert < 0, \ldots, (-1)^n \cdot \lvert H_n\rvert < 0$	$\lvert H_1\rvert, \lvert H_2\rvert, \ldots, \lvert H_n\rvert > 0$

We apply this rule to two equations that have two independent variables, so that graphs illustrate the conditions. Then we consider an example in which a consumer chooses levels of three goods subject to an income constraint.

The first example has neither a maximum nor a minimum value. Rather, the (single) extreme value of this function is at a minimum when approached from one direction and a maximum when approached from another direction. This function exhibits a "saddle point." Setting $fx = fy = 0$ and solving reveals that the saddle point is at $x = 1$ and $y = 2$

```
f(x, y):=(x-1)^2-(y-2)^2$  fx : diff(f(x,y),x)$ fy:
 diff(f(x,y),y)$ fxx:diff(f(x,y),x,2)$ fxy:diff(fx,y)$
fyx : diff(fy, x)$     fyy:diff(f(x, y), y, 2)$
matrix(["f(x,y)", "fx", "fy"], [f(x, y), fx, fy]);
matrix(["fxx","fxy","fyx","fyy"],[ fxx,fxy,fyx,fyy]);
```

$$\begin{bmatrix} f(x,y) & fx & fy \\ (x-1)^2-(y-2)^2 & 2\,(x-1) & -2\,(y-2) \end{bmatrix} \quad \begin{bmatrix} fxx & fxy & fyx & fyy \\ 2 & 0 & 0 & -2 \end{bmatrix}$$

The graph shows that if $y = 2$ and x varies, then the movement toward the saddle point from either larger or smaller x values causes $f(x, 2)$ to decrease. In contrast, if $x = 1$ and y varies, then approaching the saddle point causes $f(1, y)$ to increase.[11]

This behavior is also indicated above in $fxx = 2$, so fx is negative for $x < 1$, zero for $x = 1$ and positive for $x > 1$, indicating that a minimum value is achieved when $fx = 0$ (with y held constant). Likewise, $fyy = -2$ indicates that a maximum value is achieved when $fy = 0$ (with x held constant).

```
wxdraw3d(view = [60, 300], xlabel ="x",
explicit(f(x, y), x, -10, 10, y, -10, 10) )$
```

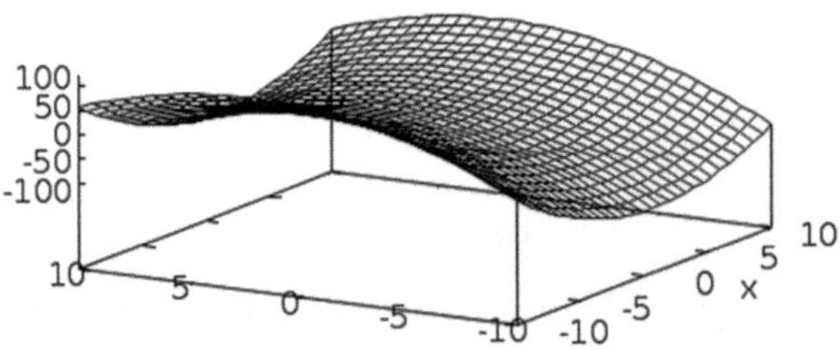

The behavior described above suggests that this function should fail the second-order conditions that are listed in the table above. It does: that $|H21| = 2$ violates the condition required for having found a maximum value, and that $|H22| = -4$ violates the condition required for having found a minimum value.

[11]The x axis is labeled. The y and z labels can be inferred.

```
subst([x=1,y=3],[gx, gy]);  H3:matrix([gxx,gxy],
[gyx,gyy]);  H31:determinant(H3[1,1])$  H32:
determinant(H3)$ matrix(["|H31|","|H32|"],[H31,H32]);
```

$$[0,0] \quad \begin{bmatrix} 2 & 0 \\ 0 & 2 \end{bmatrix} \quad \begin{bmatrix} |H31| & |H32| \\ 2 & 4 \end{bmatrix}$$

Next consider an expression that results in a minimum value for the function at its extreme point.

```
g(x, y):=x^2 +y^2-2*x-6*y+14$  gx:diff(g(x,y),x)$
gy:diff(g(x,y), y)$ gxx:diff(g(x,y),x,2)$
gxy:diff(gx,y)$  gyx :diff(gy, x)$
gyy:diff(g(x,y),y,2)$  matrix(["g(x,y)","gx","gy"],
  [g(x,y),gx,gy]);   matrix(["gxx","gxy","gyx","gyy"],
  [gxx,gxy,gyx,gyy]);
```

$$\begin{bmatrix} g(x,y) & gx & gy \\ y^2 - 6\,y + x^2 - 2\,x + 14 & 2\,x - 2 & 2\,y - 6 \end{bmatrix} \quad \begin{bmatrix} gxx & gxy & gyx & gyy \\ 2 & 0 & 0 & 2 \end{bmatrix}$$

The table above shows that $gxx > 0$ and $gyy > 0$. This implies that both gx and gy are negative for small values of x and y ($x < 1$ and $y < 3$, to be precise) and positive for larger values. Therefore, approaching the critical point (1, 3) from either direction causes $g(x, y)$ to decrease. The graph below confirms that the figure is bowl-shaped and has a minimum as its extreme value.

```
wxdraw3d(view=[70,310],xlabel="x",palette=gray,
  colorbox=false,enhanced3d=true,
  explicit(g(x,y),x,0,2,y,2,4)  )$
```

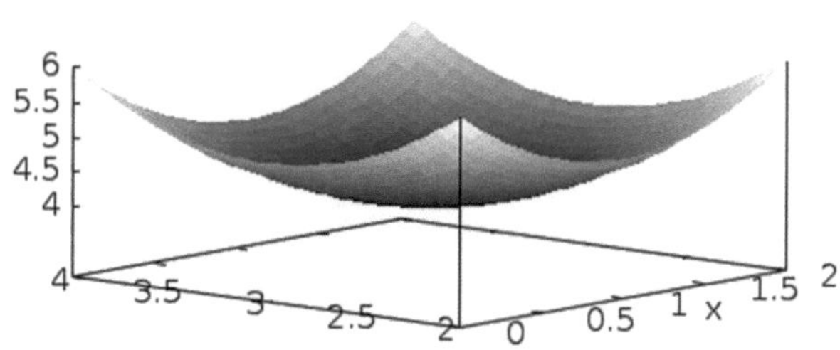

Finally, the Hessian H3 satisfies the conditions for a minimum value to have been identified. The first-order conditions are met: $gx = gy = 0$. Also, the second-order conditions are met: $|H31| > 0$ and $|H32 > 0|$.

```
subst([x=1,y=3],[gx, gy]);   H3:matrix([gxx,gxy],
 [gyx,gyy]); H31: determinant(H3[1,1])$
 H32: determinant(H3)$
matrix(["|H31|", "|H32|"], [H31, H32]);
```

$$[0,0] \quad \begin{bmatrix} 2 & 0 \\ 0 & 2 \end{bmatrix} \quad \begin{bmatrix} |H31| & |H32| \\ 2 & 4 \end{bmatrix}$$

4.4.2.3 Application to Optimization

We illustrate a constrained optimization problem using a Cobb–Douglas utility function of three goods and a corresponding budget constraint. The first line defines a budget expression G, a utility expression u, and the implied Lagrangian expression L.

We extract the requisite first and second derivatives. We use Young's theorem to keep the number of computed cross-partial derivatives to a minimum. The table shows the expressions for utility, the budget constraint, and the first partial derivatives. The second derivatives appear in the matrix below.

```
G:40-(2*x+3*y+4*z)$   u:x^.2*y^.5*z^.3$   L:u+M*(G)$
[Lx,Ly,Lz,LM]:[diff(L,x),diff(L,y),diff(L,z),
 diff(L,M)]$ [Lxx,Lyy,Lzz]:[diff(L,x,2),diff(L,y,2),
 diff(L,z,2)]$    [Lxy, Lxz, Lyz]:
 [diff(Lx, y),diff(Lx, z), diff(Lz,y)]$
[Gx, Gy, Gz]  :  [diff(G,x), diff(G, y), diff(G, z)]$
transpose(matrix(["Utility Function","---",
 "Budget Constraint","---","Lagrangian","---","Lx",
 "Ly","Lz"],[u,"---",G,"---", L,"---",Lx,Ly,Lz]));
```

$$\begin{bmatrix} \text{Utility Function} & x^{0.2}\, y^{0.5}\, z^{0.3} \\ --- & --- \\ \text{Budget Constraint} & -4\,z-3\,y-2\,x+40 \\ --- & --- \\ \text{Lagrangian} & (-4\,z-3\,y-2\,x+40)\,M + x^{0.2}\, y^{0.5}\, z^{0.3} \\ --- & --- \\ Lx & \frac{0.2\,y^{0.5}\,z^{0.3}}{x^{0.8}} - 2\,M \\ Ly & \frac{0.5\,x^{0.2}\,z^{0.3}}{y^{0.5}} - 3\,M \\ Lz & \frac{0.3\,x^{0.2}\,y^{0.5}}{z^{0.7}} - 4\,M \end{bmatrix}$$

The "bordered" Hessian matrix below contains the second-order partial derivatives of the utility function. The last row and the last column, which form the border, are the derivatives of the constraint. Alternatively, they can be determined as the second-order cross-partial derivatives. The first item in the fourth column is $\frac{\partial^2 L}{\partial x \partial M}$.

The first item in the fourth row has the same value, by virtue of Young's theorem. The other items are determined in like fashion, with $\frac{\partial^2 L}{\partial M^2} = 0$ being the final item.

```
HB: matrix([Lxx,Lxy,Lxz,Gx],[Lxy,Lyy,Lyz,Gy],
    [Lxz,Lyz,Lzz,Gz],[Gx,Gy,Gz,0] );
```

$$\begin{bmatrix} -\frac{0.16\,y^{0.5}\,z^{0.3}}{x^{1.8}} & \frac{0.1\,z^{0.3}}{x^{0.8}\,y^{0.5}} & \frac{0.06\,y^{0.5}}{x^{0.8}\,z^{0.7}} & -2 \\ \frac{0.1\,z^{0.3}}{x^{0.8}\,y^{0.5}} & -\frac{0.25\,x^{0.2}\,z^{0.3}}{y^{1.5}} & \frac{0.15\,x^{0.2}}{y^{0.5}\,z^{0.7}} & -3 \\ \frac{0.06\,y^{0.5}}{x^{0.8}\,z^{0.7}} & \frac{0.15\,x^{0.2}}{y^{0.5}\,z^{0.7}} & -\frac{0.21\,x^{0.2}\,y^{0.5}}{z^{1.7}} & -4 \\ -2 & -3 & -4 & 0 \end{bmatrix}$$

Next, mnewton solves for the optimum values of x, y, and z. The resulting utility level is reported. An exercise: Confirm that the consumer's expenditures are as follows: 20 % on x, 50 % on y, and 30 % on z, values that correspond to the coefficients of the utility function.

```
load(mnewton)$ soln:mnewton([Lx,Ly,Lz,LM],[x,y,z,M],
  [.1,.1,.1,.1])$  [x0,y0,z0,u0]:[rhs(soln[1][1]),
  rhs(soln[1][2]),rhs(soln[1][3]),subst(soln[1],u)]$
  matrix(["x0","y0","z0","u0"],[x0,y0,z0,u0]);
```

$$\begin{bmatrix} x0 & y0 & z0 & u0 \\ 4.0 & 6.6667 & 3.0 & 4.737 \end{bmatrix}$$

Next, evaluate the terms in the bordered Hessian, using the values determined in the cell above.

```
HB0: subst(soln[1],HB);
```

$$\begin{bmatrix} -0.0474 & 0.0178 & 0.0237 & -2 \\ 0.0178 & -0.0266 & 0.0355 & -3 \\ 0.0237 & 0.0355 & -0.111 & -4 \\ -2 & -3 & -4 & 0 \end{bmatrix}$$

Like extreme values in the absence of a constraint, the second-order conditions relate to the principal minors of the bordered Hessian matrix. With the bordered Hessian, however, only *border-preserving* principal minors are considered. A border-preserving principal minor of order r is the determinant of the matrix obtained by deleting $n - r$ rows and the corresponding columns. The border row and the border column cannot be deleted, though the elements of the border that correspond to the deleted rows and columns are deleted.

First, consider the border-preserving principal minor with $r = 1$ $(n - r) = 2$, because $n = 3$ in this illustration. As always, the principal minor for $r = 1$ is a four-element matrix with a zero in the lower right-hand corner and identical terms as the off-diagonal elements. The determinant must be negative, as it is in this case: $0 - (-4)^2 = -16$.

The first appearance of `determinant` in the next cell is preceded by the single-quote operator. This operator suppresses evaluation and shows the matrix for which the determinant is to be determined. In the next appearance of the command, the determinant is computed.[12]

[12]Be aware that the `matrix` command is being used in two different ways. The outermost command is used to build a table. The interior commands are used analytically.

```
matrix(['determinant(matrix([HB0[3][3],HB0[3][4]],
  [HB0[4][3],HB0[4][4]])),"=",determinant(matrix(
  [HB0[3][3],HB0[3][4]],[HB0[4][3],HB0[4][4]]))]);
```

$$\left[\text{determinant}\left(\begin{bmatrix}-0.111 & -4\\ -4 & 0\end{bmatrix}\right) = -16\right]$$

The second-order condition to establish a maximum is this: Given that the first-order conditions are satisfied, the border-preserving principal minors of order r must have the sign $(-1)^r$, for $r = 2, \ldots, n$. That is, the border-preserving principal minors must alternate in sign, beginning with a positive second-order minor.

In this example, $n = 3$, so we consider $r = 2$ $(n-r = 1)$ and $r = 3$ $(n - r = 0)$, the latter being the determinant of the original matrix. Thus, we delete the two determinants below, with the attached number indicating the value of r. The first term $(r = 2)$ is positive, as it should be: $(-1)^2 > 0$. The third term $(r = 3)$ is negative, as it should be: $(-1)^3 < 0$.

```
minor(HB0,1,1); [HB2:determinant(minor(HB0,1,1)),
       HB3 : determinant(HB0)];
```

$$\begin{bmatrix}-0.0266 & 0.0355 & -3\\ 0.0355 & -0.111 & -4\\ -3 & -4 & 0\end{bmatrix} \qquad [2.2738, -0.168]$$

The preceding material sketches the development and application of Hessians in determining the nature of the second-order terms. If computation alone is required, then *Maxima* provides the information directly. The first command below produces the general Hessian for the example above. The second command produces the Hessian given the specific values in `soln`.

```
HB:hessian(L, [x,y,z,M]); HB0:subst(soln,HB);
```

$$\begin{bmatrix}-\frac{0.16\,y^{0.5}\,z^{0.3}}{x^{1.8}} & \frac{0.1\,z^{0.3}}{x^{0.8}\,y^{0.5}} & \frac{0.06\,y^{0.5}}{x^{0.8}\,z^{0.7}} & -2\\ \frac{0.1\,z^{0.3}}{x^{0.8}\,y^{0.5}} & -\frac{0.25\,x^{0.2}\,z^{0.3}}{y^{1.5}} & \frac{0.15\,x^{0.2}}{y^{0.5}\,z^{0.7}} & -3\\ \frac{0.06\,y^{0.5}}{x^{0.8}\,z^{0.7}} & \frac{0.15\,x^{0.2}}{y^{0.5}\,z^{0.7}} & -\frac{0.21\,x^{0.2}\,y^{0.5}}{z^{1.7}} & -4\\ -2 & -3 & -4 & 0\end{bmatrix}$$

$$\begin{bmatrix}-0.0474 & 0.0178 & 0.0237 & -2\\ 0.0178 & -0.0266 & 0.0355 & -3\\ 0.0237 & 0.0355 & -0.111 & -4\\ -2 & -3 & -4 & 0\end{bmatrix}$$

The illustrative example above confirms that the values determined by satisfying the first-order conditions yield maximum utility, given a Cobb–Douglas utility function and a set of product prices. We could have stated the problem as an expenditures minimization problem, where the consumer wishes to minimize G for a given value of u. The resulting values of x, y, and z would have been the same, but the second-order condition would have been different.

The second-order condition to establish a minimum is this: Given that the first-order conditions are satisfied, the border-preserving principal minors of all orders must be negative.

4.4.2.4 Multiple Constraints

The technique illustrated above can be extended to incorporate multiple constraints, as long as the number of constraints is less than the number of choice variables. As one might expect, the resulting bordered Hessian contains an additional row and column for each added constraint. Applications of the Lagrangian technique with two constraints are in [1], Chap. 10.

References

1. Baldani J, Bradfield J, Turner RW (1996) Mathematical economics. Dryden, Fort Worth
2. Doi J, Iwasa K, Shimomura K (2009) Giffen behavior independent of the wealth level. Econ Theory 41:247–267
3. Nicholson W, Snyder C (2008) Microeconomic theory: basic principles and extensions, 11th edn. South-Western, Mason

Chapter 5
Preferences and Demand

Abstract Chapter 4 analyzes optimization and its implications for utility maximization subject to one or more constraints, and it develops the relationship between money income and the quantity consumed of a good. This chapter addresses the effects of changes in the good's price. Such a change causes both the relative prices of the goods consumed and the consumer's real income to change. Initially, analysis is based on the CES utility function. Then the analysis turns to the case of a Giffen good. A large amount of time is spent on the latter, not because of its inherent importance, but because it provides added insights into the nature of the consumer's reaction to price and income changes.

5.1 The CES Utility Function

We begin with a review of the basics. The first input group restates the CES utility function for use below along with the function for the marginal rate of substitution, mrs. This section follows [2].

```
kill(all)$  u(x,y,a,b):=(a*x^-b+(1-a)*y^-b)^-(1/b);
mrs(x,y,a,b):=''(diff(u(x,y,a,b),x)/
    diff(u(x,y,a,b),y));
```

$$\mathrm{u}(x,y,a,b) := \left(a\,x^{-b} + (1-a)\,y^{-b}\right)^{-\frac{1}{b}}$$

$$\mathrm{mrs}(x,y,a,b) := \frac{a\,x^{-b-1}\,y^{b+1}}{1-a}$$

The mrs expression above can be rephrased in terms of y/x, named `yxratio` below:

$$mrs(x,y,a,b) = \frac{a}{1-a} \cdot x^{-(b+1)} \cdot y^{b+1} = \frac{a}{1-a} \cdot \left(\frac{y}{x}\right)^{b+1}.$$

M.R. Hammock and J.W. Mixon, *Microeconomic Theory and Computation: Applying the Maxima Open-Source Computer Algebra System*,
DOI 10.1007/978-1-4614-9417-1_5, © Springer Science+Business Media New York 2013

Set $mrs = px/py$ and solve for $\frac{y}{x}$, write the budget line as $m = px \cdot x + py \cdot yxratio \cdot x$, and solve for x. The result, the single output line below, is the function for the optimal quantity of x given values for px, py, m, a, and b.

The last two input lines generate the counterpart for the optimal quantity of y. This function is used below for graphing but does not directly pertain to the text material, so it is not printed.

```
yxratio(px,py,a,b):= ''(((px/py)*(1-a)/a)^(1/(1+b)))$
xopt(px,py, m,a,b):=''(m/(px+py*yxratio(px,py,a,b)));
yopt(px,py,m,a,b):=''(xopt(px,py,m,a,b)*
    yxratio(px,py,a,b))$
```

$$\mathrm{xopt}(px, py, m, a, b) := \frac{m}{\left(\frac{(1-a)\,px}{a\,py}\right)^{\frac{1}{b+1}} py + px}$$

The next input group defines the utility function's parameters and the initial values of the three pertinent variables.

```
[a0,b0]:[0.3,0.8]$ [px0,py0,m0]:[50,12,1200]$
```

We now define the indirect utility function, expressed as a function of the three variables above and the two parameters of the original utility function. The function is printed only to allow a visual check regarding its arguments. Therefore, no attempt is made to simplify it.

The last two output lines confirm that placing a given pair of x and y values into both the original utility function and the indirect utility function yields the same numerical value for utility. That value is assigned the name `u0`.

```
u(xopt(px,py,m,a,b), yopt(px,py, m,a,b),a,b);
v(px,py,m,a,b):=''%$ [u0:float(v(px0,py0,m0,a0,b0)),
float(u(xopt(px0,py0,m0,a0,b0),
        yopt(px0,py0,m0,a0,b0),a0,b0))];
```

$$\frac{1}{\left(\frac{1-a}{\left(\frac{m\left(\frac{(1-a)\,px}{a\,py}\right)^{\frac{1}{b+1}}}{\left(\frac{(1-a)\,px}{a\,py}\right)^{\frac{1}{b+1}} py+px}\right)^{b}}+\frac{a}{\left(\frac{m}{\left(\frac{(1-a)\,px}{a\,py}\right)^{\frac{1}{b+1}} py+px}\right)^{b}}\right)^{\frac{1}{b}}}$$

$$[27.11, 27.11]$$

The next input/output group assigns names to the initial values of `xopt` and `yopt`, and utility. These values are used in graphs that follow.

```
[[xopt0,yopt0]:[xopt(px0,py0,m0,a0,b0),yopt(
   px0,py0,m0,a0,b0)], u0:u(xopt0,yopt0,a0,b0)];
```

$$[[12.979, 45.92], 27.11]$$

5.1.1 The Uncompensated Demand Curve

We now derive a demand curve for good x, given a set of preferences, a money income level, and a price of good y. We select two prices of x, one higher than the initial price and one lower than the initial price, and determine the resulting quantities of x. This provides three points on the "uncompensated" demand curve. Here "uncompensated" means that the consumer's utility level is allowed to change.

First set $px1 < px0$ and then set $px2 > px0$. The order matters only for purposes of graphing. Then define corresponding values for `xopt`, `yopt`, and `u`. The final command collects the x, y pairs, including the initial pair, for use in the graph below.

```
[px1,px2]:[25, 80]$ [xopt1,yopt1,u1]:[xopt(
  px1,py0,m0,a0,b0), yopt(px1,py0,m0,a0,b0),
  u(xopt1,yopt1, a0, b0)]$ [xopt2,yopt2,u2]:
float([xopt(px2,py0,m0,a0,b0),yopt(px2,py0,m0,a0,b0),
u(xopt2,yopt2,a0,b0)])$ matrix(["x","y","u"],
 [xopt1,yopt1,u1],[xopt0,yopt0,u0],[xopt2,yopt2,u2]);
 matrix(
  ["x","p"],[xopt1,px1],[xopt0,px0],[xopt2,px2]);
```

$$\begin{bmatrix} x & y & u \\ 22.269 & 53.607 & \frac{1}{\left(\frac{0.7}{yopt1^{0.8}}+\frac{0.3}{xopt1^{0.8}}\right)^{1.25}} \\ 12.979 & 45.92 & 27.11 \\ 8.8807 & 40.795 & \frac{1}{\left(\frac{0.7}{yopt2^{0.8}}+\frac{0.3}{xopt2^{0.8}}\right)^{1.25}} \end{bmatrix} \qquad \begin{bmatrix} x & p \\ 22.269 & 25 \\ 12.979 & 50 \\ 8.8807 & 80 \end{bmatrix}$$

A lower price decreases consumption of x and utility, and a higher price does the opposite, as the consumer moves along the (uncompensated) demand curve.

We develop the graph of the demand curve in three stages. The first stage produces a "scenario" that graphs the relevant indifference curves and budget lines. The scenario is named `indiff`. The next stage produces a scenario, `uncompdem` that shows the uncompensated demand curve. The third stage draws the graphs implied by the scenarios.

For the indifference curves and budget lines, we set a range of y values that begin at zero and end just above the value at which all income is spent on good y, $m0/py0$. The commands that produce the curves also determine the range of x values.

We do not derive an explicit function for the indifference curves. Rather, we use `implicit` to plot values of x and y that are consistent with each of the three utility levels that have been defined above. Each indifference curve is accompanied by a budget line that is stated as an explicit function. The first indifference curve/budget line pair is the one that pertains when $px = px0$, as indicated by the label. Thus, each label applies to two curves. Note the use of `key=""` to suppress a second printing of the legend entry.

```
indiff: gr2d( ... Some commands omitted ...
 key="px=px0",xtics={xopt1,xopt0,xopt2},
 ytics={20,80},implicit(u(x,y,a0,b0)-u0,x,1,m0/
  px1,y,1,m0/py0),key = "", explicit(m0/py0 -
  px0*x/py0,x,0,m0/px0), ... Two sets of implicit &
  explicit commands omitted ...  )$
```

The demand curve representation consists of two parts. The demand curve is graphed using the `implicit` command. One cannot derive an explicit expression for px as a function of x, so we use the implicit expression below. The second part of the graph consists of the three points on this curve that we derived above.

```
uncompdem:gr2d( ... Some commands omitted ...
 xtics={xopt1,xopt0,xopt2},ytics={px1,px0,px2},
 ... key= "Uncomp. demand",implicit(x-xopt(
 px,py0,m0,a0,b0),x,0,1.5*xopt1,px,0,100), key="",
 points( [[xopt1,px1],[xopt0,px0],[xopt2,px2]]))$
```

Finally, `wxdraw` produces the graphs that these two scenarios have generated. The demand curve is directly below the indifference curve/budget line graph, so the x values can be compared. The horizontal axes on both graphs are identical, but the vertical axes are completely different. The vertical axis on the top graph measures a number of physical units of good y. The vertical axis on the bottom graph measures the number of dollars spent per unit of good x. The similarity of the numbers is an artifact of the particular specification of this illustration.

```
wxdraw(indiff,uncompdem,dimensions=[480,360])$
```

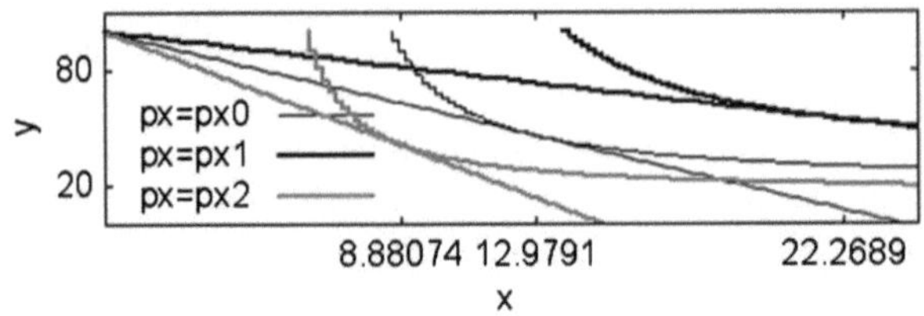

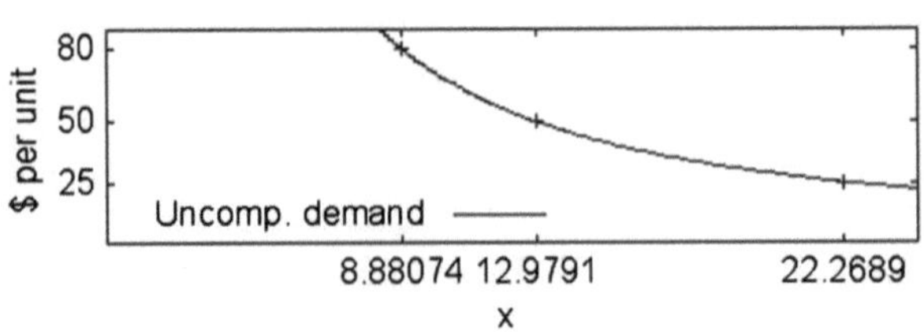

5.1.2 Compensation

We next compensate the consumer for the price level change. That is, we change the income level by enough to restore the consumer's initial utility level. First, we generate two sets of commands that determine the money income change required to compensate the consumer. We must simultaneously determine the income level change and the change in the quantity of x. Therefore, we apply `mnewton`, to a pair of expressions. We enter three lists: a list of expressions, a list of variables, and a list of guesses. The first expression sets $u(x, y, a0, b0) = u0$. The second expression sets the marginal rate of substitution equal to the price ratio.

We first reduce the price of x from 50 to 25. The compensating money income level changes from $m = 1200$ to $m = 847.13$, a reduction of 352.87.

```
soln:mnewton([u(x,y,a0,b0)-u0,mrs(x,y,a0,b0)-px1/py0],
[x,y],[xopt0,yopt0])$ [xcomp1,ycomp1]:
 [rhs(soln[1][1]),rhs(soln[1][2])]$
 m1:px1*xcomp1+py0*ycomp1$    matrix(["xcomp",
 "ycomp","m1","m1 - m0"],[xcomp1,ycomp1,m1,m1-m0]);
```

$$\begin{bmatrix} xcomp & ycomp & m1 & m1 - m0 \\ 15.721 & 37.843 & 847.13 & -352.87 \end{bmatrix}$$

Next, we raise x's price from 50 to 80. To compensate for this price increase, money income must increase by 366.12, from 1200 to 1566.12.

```
soln:mnewton([u(x,y,a0,b0)-u0,mrs(x,y,a0,b0)-px2/py0],
  [x,y],[xopt0,yopt0])[1]$      [xcomp2, ycomp2]:
  [rhs(soln[1]),rhs(soln[2])]$ m2:px2*xcomp2+
 py0*ycomp2$    matrix(["xcomp","ycomp","m2",
 "m2 - m0"],  [xcomp2,ycomp2,m2,m2-m0]);
```

$$\begin{bmatrix} xcomp & ycomp & m2 & m2 - m0 \\ 11.59 & 53.241 & 1566.1 & 366.12 \end{bmatrix}$$

We now determine the nature of the compensated demand curve. First, we determine the value of y given a utility level u and the utility function. That solution appears as the single item in an output list named `solny`. We use that solution just below. Before doing so, we discuss the dialog with *Maxima* that is required before the solution is achieved.

Maxima must know the answer to two questions.[1] The answer to the first question, `Is 1/b an integer?`, is no. The answer to the second question

[1]More questions would have to be answered, but we use `assume` and `declare` to anticipate two questions.

is not obvious, but we can show that the term $(a - 1) \cdot (a \cdot u^b - x^b)$ is positive.[2]

```
assume(u>0)$ declare(b,noninteger)$
solny:solve(u(x,y,a,b)-u,y);
```

Is $\frac{1}{b}$an integer? n; Is $(a-1)\ x\ \left(a\ u^b - x^b\right)$ positive, negative, or zero? p;

$$[y = \left(\frac{a\,u^b\,x^b}{a\,u^b-x^b} - \frac{u^b\,x^b}{a\,u^b-x^b}\right)^{\frac{1}{b}}]$$

To derive an expression for the inverse compensated demand curve for x, substitute the output above for y in the *mrs* function. The resulting function, named `pxc(u, x, py, a, b)` is defined below. The output shows the right-hand side of this equation, confirming the variables and parameters that appear in the equation's name.

```
pxc(u,x,py,a,b):=''(rhs(solve(mrs(
x,rhs(solny[1]),a,b)-px/py,px)[1]));
```

$$\mathrm{pxc}\left(u,x,py,a,b\right) := -\frac{a\,py\,x^{-b-1}\left(\left(\frac{a\,u^b\,x^b}{a\,u^b-x^b}-\frac{u^b\,x^b}{a\,u^b-x^b}\right)^{\frac{1}{b}}\right)^{b+1}}{a-1}$$

The demand curve is defined for values of x such that $x^b > a \cdot u^b$. To restrict the graph to that range, we create an expression `xcrit` below. Also, we create a list of the three x, px pairs derived above. The graph contains the uncompensated demand curve shown earlier along with its counterpart when the consumer is compensated for any change in the price of x.

```
xcrit:(a0*u0^b0)^(1/b0)$
print("The critical value of x is ",xcrit)$
wxdraw2d( ... Some commands omitted ... key=
 "Uncomp. demand",implicit(x-xopt(px,py0,m0,a0,b0),
 x, xcrit+.1,1.5*xopt1,px,0,100),key="",
 points([[xopt1,px1],[xopt0,px0],  [xopt2,px2]]),
 line_width=2,key= "Comp. demand",explicit(
 pxc(u0,x,py0,a0,b0),x,xcrit+0.1,20),key="",
 points([[xcomp1,px1],[xopt0,px0],[xcomp2,px2]]),
 ... line_type=dots,points([0,xopt0],[px0,px0]),
 points([0,xopt1],[px1,px1]),points([0,xopt2],
 [px2,px2]), points([xopt0,xopt0], [0,px0]) )$
```

[2]The utility function can be written as $u^b = \frac{1}{\frac{a}{x^b} + \frac{1-a}{y^b}}$.

Therefore, $a \cdot u^b - x = \frac{a}{\frac{a}{x^b} + \frac{1-a}{y^b}} - x$. The right-hand side of this term equals $\{a \cdot x^b \cdot y^b - x^b \cdot [(1-a) \cdot x^b + a \cdot y^b]\}/D$, where $D = (1-a) \cdot x^b + a \cdot y^b$. The denominator D is positive, so the sign of the numerator determines the sign of $a \cdot u^b - x^b$. Multiplying through by x^b and collecting terms shows that the numerator is $-(1-a) \cdot x^{2 \cdot b}$, which must be negative because $a < 1$. This negative term is multiplied by the negative value $(a - 1) \cdot x$, resulting in a positive expression.

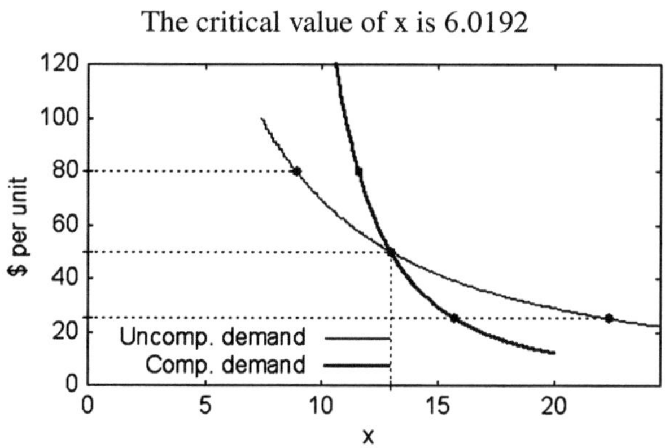

The dashed lines are added to the graph to show the area that corresponds to the change in consumer surplus that a price level change causes. When the price of x falls from 50 to 25, the consumer gains for two reasons. The 12.979 units that were initially consumed now cost less. This gain is a pure transfer, not related to the consumer's behavior. The other reason is that the consumer now consumes additional units.[3]

Look at the second part of this gain first. Integrate the inverse compensated demand curve from `xopt0`, 12.979, to `xcomp1`, 15.721. *Maxima*'s symbolic integration command `integrate` cannot be used here, but we can use the `romberg` numerical method to determine the area. The area to be calculated is the height of the demand curve (marginal willingness to pay) *less* $px1$ over the range indicated just above. The result, shown below, is 28.289. Adding this value to the area that corresponds to the transfer results in 352.87, the same value we determined above.

```
comptriangle:romberg(pxc(u0,x,py0,a0,b0)-px1,x,
 xopt0,xcomp1)$  compcschange:comptriangle+(xopt0*
 (px0-px1))$  matrix(["Comp Triangle",
 "Comp CS change"],[comptriangle,compcschange]);
```

$$\begin{bmatrix} \text{Comp Triangle} & \text{Comp CS change} \\ 28.389 & 352.87 \end{bmatrix}$$

In general one cannot expect to see a compensated demand curve. In contrast, it might be possible to estimate the effect that a price change has on consumers' behavior when money income is held constant. That is, it might be possible to estimate the compensated demand curve. The input/output cell below shows the result of this calculation given the parameters in this illustrative case.

[3]The price increase to 80 has symmetrical effects: Each of the 8.8 units still bought costs more, and the consumer loses the net gain from buying 4.2 (13–8.8) units at the lower initial price.

Again, we use the `romberg` numerical method for integration. We now have x stated as a function of px, so the integration relates to the vertical axis. That is, we calculate the expression shown in the first output line below. The resulting area is the roughly trapezoidal area defined in the graph. The calculated result, 417.86, exceeds its counterpart for the compensated demand curve, as the graph shows. Thus, the (possibly observable) compensated demand curve tends to overstate the increase in consumer surplus that results from a price decrease.

```
[integrate(xopt(px,py,m0,a0,b0),px, px1,px0),
uncompcschange:romberg(xopt(px,py0,m0,a0,b0),px1,
 px0),uncomptriangle:uncompcschange-(px0-px1)*xopt0];
```

$$[1200 \quad \int_{25}^{50} \frac{1}{1.6011\left(\frac{px}{py}\right)^{0.556} py+px} dpx, \quad 417.86, \quad 93.386]$$

An important question regards the size of the discrepancy between the theoretically appropriate value and its observable counterpart. The two computations below show that for this example, the discrepancy is 18.4 % for the entire area. This comparison tends to give a small discrepancy, however, because both values share a large common value, the transfer inherent in the price change. Focusing on the change due to the consumer's behavior shows a quite different value, with the uncompensated demand curve yielding a value 228.95 % larger than its counterpart for the uncompensated demand curve.

In large part, this great discrepancy reflects the large income effect of the changed price of x, which results from the fact that our analysis uses two goods. In real cases, the income effect of a price level change will be much smaller than in this illustration, and the difference between the observable and appropriate values will be much smaller. See [1].

```
[(uncompcschange-compcschange)/compcschange,
(uncomptriangle-comptriangle)/comptriangle];
```

$$[0.184, 2.2895]$$

5.1.3 An Alternative Measure

We earlier referred to the compensation required to restore the consumer to the initial utility level given a price change as *the* theoretically proper measure of the change in consumer surplus due to the price change. A better characterization is that such a measure is *one* theoretically proper measure. An alternative is this: Determine the size of the income change that would be sufficient to compensate the consumer for *forgoing* a price change. That is, the new utility level, not the initial one, becomes the reference point, and the payment in lieu of a price change is considered an *equivalent variation*. The next cell determines the size of this variation for the consumer being considered here.

For the price reduction being considered, the consumer would have to be awarded $499.85 to compensate for not being allowed to enjoy the reduction.

```
soln:mnewton([u(x,y,a0,b0)-u1,mrs(x,y,a0,b0)-
 px0/py0], [x,y],[xopt0,yopt0])$ [[xequiv1:
 rhs(soln[1][1]), yequiv1: rhs(soln[1][2])],
m3: px0*xequiv1+py0*yequiv1,m3-m0];
```

$$[[18.386, \quad 65.048], \qquad 1699.\ 9, 499.85]$$

The next cell reproduces the preceding graph and adds a second demand curve along which utility is held constant, the "equivalent demand curve." The graph shows that the change in consumer surplus when the equivalent measure is used exceeds its counterparts when the compensating measure is used and when the uncompensated demand curve is used. Indeed, the graph shows that the measure of the change in consumer surplus using the uncompensated demand curve is bracketed by the two demand curves for which utility is held constant at one level or the other.

```
wxdraw2d( ... Some commands omitted ...
 key = "Uncomp. demand",implicit(x - xopt(
 px,py0,m0,a0,b0),x,xcrit+.1,1.5*xopt1,px,0,100),
 key = "", points( [[xopt1,px1],[xopt0,px0],
 [xopt2,px2]] ), points([0, xequiv1],[px0,px0]),
 ... explicit(pxc(u0,x,py0,a0,b0),x,xcrit+0.1,
 1.1*xopt1),key = "Equiv. demand, u = u1", ...
 explicit(pxc(u1,x,py0,a0,b0),x,xcrit+0.1,
 1.1*xopt1), ... points([[xcomp1,px1],[xopt0,px0],
 [xcomp2,px2]]), ... label(["A",xopt0-0.5,
 px0-3]), label(["B", xcomp1, px1-3]), ...)$
```

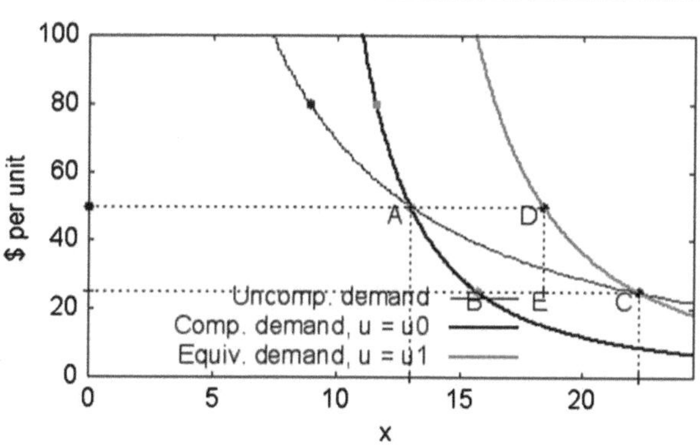

The computation below confirms the information in the graph. To determine the area that corresponds to the change in consumer surplus, we calculate the area *xequiv*1 $\cdot$ $(p0 - p1)$ plus the area ECD. The latter area is determined by the `romberg` method. We see that the area is $499.85, which exceeds the area defined by the uncompensated demand curve, $417.86.

```
equivtrapezoid: (px0-px1)*xequiv1+romberg(
pxc(u1,x,py0,a0,b0)-px1,x,xequiv1,xopt1);
```

499.85

5.2 A Giffen Good

This section extends Chap. 4's treatment of an inferior good. It largely parallels the treatment of the CES case and can be omitted without loss of continuity.

The utility function below can represent cases in which both goods are normal goods, one good is a non-Giffen inferior good, and one good is a Giffen good, depending on the parameter values. We select values that produce the third case.

```
u(x,y,a,b,c):=a*log(x)+b*log(y)+c*x*y;
[a0,b0,c0]:[40,50,-0.1]$
```

$$\mathrm{u}(x, y, a, b, c) := a \log(x) + b \log(y) + c\, x\, y$$

A typical indifference curve passes through four areas: one in which both goods are normal goods, one in which Good x is an inferior good, and one within which Good x is a Giffen good (actually, a subset of the inferior good area), and an area in which the marginal utility of x is negative. The boundary conditions that define these areas are stated in terms of $x \cdot y$. For `xy1`, for example, $x \cdot y = -\frac{a0}{c0}$. Therefore, for this function, y = `xy1/x`. Likewise for the other two. The second boundary condition divides the graph into Giffen/non-Giffen areas, and the third boundary condition divides the graph into inferior/normal regions.

```
[xy1,xy2,xy3]:[-a0/c0, -(b0-sqrt(b0*(b0-a0)/2))/c0,
    -(b0 - sqrt(b0*(b0-a0)))/c0]$
```

The graph below shows a representative indifference curve, for $u = 230$ and the boundary conditions defined above. Points to the northeast of the curve labeled `mux = 0` in the graph below show x, y pairs for which the marginal utility of x is negative. Between this curve and the one labeled `Giffen`, x is a Giffen good. Between `xy2: Giffen` and `xy3: Inferior`, x is a non-Giffen inferior good. To the southwest of `xy3: Inferior`, both x and y are normal goods.

```
wxdraw2d ... Some commands omitted ...
 implicit(u(x,y,a0,b0,c0)-230,x,0.01,30,y,0.1,35),
 ... key="xy1:mux=0",explicit(xy1/x,x,0.1,30), ...
 key="xy2: Giffen",explicit(xy2/x,x,0.1,30), ...
 key="xy3: Inferior",explicit(xy3/x,x,0.1,30))$
```

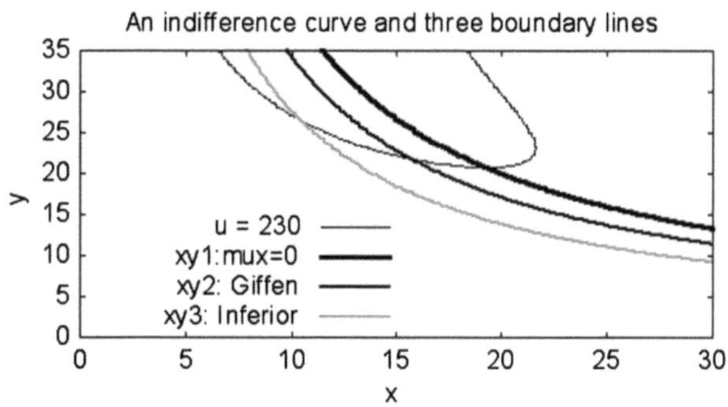

A consumer would never move beyond the line labeled `xy1: mux = 0` unless forced or paid to do so. A consumer would not give up any y for the unit of x that yields zero additional utility. In the area bounded by `xy3:Inferior` and `xy2:Giffen`, x is an inferior good but not a Giffen good. In the area bounded by `xy2:Giffen` and `xy1:mux=0`, x is a Giffen good. In the area to the left of `xy3:Inferior` (the largest part of the graph for this set of parameters), x is a normal good.

Determining the optimal consumption pattern requires numerical methods. We determine the optimal outcome by setting the slope of the indifference curve equal to that of the budget line. The cell below shows the equation for the marginal rate of substitution.

```
mrs(x,y,a,b,c):=
''(radcan(diff(u(x,y,a,b,c),x)/
diff(u(x,y,a,b,c),y)));
```

$$\text{mrs}(x, y, a, b, c) := \frac{c\,x\,y^2 + a\,y}{c\,x^2\,y + b\,x}$$

To rephrase the mrs function in terms of m and p, insert the budget constraint directly into the mrs function, generating an "augmented" mrs function below.[4] This function restates the marginal rate of substitution in terms of m, x, and p, along with the original parameters.

```
mrs_aug(a,b,c,m,x,p):=''(mrs(x,m-p*x,a,b,c));
```

$$\text{mrs\_aug}(a, b, c, m, x, p) := \frac{c\,x\,(m-p\,x)^2 + a\,(m-p\,x)}{c\,x^2\,(m-p\,x) + b\,x}$$

To guide the numerical analysis, we produce a graph of the difference between mrs and p. We look at changes in either m or p, but not both, so three curves are considered. All intersect the horizontal axis at values below $x = 30$. The behavior for larger x values can be ignored.

Two of the three cases result in undefined functions at specific values. We use these values to guide the selection of endpoints for `find_root` below, ensuring that both ends are within the first, relevant segment of the curve.

[4]From the budget constraint, $m = y + p \cdot x$, with y units defined so that its price is 1. Thus, $y = m - p \cdot x$.

```
[m0, m1]:[22.5, 25]$ [p0,p1]:[0.2,0.3]$  wxdraw2d(
  ... Some material omitted ... explicit(mrs_aug(
  a0,b0,c0,m0,x,p0)-p0,x,.11,50),...
  key="m1,p0",explicit(mrs_aug(a0,b0,c0,m1,x,p0)-p0,
  x,.11,50),color=gray,key="m0,p1",
 explicit(mrs_aug(a0,b0,c0,m0,x,p1)-p1,x,.11,50))$
```

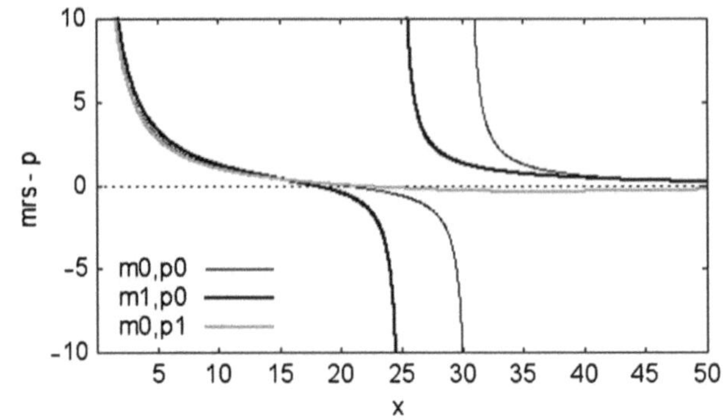

5.2.1 Analyzing Consumer Behavior

Maxima's `find_root` function determines the numerical values that appear below. In each case, a level of m is specified and the augmented mrs is set equal to a price.[5] The expression to be evaluated is entered as an implicit function, $mrs\_aug - p$.

The results of the computation appear in the table. The first entry is the value x, and the second is that of y. The third item is expenditures, which is compared to the specified value of m, the fourth item. The final item is the utility level. Once these base values are determined, we raise money income, price constant, and then we raise price, money income constant. We leave the analysis of income and price reductions as exercises.

```
[m0,m1]:[22.5,25]$ [p0,p1]:[0.2,0.3]$   x0:find_root(
 mrs_aug(a0,b0,c0,m0,x,p0)-p0,x,.1,25)$  y0:m0-p0*x0$
u0:u(x0,y0,a0,b0,c0)$      matrix(["x0","y0",
 "spending","m0","u0"],[x0,y0,y0+p0*x0,m0,u0]);
```

[5]We selected the values for m and p, m0 and po as follows. We started with $x = 10$ and used the xy1 and xy2 functions to find a range of y values. We selected a value near the middle of this range and determined the mrs as this point. The initial price is approximately that mrs value, and the resulting income level is the one required to fit a budget line through the selected point.

$$\begin{bmatrix} x0 & y0 & \text{spending} & m0 & u0 \\ 20.145 & 18.471 & 22.5 & 22.5 & 228.72 \end{bmatrix}$$

5.2.2 *The Income Effect*

First, consider the effects of increased income, holding price constant. The cell below increases income. As a result, the consumption of the inferior good x decreases and that of the normal good y increases. The utility level increases. The table reproduces the base values for comparison.

```
x1:find_root(mrs_aug(a0,b0,c0,m1,x,p0)-p0,x,.1,22)$
y1: m1-p0*x1$  u1:u(x1,y1,a0,b0,c0)$   matrix(
 ["m","x","y","spending","u"],[m0,x0,y0,
 y0+p0*x0,u0], [m1,x1,y1,y1+p0*x1,u1]  );
```

$$\begin{bmatrix} m & x & y & \text{spending} & u \\ 22.5 & 20.145 & 18.471 & 22.5 & 228.72 \\ 25 & 17.721 & 21.456 & 25.0 & 230.27 \end{bmatrix}$$

Over the range examined, the income elasticity of demand for x is approximately -1.2: a 1 % income change causes the quantity to change by about 1.2 % *in the opposite direction.*

```
((x1-x0)/(x1+x0))  /  ((m1-m0)/(m1+m0));
```

-1.2165

The graph below illustrates these results. The thin gray lines represent the initial budget line and indifference curve. The thick black lines represent the second budget line and indifference curve. The thick gray line segment that connects the optimal consumption points approximates part of the consumer's Income Consumption Curve (ICC).

```
[xlow,xhigh]:[min(x0,x1),max(x0,x1)]$ [ylow, yhigh]:
 [min(y0,y1),max(y0,y1)]$ wxdraw2d( ... Some
 commands omitted ... key="u = u0",implicit(
 u(x,y,a0,b0,c0)-u0,x,.75*xlow,1.5*xhigh,y,
 0.8*ylow,1.2*yhigh),key="m = m0",explicit(m0-p0*x,
 x,0,1.5*xhigh), ... key="u = u1l",implicit(
 u(x,y,a0,b0,c0)-u1,x,.75*xlow,1.5*xhigh,y,
 0.8*ylow,1.2*yhigh),key="m = m1",explicit(m1-p0*x,
 x,0,1.5*xhigh),key= "ICC",color= gray50, ...
 points_joined=true, points([x0, x1], [y0, y1]) )$
```

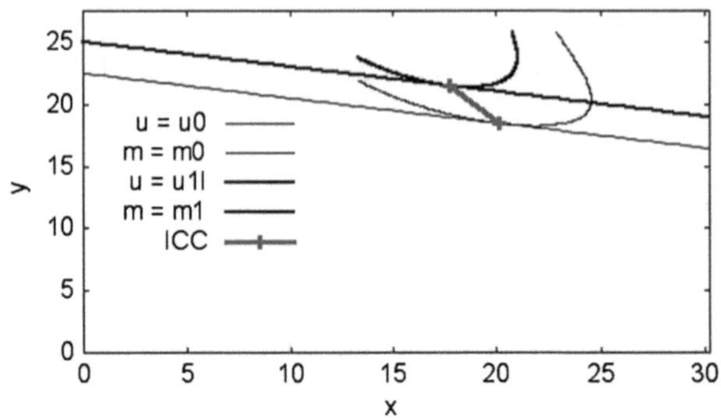

5.2.3 *The Price Effect*

The output below shows the effects of changing x's price with income held constant. The income level is returned to $m = m0$. For ease of comparison, we gather the values into a matrix. The top row shows the initial values. The bottom row shows the effect of increasing x's price: x consumption increases, and y consumption decreases, as does utility. These results confirm that the uncompensated demand curve for this good slopes upward over the price range considered.

```
x2:find_root(mrs_aug(a0,b0,c0,m0,x,p1)-p1,x,.1,25)$
y2:m0-p1*x2$       u2:u(x2,y2,a0,b0,c0)$    matrix(
   ["p","x","y","u","spending =  m"],
   [p0,x0,y0,u0,y0+p0*x0],[p1,x2,y2,u2,y2+p1*x2]);
```

$$\begin{bmatrix} p & x & y & u & \text{spending} = m \\ 0.2 & 20.145 & 18.471 & 228.72 & 22.5 \\ 0.3 & 20.771 & 16.269 & 227.01 & 22.5 \end{bmatrix}$$

The uncompensated demand curve for a Giffen good slopes upward, so the price elasticity is positive. The computation below confirms that, over the selected range of price and for the selected money income level, the price elasticity value is a small positive number.

```
((x2-x0)/(x2+x0))/((p1-p0)/(p1+p0));
```

0.0765

As with the income effect, the graph illustrates the effects defined above. The line segments are part of the Price Consumption Curve (PCC).

```
[xlow, xhigh]: [min(x0, x2), max(x0, x2)]$
[ylow, yhigh]: [min(y0, y2), max(y0, y2)]$
wxdraw2d( ... Some commands omitted ...
 implicit(u(x,y,a0,b0,c0)-u0,x,.75*xlow,1.5*xhigh,
 y,0.8*ylow,1.2*yhigh),key=concat("p = p0, x =
 ",string(x0)),explicit(m0-p0*x,x,0,1.5*xhigh),
 ... key="u = u2",implicit(u(x,y,a0,b0,c0)-u2,x,
 .75*xlow,1.5*xhigh,y,0.8*ylow,1.2*yhigh),key=concat(
 "p = p1, x = ",string(x2)),explicit(m0-p1*x,x,0,
 1.5*xhigh),key="PCC",points_joined=true, ...
 points([x0, x2], [y0, y2])    )$
```

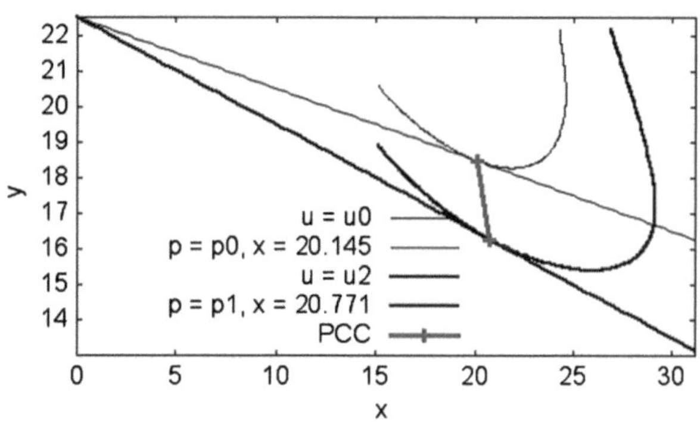

5.2.4 *Income and Substitution Effects of a Price Change*

The quantity response to a price change involves both the income effect and the substitution effect. Increasing the price from 0.16 to 0.24 reduces the consumer's real income. For this inferior good, the result of the implied income reduction would be an increased quantity. The price rise also makes x relatively more costly than before, which works in the opposite direction. To determine the size of these two effects, we create an augmented utility function, replacing y with $m - p \cdot x$, and set that function equal to the initial utility level. We also set $mrs\_aug = p2$. Now, two values can vary: those of x and m. We use `mnewton` to find the approximate solution to this pair of equations. The solution is bound to the name `solsubeff` (for `sol`ution, `sub`stitution `eff`ect). The `mnewton` command has three arguments, each of which is a list. The first is a list of expressions to be evaluated.The second is a list of variables whose values are to be determined. The third is the initial guesses. We use the initial values of x and m as the initial guesses.

```
u_aug(a,b,c,m,x,p):=''(u(x,m-p*x,a,b,c))$  solsubeff:
 mnewton([u_aug(a0,b0,c0,m,x,p1)=u0,mrs_aug(
 a0,b0,c0,m,x,p1)=p1],[x,m],[x0,m0]);
```

$$[[x = 18.997, m = 24.458]]$$

If the consumer is returned to the initial utility level, $u \approx 229$, the compensating income change must be increased from $m = 22.5$ to $m = 24.458$. At this income level, with $p = p2$, the consumer would consume 18.997 units of x. The utility level named u3 is the same as u0.

The input group below uses rhs to extract the right-hand side term of the indicated item from the list that is created in the preceding group. It assigns the names x3 and m3 to these values. The second line determines the utility level and the value of y given the values of x and m. The last line creates the table that contains these values, along with u0 to confirm that utility remains unchanged.

```
[x3, m3]:[rhs(solsubeff[1][1]),rhs(solsubeff[1][2])]$
u3: u_aug(a0, b0, c0, m3, x3, p1)$    y3:m3-p1*x3$
matrix(["x3","y3","m3","u3","u0"],[x3,y3,m3,u3,u0]);
```

$$\begin{bmatrix} \text{x3} & \text{y3} & \text{m3} & \text{u3} & \text{u0} \\ 18.997 & 18.759 & 24.458 & 228.72 & 228.72 \end{bmatrix}$$

The graph shows the substitution effect. The quantity falls from $x0 = 9.4981$ to $x5 = 9.3125$. Of course, reducing the income level to its initial value causes the quantity to increase by more than the decrease in quantity caused by the substitution effect, so that the net effect is to increase x.

```
[xlow,xhigh]:[min(x0,x3),max(x0,x3)]$
[ylow,yhigh]:[min(y0,y3),max(y0,y3)]$
wxdraw2d(... Some commands omitted ...
 points([[x0,y0]]), key="u0, p = p0, m = m0", ...,
 implicit(u(x,y,a0,b0,c0)-u0,x,.5*xlow,1.25*xhigh,
 y,.5*ylow,1.5*yhigh),..., explicit(m0-p0*x,x,0,
 1.1*xhigh), ...,key ="Budget line, p = p1",
 explicit(m3-p1*x,x,0,1.1*xhigh),
 point_type=filled_circle,points([[x3,y3]]))$
```

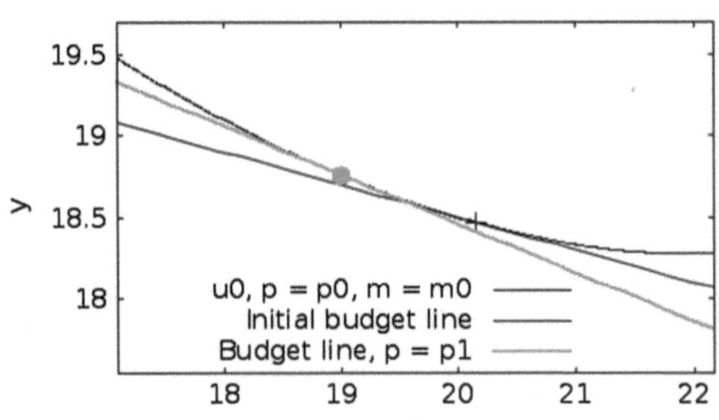

The table below summarizes the results of the analysis in this section. It shows the four values of x that correspond to the scenarios developed above. Values of money income appear at the left, prices appear on the top row, and the remaining cells report the quantities of good x.

```
matrix(["-",  "-","|","p0=",p0,"p1=",p1],
["--","--","--","--","--","--","--"],
 ["m0=",m0,"|","x0=",x0,"x2=",x2],
 ["m1=",m1,"|","x1=",x1,"-","-"],
 ["m3=",m3,"|","-","-","x3=",x3] );
```

$$\begin{bmatrix} - & - & | & p0= & 0.2 & p1= & 0.3 \\ -- & -- & -- & -- & -- & -- & -- \\ m0= & 22.5 & | & x0= & 20.145 & x2= & 20.771 \\ m1= & 25 & | & x1= & 17.721 & - & - \\ m3= & 24.458 & | & - & - & x3= & 18.997 \end{bmatrix}$$

5.2.5 *The Price Consumption Curve and the Income Consumption Curve*

This section generalizes material developed above. It can be omitted without a loss of continuity.

We first initialize the model. Doing so involves choosing an initial value for x and then determining initial values for four other variables: y, p, m, and u. The value of y is determined by evaluating the first two boundary expressions. The expression `xy1` says that $x \cdot y$ equals a constant. Thus y is a constant divided by $x0$, the initial value of x. Likewise for $xy2$. The initial value of y lies halfway between these two boundary lines.[6] Once x and y values are determined, the marginal rate of substitution is established. We set $p = mrs$ because that condition must hold if the consumer is to be at a point consistent with optimization. That is, it must hold if the point is to be one on the consumer's PCC.

Likewise, we establish m's value by forcing a budget line with slope $-p0$ through the point. Such a line must pass through the point if that point is to be a point on the ICC. Finally, given the values of x and y, we determine the implied utility level. This value, along with the others, is required to draw the graphs below.

[6]Selecting y values closer to the `xy1` boundary can cause the graphic representations below to misbehave. Doi et al. modify this function to remove this difficulty, but doing so makes using it for our illustrative purposes more difficult.

```
x0:15$  y0:float((xy1/x+xy2/x)/2),x=x0$
p0:float(mrs(x0,y0,a0,b0,c0))$  m0:x0*p0+y0$
u0:float(u(x0,y0,a0,b0,c0))$    matrix(
   ["x0","y0","p0","m0","u0"],[x0,y0,p0,m0,u0]);
```

$$\begin{bmatrix} x0 & y0 & p0 & m0 & u0 \\ 15 & 24.73 & 0.371 & 30.297 & 231.63 \end{bmatrix}$$

Before using these values, we define functions for the optimal values of x and y. To do this, we use the augmented mrs function, setting its value equal to p, and we solve for x. The result is a quite long and messy set of expressions, not shown here. The first two expressions contain complex numbers; only the third solution is real. The real solution is extracted and used below.

```
temp:solve(mrs_aug(a, b, c, m, x, p)-p, x)$
```

The first command below extracts the right-hand side of the output above and assigns that expression to the name xopt. The second command defines the optimal value of y, yopt, given the optimal value of x, xopt, and the equation for the budget constraint. Both of these are long and messy expressions that involve the parameters a, b, and c, along with the variables m and p.

```
xopt: rhs(temp[3])$ yopt  : m-p*xopt$
```

We now define expressions for optimal values of x and y in terms of the utility function's parameters and either m or p. The first two expressions make x and y functions of m, given $p = p0$. The second two expressions make x and y functions of p, given $m = m0$.

```
[xopt_p0,yopt_p0]:subst([a=a0,b=b0,c=c0,p=p0],
 [xopt,yopt])$  [xopt_m0,yopt_m0]:subst(
 [a=a0,b=b0,c=c0,m=m0],[xopt,yopt])$
```

5.2.6 Graphing the ICC and PCC

This section develops a general representation of the effects that income change and price changes have on the quantity consumed of good x. The first of three graphs shows both the ICC and the PCC. The second graph focuses on the ICC. The third graph focuses on the PCC.

The first graph shows three sets of information.

- The first set consists of the boundary lines that define the regions over which x is a normal good, then an inferior good, then a Giffen good, and finally a "bad" as we move to the northeast.
- The second set draws the ICC and the PCC using the four equations in the preceding input group. This pair of graphs uses parametric functions. For such

functions, x and y are separate functions of a third variable (either m or p here). This variable is treated as a parameter in these functions. For each value of the variable, over some range, the corresponding values of x and y are plotted. Thus the graph consists of a large number of points, but appears to be continuous.[7]

- The third set of commands generates the indifference curve and budget line that define the terminal point on our representation of the ICC and the PCC. The point of tangency must be on both due to the way these curves are defined.

```
wxdraw2d(user_preamble="set key bottom left",
yrange=[10,1.1*y0],line_width=2,color=gray,
key="Boundary lines",explicit(xy1/x,x,0.1,1.5*x0),
key ="",explicit(xy2/x,x,0.1,1.5*x0),
explicit(xy3/x,x,0.1,1.5*x0),color=black,line_width=2,
key = "ICC",parametric(xopt_p0, yopt_p0, m, 0.5, m0),
key="PCC",line_width=3,parametric(xopt_m0,
 yopt_m0, p, p0, 5*p0),line_width = 1,
key = "Indiff Curve and Budget Line",
implicit(u(x,y,a0,b0,c0)-u0,x,.1,1.5*x0,y,.1,1.5*y0),
key = "", explicit(m0 - p0*x, x, 0, 1.5*x0) )$
```

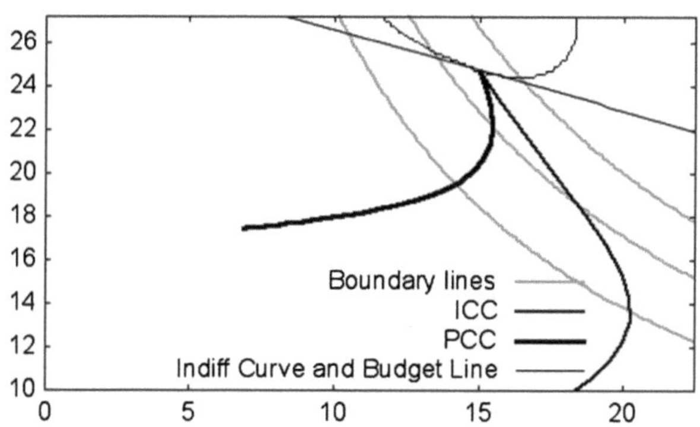

The ICC and PCC behave as expected. The ICC slopes upward over the range for which x is a normal good and bends backward in the range for which x is an inferior good. The ICC must never reach the right-most boundary line, no matter how much income increases, as long as its price is positive. Thus as income increases beyond the one represented, the ICC will continue in the direction shown in the graph. The PCC is defined for a range of prices between $p = p0$ and a price that is five times that value. Extending the PCC farther to the west by allowing higher prices

---

[7]The requisite number of points can vary. The command has a default value that determines how finely the curve is represented. If the resulting graph is insufficiently smooth, this value can be changed using either of two commands, `ip_grid` or `ip_grid_in`. Increasing the number of values of the parameter values for which the functions are to be evaluated can slow processing appreciably.

compromises the quality with which the curve is represented, given the default fineness of the search. This range is large enough to extend the PCC through the three relevant ranges.

The PCC bends backward only when it reaches the Giffen range. Over the range for which x is a normal good, the income and substitution effects work together, so that reducing p causes the consumption of more x. Over the non-Giffen inferior range, the two work opposite each other, but the positive substitution effect of a price reduction dominates the negative income effect. Over the Giffen range, however, this reverses with the income effect dominating. As the price falls, the PCC moves ever closer to the northeast boundary of the Giffen range, but it does not reach that boundary as long as $p > 0$.

References

1. Hausman JA (1981) Exact consumer's surplus and deadweight loss. Am Econ Rev 71:662–676
2. Rutherford TF (2008) Calibrated CES utility functions: a worked example, mpsge.org/calibration.pdf

Chapter 6
Production Theory

Abstract This chapter contains six sections. The first considers the nature of production and a production function. The second introduces the equation that we use to illustrate production theory. The third section looks at a specific aspect of that theory, returns to scale. In this section the inputs are used in a fixed proportion, so that the output can be phrased in terms of the employment of a single input. The fourth section shows aspects of production when employment of either input can be varied over wide ranges. The fifth section considers "the short run," a period during which the employment of only one input can be varied. The final section examines substitution between inputs.

6.1 Introduction

A firm employs a set of inputs to produce one or more types of output. The relationship between the inputs and the firm's output is governed by engineering technology and by issues involving human interactions. We cannot explore the details of the relationship for any single firm. Doing so would involve acquiring a daunting amount of information, much of it known only by individuals within the firm, and much of it not known by anyone.

Furthermore, acquiring and assimilating such information would be of limited value to the larger purpose of exploring general principles that underlie production. Rather than addressing the details of production and their implications for cost, we posit a general function that governs production, $x = f(K, L, \ldots)$, where x indicates the quantity of a good that the firm produces, $f(\ldots)$ indicates that the number on the left is a function of arguments of the function: K is the number of units of capital services employed, L is the number of units of labor services employed, and ... indicates that other arguments can be included.

M.R. Hammock and J.W. Mixon, *Microeconomic Theory and Computation: Applying the Maxima Open-Source Computer Algebra System*,
DOI 10.1007/978-1-4614-9417-1_6,

6.2 An Illustrative Function

The first input cell below contains a relatively simple production function that involves two inputs and four parameters. We use this function through the remainder of this chapter. The function used is this:

$$x = a \cdot K^e \cdot L^{1-e} + b \cdot K^{2 \cdot e} \cdot L^{2 \cdot (1-e)} + c \cdot K^{3 \cdot e} \cdot L^{3 \cdot (1-e)}.$$

This function is based on the Cobb–Douglas function, but uses higher powers to allow for varying effects when the scale of operation changes.[1]

```
f(K,L,a,b,c,e):=a*K^e*L^(1-e)+
        b*K^(2*e)*L^(2*(1-e))+c*K^(3*e)*L^(3*(1-e))$
```

We derive two pairs of functions from the production function. These are the average product of each input and the marginal product of each input. The average product functions (denoted as APK and APL) are the original function divided by the employment levels. APK is a function of K, and L as is APL.

The marginal product functions MPK and MPL are the partial derivatives of the original function with respect to K and L, respectively. Again, each of these is a function of both K and L.

The expression for the marginal product of capital appears below, to show the general nature of these functions. The other three functions are not printed because printing them does not provide useful information. We use the information to develop graphical representations below. The graphs provide insight into the nature of these relationships.

```
APK(K,L,a,b,c,e):=''(f(K,L,a,b,c,e)/K)$
APL(K,L,a,b,c,e):=''(f(K,L,a,b,c,e)/L)$
MPK(K,L,a,b,c,e):=''(diff(f(K,L,a,b,c,e),K));
MPL(K,L,a,b,c,e):=''(diff(f(K,L,a,b,c,e),L))$
```

$$\mathrm{MPK}(K, L, a, b, c, e) := \\ 3\, c\, e\, K^{3e-1}\, L^{3\,(1-e)} + 2\, b\, e\, K^{2e-1}\, L^{2\,(1-e)} + a\, e\, K^{e-1}\, L^{1-e}$$

Before developing the graphs, we determine the values of K and L that lead to maximum output, given a set of parameters. The parameter values appear below.

```
[a0, b0, c0, e0] : [2, 10, - 0.2, 0.3]$
```

Examining the production function reveals that at some set of values, production will decrease as input employment increases (the coefficient of the terms with the largest exponent is negative). Before we examine the production function's other characteristics, we consider the question of *returns to scale*.

[1]When L and K vary proportionately, the function reduces to a cubic function in one variable. This important special case is examined below.

6.3 Returns to Scale

An important aspect of any firm is its size. Presumably, over some range increasing a firm's size can lead to *increasing returns to scale*. That is, over some range, increasing the employment of all inputs proportionately leads to a proportionately larger increase in output. Beyond some scale, however, either engineering relationships or issues regarding coordination are likely to bring an end to this gain from size. Indeed, it is possible that a firm can become so large that further increases in size reduce total output (the firm experiences *negative* returns to scale).

6.3.1 Ranges of Positive and Negative Returns to Scale

For the function used here, increasing employment levels eventually leads to reduced total output. We formalize *returns to scale* and determine the employment levels for which returns to scale are positive. We derive an expression for the production function given that $m = \frac{K}{L}$, where m is a positive constant. The firm's output is now expressed in terms of m and L.

We next take the derivative of that expression and set it equal to zero. We then solve for m.[2] Our purpose here is to determine the values of L for which x is maximized, given m, and the expression below serves this purpose. The name `solnm` is bound to the result for subsequent use.

The first output line below shows the production function, given that K changes in direct proportion to L. The second line shows values m and L that define the border between positive and negative returns to scale.

```
radcan(f(m*L,L,a0,b0,c0,e0));
solnm:solve(diff(f(m*L,L,a0,b0,c0,e0), L)=0, m);
```

$$-\frac{m^{\frac{9}{10}}\,L^3-50\,m^{\frac{3}{5}}\,L^2-10\,m^{\frac{3}{10}}\,L}{5}$$

$$[m = -\frac{100\,m^{\frac{3}{5}}\,L^{\frac{13}{10}}\,(m\,L)^{\frac{7}{10}}-3\,m^{\frac{8}{5}}\,L^3}{10\,L}]$$

The graph shows an inverse relationship between $\frac{K}{L}$ and the value of L at which returns to scale become negative.

```
wxdraw2d( xlabel= "K/L = m",ylabel="L",
 implicit(solnm[1], m, 0.01, 5, L, 0.1, 75),
 label(["Positive returns to scale", 2.5, 10]),
 label(["Negative returns to scale", 3, 50]),
 title="L, m, and returns to scale"     )$
```

[2]We solve for m rather than L, because the latter results in a list of expressions that are hard to manage.

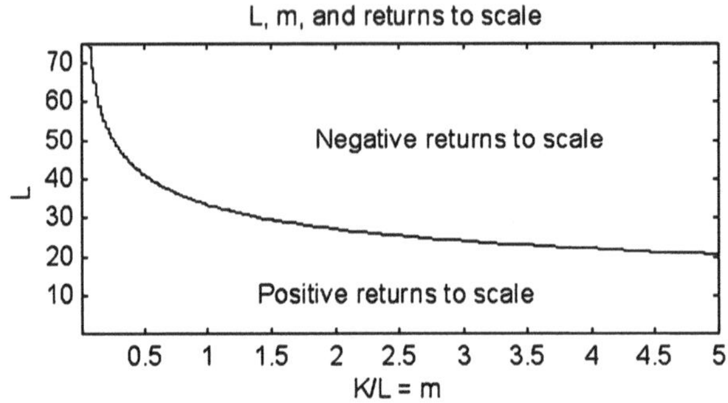

As an example, the value of L for $m = 1$ appears below. When $K = L$, maximum output occurs when $L = K = 33.433$. We assign the name `solnm0` to this list of solutions. Also, we assign the name `imp` (for implicit) to the solution that is derived above.

```
imp:solnm[1]; solnm0:float(solve(subst(m=1,imp),L));
```

$$m = -\frac{100\, m^{\frac{3}{5}}\, L^{\frac{13}{10}}\, (m\, L)^{\frac{7}{10}} - 3\, m^{\frac{8}{5}}\, L^3}{10\, L} \qquad [L = -.099702, L = 33.433]$$

The next cell shows the output, given that L = K = 33.433. At these employment levels, $x = 3770.5$, its maximum value for the selected $\frac{K}{L}$.

```
L0:rhs(solnm0[2])$ K0:L0$
x0:f(K0,L0,a0,b0,c0,e0)$
matrix(["L0","K0","x0"],
[L0,K0,x0]);
```

$$\begin{bmatrix} L0 & K0 & x0 \\ 33.433 & 33.433 & 3770.5 \end{bmatrix}$$

6.3.2 *Increasing and Decreasing Returns: Fixed Proportions*

So far, we have divided the firm's employment ranges into those that result in either positive or negative returns to scale. We can divide the range of positive returns to scale into the range over which the returns to scale are increasing and the range over which they decrease. For illustration, let $m = 0.5$. The resulting expression, stated in terms of L, is a cubic function.

The first derivative of that function with respect to L yields insight into its behavior. Over some range the slope increases as L increases, but eventually the term $-0.332L^{2.0}$ reduces the slope's growth rate and then turns the slope negative.

The next cell provides an expression for the production function, given that $m = m1 = 0.5$, and the derivative of that expression. The name `xm1` is bound to the expression for x given that $K = m1 \cdot L$. This expression is used later to discover the value of L for which the slope begins to decrease.

```
m1:.5$ xm1:f(m1*L,L,a0,b0,c0,e0);
diff(f(m1*L,L,a0,b0,c0,e0),L);
```

$$-.10718\,L^{3.0} + 6.5975\,L^{2.0} + 1.6245\,L$$
$$-.32153\,L^{2.0} + 13.195\,L + 1.6245$$

We use the imp expression to determine where the x function reaches its maximum value. That value is $L = L1 = 41.161$. Thus, the range over which returns to scale are positive is from $L = 0$ to $L = 41.161$.

The name solnimpm0 is bound to the list that contains the solutions. The last two lines of the cell below bind the names L1 to the value of L that generates maximum x when $m = m0$, and x1 to the corresponding output level.

```
matrix(["L values","L","K","x"], [solnimpm1:float(
solve(subst(m1,m,imp), L)), L1:rhs(solnimpm1[2]),
  K1:m1*L1,   x1:f(K1,L1,a0,b0,c0,e0)]);
```

$$\begin{bmatrix} \text{L values} & L & K & x \\ [L = -.12275, L = 41.161] & 41.161 & 20.58 & 3770.5 \end{bmatrix}$$

We now use the value of L for which the *second* derivative equals zero to find the value of L for which the slope for the first derivative changes signs, i.e., the inflection point for the production function. This value is $L = L2 = 20.519$.

The cell below derives the expression for the second derivative. It also solves that expression for L and assigns the name L2 to the resulting value. Next, it computes the associated values of K and x. Finally, these values, along with those generated in the previous cell, are collected into the table.

```
solve(diff(xm1,L,2),L)$
L2:float(rhs(%[1]))$
K2:m1*L2$    x2:f(K2,L2,a0,b0,c0,e0)$
matrix(["K","L","x"] ,[K1,L1,x1],
[K2, L2, x2]  );
```

$$\begin{bmatrix} K & L & x \\ 20.58 & 41.161 & 3770.5 \\ 10.26 & 20.519 & 1885.2 \end{bmatrix}$$

The graph below represents these results. The horizontal axis shows values of L (the corresponding K values are $K = m1 \cdot L$). The vertical axis shows output. This graph represents a slice across the three-dimensional graph, as shown below. For values of L between 0 and $L2$, the quantity is increasing as L increases and doing so at an increasing rate (increasing returns to scale). For values of L between $L2$ and $L1$, the quantity is increasing as L increases but doing so at a decreasing rate (decreasing returns to scale). Beyond $L1$, increasing L causes x to decrease (negative returns to scale). Keep in mind that, while L appears on the horizontal axis, x could equally well have been graphed as a function of K, because $K = m1 \cdot L$.

```
wxdraw2d(xlabel = "L, when K = m1*L",ylabel="x",
 yrange=[0,1.1*x1],explicit(xm1,L,0.1,50),
 points_joined=true,line_type=dots,points(
 [[L2,0],[L2,x2] ]),points([[L1,0],[L1,x1]]),
 label(["Diminishing Returns",L2+.5*(L1-L2),x1/4]),
 label( ["Increasing Returns", L2/2, x2/4]))$
```

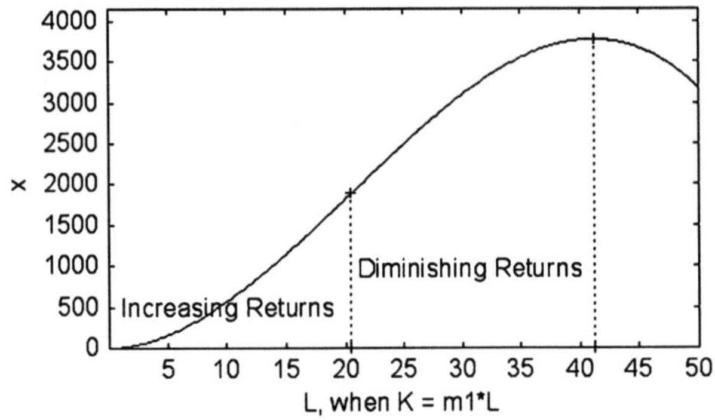

The table below shows a set of values consistent with $K = m1 \cdot L$. These values are added to the surface plot in order to illustrate how the production function with a constant input ratio relates to the production function absent this constraint.

Three lists are created in the cell below. First, `Llist` contains a set of L values that range from 0 to 1.2 times the value that yields maximum output. Then `Klist` shows $K = m1 \cdot L$ values. The values in `xlist` result from placing the K, L pairs into the production function.

```
Llist: makelist(i*L2/5.0,i,0,6)$ Klist:m1*Llist$
xlist:f(Klist,Llist,a0,b0,c0,e0)$ addrow(matrix(["K",
"L","x"]),transpose(matrix(Klist,Llist,xlist)));
```

K	L	x
0.0	0.0	0.0
2.0519	4.1038	110.37
4.1038	8.2076	398.52
6.1557	12.311	820.0
8.2076	16.415	1330.4
10.26	20.519	1885.2
12.311	24.623	2440.0

The graph below represents the production function in a three-dimensional graph. Also, the data in the preceding cell are added, appearing as a set of points.

```
wxdraw3d(view=[60,330],zrange=[0,1.1*x2],xlabel=
 "K",ylabel="L",zlabel="x",xtics={0,m1*L2},color=
 gray,ytics={0,L2},ztics={0,x0},explicit(
  f(K,L,a0,b0,c0,e0),K,.1,1.2*K2,L,.1,1.2*L2),
  color=gray30,points(Klist,Llist,xlist),title
  ="Trace across the production surface" )$
```

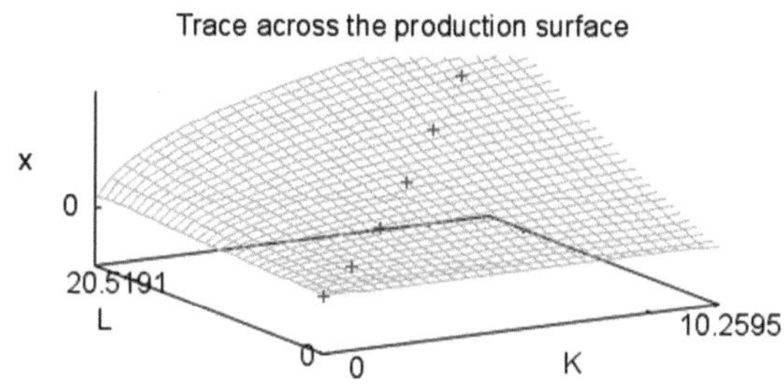

6.4 Production with Variable Proportions

Except for the last graph, the previous section treats only the case in which inputs are used in a fixed ratio. The graph below shows the same production surface as the one above, except that the ranges of K and L values are extended. The surface shows the maximum amount of x the firm can produce for each input combination. The graph shows that the maximum quantity increases over most of the input values used, but it begins to decrease at high employment levels.

The graph shows one contour line, for $x = 0.9{\cdot}x0$. Two segments of this contour line appear on the base. The segment nearest the origin is the relevant one.[3]

These contour lines are like the indifference curves for a utility function that generates "bliss points." There, beyond some quantities the consumer had so much of the two goods that both had become "bads." Here, beyond some employment levels the firm cannot manage the additional resources. The production function illustrates what *could* happen. Presumably, managers would not employ so many resources that total output falls when more are hired.

[3]If the values of K and L had been extended sufficiently, this contour line would define an enclosed area. As an exercise, execute the commands with K ranging from 0.1 to $30 \cdot$ `K0` and L ranging from 0.1 to $15 \cdot$ `L0` to confirm this assertion.

```
wxdraw3d(view = [55,330],dimensions=[480,400],
 contour=base,contour_levels={.9*x0},xlabel="K",
 ylabel="L",zlabel="x",xtics={K0/4,3*K0},ytics=
 {L0/4,1.5*K0},ztics={0,x0},color="gray",explicit(
 f(K,L,a0,b0,c0,e0),K,.1,3.0*K0,L,.1,1.5*L0))$
```

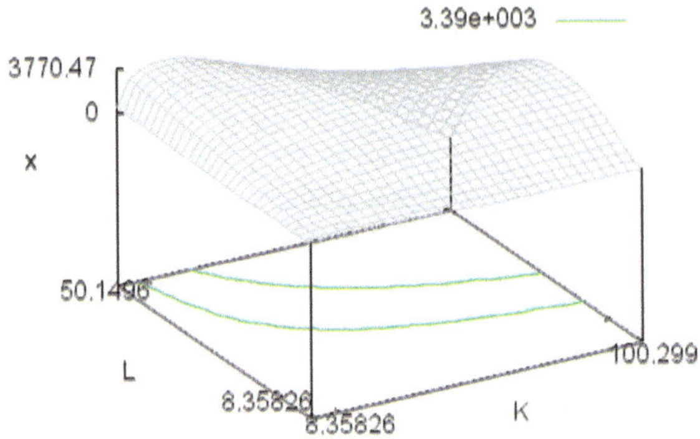

6.5 Varying a Single Input

The two curves at the top of the panel of graphs below are cross-sections of the three-dimensional production surface shown above, extended slightly beyond the range used there. They are generated by `gr2d` and given the names `Ktotalcurve` and `Ltotalcurve`. For illustration, $L = L0 = 33.433$ is used for the capital curve and $K = K0 = 33.433$ is used for the labor curve. In both cases, the range of values is sufficiently large for the total product curve to turn downward (the employment levels at which the marginal product curves intersect the horizontal axis). The total product function above can be used to generate two sets of functions that state output per unit of each of the two inputs, given the employment level of the other input. The thin lines on the bottom two graphs define the average product of capital and labor (APK and APL). Each of these is defined by dividing the production function by the relevant input, holding constant the employment level of the other input.

The marginal product functions (MPK and MPL, the dark lines on the bottom two graphs) are defined by taking the relevant derivatives of the production function, again for a specified employment level of the other input. The “quote-quote” operator is used to produce a functional expression of the relationship.

```
print("L0 =", L0, " and K0 =", K0)$
Ktotalcurve:gr2d(xlabel="K", ylabel="x",
 yrange=[0, 1.1*x0], ytics=2000,xtics=10,explicit(
 f(K, L0,a0,b0,c0,e0),K, .1,1.1*K0))$
Kunitcurves:gr2d(yrange=[-50,200],xlabel="K",xtics=
  10,ylabel="x per unit of K",xaxis=true,ytics=100,
  explicit(APK(K,L0,a0,b0,c0,e0),K,.1,1.1*K0),
  line_width=2,explicit(MPK(K,L0,a0,b0,c0,e0),
  K,.1,1.1*K0) )$
Ltotalcurve:gr2d(yrange=[0, 1.1*x0],ytics=2000,
  xtics=10,xlabel = "L", ylabel="x",
  explicit(f(K0, L, a0, b0,c0,e0),L,.1,1.1*L0) )$
Lunitcurves: gr2d( xaxis = true, yrange=[-50, 200],
  xlabel="L",ylabel="x per unit of L",xtics=10,
  ytics=100,explicit(APL(K0,L,a0,b0,c0,e0),
  L,.1,1.1*L0),line_width=2,
  explicit(MPL(K0,L,a0,b0,c0,e0),L,.1,1.1*L0))$
wxdraw(dimensions=[480,480],columns=2,Ktotalcurve,
   kLtotalcurve,Kunitcurves, Lunitcurves)$
```

L0 = 33.433 and K0 = 33.433

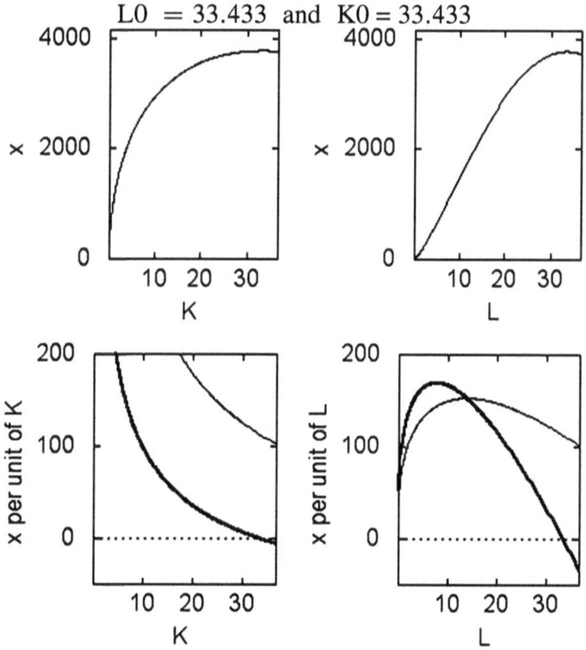

The per-unit graphs above confirm that when $L = L0$ and $K = K0$, then both marginal product curves have values of zero. Thus adding more of either input reduces output.

An important relationship between the APL and MPL curve warrants notice. When $MPL > APL$, then APL is increasing; when $MPL < APL$, then APL is decreasing; and when $MPL = APL$, then APL reaches its maximum value.[4]

6.6 Interactions Between Inputs

We now focus on the behavior of output as L changes, holding K constant. We define three scenarios: `totals` creates a graph of the total output x at each value of L for each of the two values of K, `averages` does the same thing for x/L, and `marginal` creates the corresponding graphs for the marginal product curves.

```
totals: gr2d(user_preamble="set key bottom right",
 xlabel = "L", ylabel="x",ytics=2000,key="K=K0/4",
 explicit(f(K0/4,L,a0,b0,c0,e0),L,.1,2*L0),
 line_width=2,key="K = K0/2",explicit(
 f(K0/2,L,a0,b0,c0,e0),L,.1,2*L0))$
averages: gr2d( ... Like totals above ...
  explicit(APL(K0/2,L,a0,b0,c0,e0),L, .1,2*L0))$
marginals: gr2d( ... Like totals above ...
  explicit(MPL(K0/2,L,a0,b0,c0,e0),L,.1,2*L0) )$
```

The panel below contains the figures defined above. The first graph shows that, over the range $L = 0$ to $L = 3 \cdot L0$, increasing K from `K0/2` to `K0/4` increases the total output at each value of L. Eventually, however, additional units of L take us to the back side of the production surface, so that increasing K any more decreases x. Again, one should not expect ever to observe production in this range.

The average product curves appear in the second panel. For both values of K, the average product increases over a range and then decreases.

```
wxdraw(totals,averages,marginals,
 dimensions=[480,480])$
```

[4]This is a general relationship, not limited to the illustrative example used here. When a marginal is above an average, it is pulling the average up. When a marginal is below an average, it is pulling the average down.

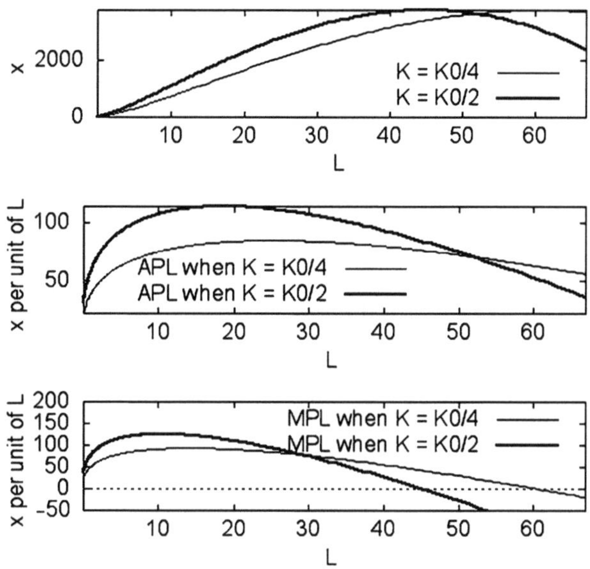

The marginal product curves appear in the bottom panel. For any K value, the MPL curve shows the slope of the corresponding x curve. When $K = K0/2$, MPL becomes negative at $L \approx 46$, where the x curve reaches its maximum value. When $K = K0/4$, MPL becomes negative for larger values of L, $L \approx 60$.

This intersection of product curves is an artifact of the production function being used. Specifically, it is due to varying returns to scale. The intersection would not occur if we replace this function with a homogeneous function like the Cobb–Douglas or CES functions.

6.7 Isoquants and Substitution

The contour line shown at the beginning of Sect. 6.4 is an *isoquant curve* (signifying equal quantity produced), or more simply an *isoquant*. An isoquant passes through any selected point in the L, K space and shows other L, K combinations that would yield the same quantity as the point selected.

Before drawing illustrative isoquants, we determine a function for the isoquant's slope, $\frac{dK}{dL}$. That slope is the ratio of the two partial derivatives. That is, the slope is the ratio $-\frac{MPL}{MPK}$. We assign the name *marginal rate of technical substitution*, or $mrts$ to this variable. The output below shows that the value of $mrts$ depends on the ratio $\frac{K}{L}$ and the exponent e. Thus, this function is *homothetic*. Note that $e - 1$ is negative (isoquants slope downward).

```
mrts(K,L,e) :=
  ''(radcan(-diff(f(K,L,a,b,c,e),L)/
    diff(f(K,L,a,b,c,e),K)));
```

$\text{mrts}(K, L, e) := \frac{(e-1)K}{eL}$

6.7.1 *Elasticity of Substitution*

Because the marginal rate of substitution and $\frac{K}{L}$ are proportional to each other, the elasticity of substitution for this function is 1, as it is for the simple Cobb–Douglas function. Recall that the elasticity of substitution is the percent change in $\frac{K}{L}$ per 1 % change in the marginal rate of substitution. Write $\frac{K}{L}$ as k. The elasticity of substitution can be written as $\sigma = \frac{dk}{d(mrts)} \cdot \frac{mrts}{k}$, which equals 1. Note: The e in the output below is the function's parameter and does not relate to taking a derivative.

```
[k:mrts*(e/(1-e)),
 diff(k,mrts)*(mrts/k)];
```

$[\frac{e\, mrts}{1-e}, 1]$

6.7.2 *Effect of L on mrts*

Given that $mrts(K, L, e) = \frac{e-1}{e} \cdot \frac{K}{L}$, the $mrts$ must approach zero as L becomes large, as we move along any isoquant, because increasing L implies reducing K, so the ratio decreases.

If K is constant, then increasing L causes movement to a higher isoquant. This rightward movement in the L, K plane will lead to a lower $mrts$ on the new isoquant than on the initial one. In the graph below, compare the slopes at the intersections of the three isoquants and the dotted line, where $K = 1.5$.

6.7.3 *Illustration*

The following input group defines the three output levels for which isoquants are drawn below. It also determines the value of $mrts$ when $\frac{K}{L}$ = `m1`. The name `slope` is bound to the expression for the marginal rate of technical substitution.

```
[L1,L2,L3]:[2,4,6]$      [K1,K2,K3]:m1*[L1,L2,L3]$
[x1,x2,x3]:[f(m1*L1,L1,a0,b0,c0,e0),f(m1*L2,
 L2,a0,b0,c0,e0), f(m1*L3,L3,a0,b0,c0,e0)]$ slope:
  mrts(m1*L1,L1,e0)$  matrix(["slope =",slope]);
matrix(["L","K","x"],[L1,K1,x1],
 [L2,K2,x2],[L3,K3,x3]);
```

$$\begin{bmatrix} \text{slope} = & -1.1667 \end{bmatrix}$$

$$\begin{bmatrix} L & K & x \\ 2 & 1.0 & 28.782 \\ 4 & 2.0 & 105.2 \\ 6 & 3.0 & 224.11 \end{bmatrix}$$

```
wxdraw2d( ... Some commands omitted ...
 key="Isoquants and tangents", implicit(
 f(K,L,a0,b0,c0,e0)=x1,L,.1,1.2*L3,K,.1,1.5*K3),
 ...explicit(K1-L1*slope+slope*L,L,L1-1,1.5*L1) ,
 implicit(f(K,L,a0,b0,c0,e0)=x2,L,.1,1.2*L3,
 K,.1,1.5*K3),explicit(K2-L2*slope+slope*L,
 L,L2-1, 1.25*L2),implicit(f(K,L,a0,b0,c0,e0)=x3,
 L,.1,1.2*L3,K,.1,2*K3),explicit(K3-L3*slope +
 slope*L, L, L3-2, 1.2*L3),line_width=2,key=
 "K = m1*L",explicit(m1*L,L, 0, 1.1*L3))$
```

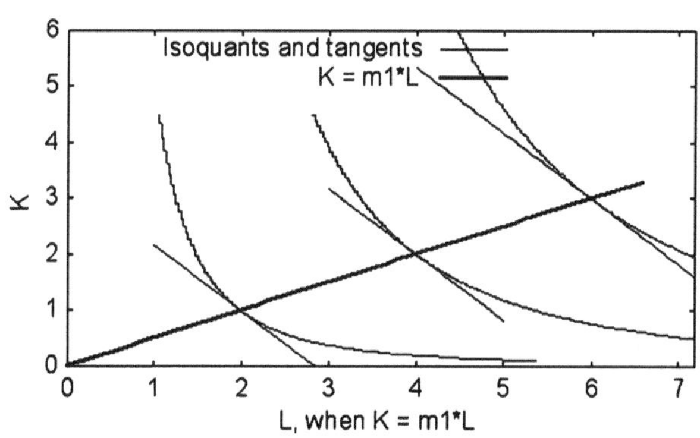

6.8 Homotheticity Revisited

The function being used here is homothetic. That is, the marginal rate of technical substitution is determined by the input ratio and not affected by the scale of production. The graph above shows three isoquants. An isocost line is drawn tangent to each of the three isoquants. All have the same slope and the points of tangency all lie on the expansion path, along which $\frac{K}{L} = \texttt{m1} = 0.5$. The common slope is $(-0.7/0.3) \cdot 0.5$.

One should not expect homothetic production functions in general. We use such functions because they are mathematically tractable and because they provide insights into the behavior of actual production.

Chapter 7
Cost Theory

Abstract Building on the production theory developed and illustrated in Chap. 6, this chapter analyzes and illustrates the relationship between the quantity produced and production cost. The illustration is based on the production function and the assumption that each of the two inputs can be purchased in competitive resource markets. Also, the firm is assumed to employ resources in a cost-minimizing combination.

7.1 Overview

The analysis of cost is divided into a stylized long run in which the firm can select any quantities of both inputs, and a stylized short run in which only the employment of L can be varied. Any real firm employs more than two inputs and will encounter many "runs." Even so, much can be learned from our simple dichotomy. In particular, we address the crucial distinction between per-period costs which are affected by the quantity of x that the firm produces and those that are independent of the production level.

We begin by recalling the production function and some relevant parameter values. The first cell defines the production function, and it recalls the parameter values for that function that applied in the preceding chapter. The other two parameters, $w0$ and $r0$, are the per-unit costs of employing labor (L) and capital (K).

```
x(K, L, a, b, c, e) := a*K^e*L^(1-e) +
    b*K^(2*e)*L^(2*(1-e)) + c*K^(3*e)*L^(3*(1-e))$
[a0,b0,c0,e0]:[2,30,-0.25,0.3]$   [w0,r0]:[50,80]$
```

M.R. Hammock and J.W. Mixon, *Microeconomic Theory and Computation: Applying the Maxima Open-Source Computer Algebra System*,
DOI 10.1007/978-1-4614-9417-1_7,

The next cell recreates the functional expression for the marginal rate of technical substitution. The $mrts$ is a function of the single parameter e and the capital/labor ratio.

```
mrts(K,L,e):=''(ratsimp(
-diff(x(K,L,a,b,c,e),L)/
diff(x(K,L,a,b,c,e),K)));
```

$$\text{mrts}(K, L, e) := \frac{(e-1)\,K}{e\,L}$$

7.2 The Optimal Input Mix

The analysis of consumer choice shows optimization (maximizing utility subject to a budget constraint or, equivalently, minimizing the expenditures required to achieve a specific utility level) requires selecting the consumption bundle such that the (negative of the) slope of the consumer's indifference curve equals the ratio of the prices of the goods consumed. Likewise, the firm must find the input mix that results in $mrts = w/r$. To see this, consider a general production function $y(K, L)$.

Define a Lagrangian function $W = y(K, L) - \mu \cdot (ltc - r \cdot K - w \cdot L)$, where ltc represents long run total costs. The first two first-order conditions are that $\mu = \frac{MPK}{r}$ and $\mu = \frac{MPL}{w}$. Taking the ratio of these two conditions and cross-multiplying shows that output is maximized, given a cost level, when $\frac{MPL}{MPK} = \frac{w}{r}$.

```
W:y(K,L)-%mu*(ltc-r*K-w*L)$ [WK,WL]:[diff(W, K),
diff(W, L)]$    [expK: solve(WK,%mu)[1],
 expL:solve(WL,%mu)[1],    w/r*expL/expK ];
```

$$[\mu = -\frac{\frac{d}{dK}\,y(K,L)}{r},\quad \mu = -\frac{\frac{d}{dL}\,y(K,L)}{w},\quad \frac{w}{r} = \frac{\frac{d}{dL}\,y(K,L)}{\frac{d}{dK}\,y(K,L)}]$$

We can confirm this result graphically. The graph below contains a single isoquant. One point on that isoquant consists of the input pair $K = 20$ and $L = 20$. The resulting output, $x = 10{,}040$, is defined inside the `key` command. A single isocost line is also defined, one that begins at $K = 30$ and whose slope is determined by the parameters above, $w = 50$ and $r = 80$.

An *isocost line* is a set of input combinations that result in the same level of cost. Unlike the consumer's budget line, the firm's isocost line is endogenous. That is, the firm may choose an output level and the least-cost input mix for that level, which implies a cost level. Alternatively, the firm may choose a total cost level and then select the input mix that yields the maximum output for that cost. Either way, it may select values for three of these four variables: K, L, x, and total cost.

The graph shows a situation in which the firm can produce more than 10,040 units of output at the same cost level by finding the input combination along the isocost line at which $mrts = 50/80$. Alternatively, it can reduce costs below the level depicted by the isocost line ($cost = 30 \cdot \$80 = \$240$) by finding a point on the current isoquant for which $mrts = 50/80$.

```
wxdraw2d( ... Some commands omitted ... key=
 concat("The isoquant for x ",string(x(20,20,a0,
 b0,c0,e0))), implicit(x(K,L,a0,b0,c0,e0)=
 x(20,20,a0,b0,c0,e0),L,.1,40, K,.1,40), ...
 explicit(30 - (w0/r0)*L, L, 0, 30*r0/w0) )$
```

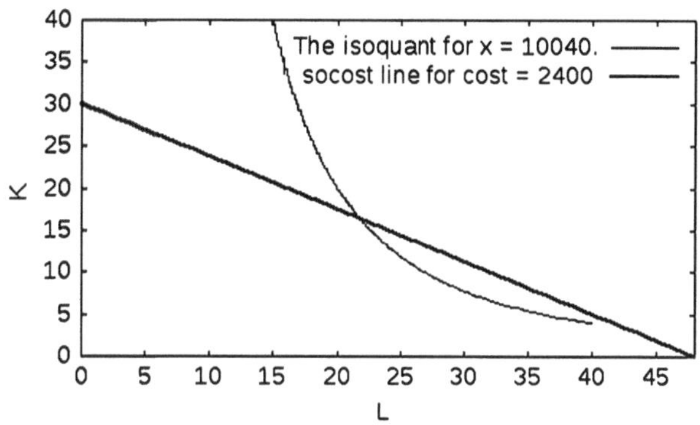

Setting $-mrts = \frac{w}{r}$ and solving reveals that for a given w, r pair, the marginal rate of technical substitution does not depend on the amount of x that the firm produces. We use this notation below: $K = m \cdot L$, or $\frac{K}{L} = m$, where $m = \frac{e}{1-e} \cdot \frac{w}{r}$.

```
[ksoln: solve(-mrts(K, L, e) = w/r, K),
m(e, w, r):= ''(rhs(ksoln[1]/L))];
```

$$[[K = -\frac{e\,w\,L}{(e-1)\,r}],\ \mathrm{m}(e, w, r) := -\frac{e\,w}{(e-1)\,r}]$$

7.3 Long-Run Cost Curves

The long run is a sufficiently long period for the firm to be able to implement the optimal input mix. The long-run is more conceptual than real. That is, production must occur when some input levels cannot be varied. Even so, it is important to know the characteristics of the mix toward which the firm will tend, given time to adjust to conditions.

7.3.1 The Total Cost Curve in the Long Run

The production function used here does not generate a tractable explicit function in which cost is stated as a function of quantity. Rather, we derive both the output level and cost as functions of L and examine the implicit relationship between the two.

First, we state output in terms of L. Substitute the expression $K = -\frac{e \cdot w \cdot L}{(e-1) \cdot r}$, which is `ksoln[1]`, into the production function to generate the augmented production function $xaug$. The result is an expression in terms of w and r, along with the original function's parameters and L.

Next, create an augmented total cost function, with cost as an explicit function of L. The firm's total cost is $ltc = r * K + w * L$. Substitute `ksoln[1]` into that expression to create the function $ltcaug$, expressed in terms of L, w, r, and e.

```
xaug(L, w, r, a, b, c, e) := ''(radcan(
x(rhs(ksoln[1]), L,a,b,c,e)));
 ltcaug(L,w,r,e):=w*L-(e*w*L)/(e-1);
```

$$\text{xaug}\,(L, w, r, a, b, c, e) := \frac{c\, e^{3e}\, w^{3e}\, L^3 + b\,(1-e)^e\, e^{2e}\, r^e\, w^{2e}\, L^2 + a\,(1-e)^{2e}\, e^e\, r^{2e}\, w^e\, L}{(1-e)^{3e}\, r^{3e}}$$

$$\text{ltcaug}\,(L, w, r, e) := w\,L - \frac{e\,w\,L}{e-1}$$

The two functions of L can be combined to generate a parametric graph of the firm's long-run total cost (ltc) curve.

Before drawing the long-run total cost curve, we define some values that to use in building the graph. These are $m0$, $L2$, $K2$, $x2$, and $ltc2$. The value of $x2$ is the largest amount the firm can produce given $m0$, and $L2$ and $K2$ are the employment levels consistent with that output. Finally, $ltc2$ is the cost of employing $K2$ and $L2$ units of the inputs.

The first input line solves for the L value at which output is maximized, given m. The second line sets a value of m, based on the relevant argument values. The third line solves the expression generated by the first line, given the m value. The next two lines define the value of L that yields the maximum x, the maximum value of x, and the associated cost level. These values are passed to the `matrix` command for printing.

```
diff(x(m*L,L,a0,b0,c0,e0),L)$  expr:solve(%, m)$
m0:m(e0,w0,r0)$
float(solve(subst(m=m0,rhs(expr[1])), L) )$
L2:rhs(%[1])$  K2:m0*L2$  ltc2:r0*K2+w0*L2$
x2:x(m0*L2,L2,a0,b0,c0,e0)$        matrix(
  ["m0","K2","L2","x2","ltc2"],[m0,K2,L2,x2,ltc2]);
```

$$\begin{bmatrix} \text{m0} & \text{K2} & \text{L2} & \text{x2} & \text{ltc2} \\ .26786 & 31.814 & 118.77 & 64160. & 8483.8 \end{bmatrix}$$

The next graph shows the cost of producing output given that the firm has selected the least-cost input combination. Thus, the cost curve represents a frontier below which the firm cannot operate. We include values of L (and implicitly, values of K) that are beyond the range of positive returns to scale.

The relevant, upward-sloping segment of the cost curve has a backward "S" shape, reflecting the "S"-shaped production that the preceding chapter developed.

```
wxdraw2d(... Some commands omitted ...
 parametric(xaug(L,w0,r0,a0,b0,c0,e0),
 ltcaug(L,w0,r0,e0),L,0.1,1.25*L2), ...
 points( [[0,ltc2],[x2,ltc2],[x2,0]]))$
```

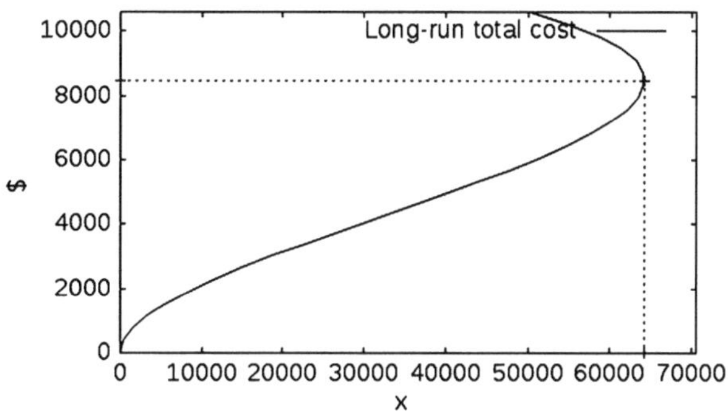

7.3.2 *Per-Unit Curves in the Long Run*

For many purposes, it helps to rephrase the information in the long-run total cost curve in terms of per-unit curves. Two such curves can be derived: the average cost curve and the marginal cost curve. The average cost curve shows the total cost divided by the quantity being produced, and the marginal cost curve shows the change in total cost per one-unit change in the quantity being produced.

The table below shows ten values of x and the associated total cost curve (ltc) and per-unit cost curves (lac and lmc). The table shows that average cost falls until $x = 2950.4$ and rises beyond $x = 3371.9$. Thus, it reaches its minimum value between these two quantities. Also note that for $x \leq 2950.4$, $lmc < lac$, and for $x \geq 3371.9$, $lmc > lac$. That is, when the marginal value is below (above) the average value, the average value decreases (increases). The graph below the table shows this relationship more directly.

```
diff(ltcaug(L,w,r,e),L)/diff(xaug(L,w,r,a,
 b,c,e),L)$  lmc(L,w,r,a,b,c,e):=''%$  Llist:
 makelist(L2*i/10,i,1,10)$  Klist: m0*Llist$
xlist : x(Klist, Llist, a0, b0, c0, e0)$ltclist:
 ltcaug(Llist,w0,r0,e0)$ laclist:ltclist/xlist$
lmclist:lmc(Llist,w0,r0,a0,b0,c0,e0)$
names : matrix(["x", "ltc", "lac", "lmc"])$
costs: transpose(matrix(xlist,ltclist,laclist,
   lmclist) )$      addrow(names, costs);
```

$$\begin{bmatrix} x & ltc & lac & lmc \\ 1808.0 & 848.38 & .46924 & .24435 \\ 6688.0 & 1696.8 & 0.2537 & .13772 \\ 13872. & 2545.1 & .18347 & 0.105 \\ 22592. & 3393.5 & .15021 & .091896 \\ 32080. & 4241.9 & .13223 & .088226 \\ 41568. & 5090.3 & .12246 & .091896 \\ 50288. & 5938.7 & .11809 & 0.105 \\ 57472. & 6787.0 & .11809 & .13772 \\ 62352. & 7635.4 & .12246 & .24435 \\ 64160. & 8483.8 & .13223 & 53.024 \end{bmatrix}$$

Before looking at a graph of the per-unit curves consider the relationship between an arc marginal cost value, calculated over a range of x values, and point marginal cost values. Look at the fourth and fifth lines above. The cell below contains two point lmc values and, between them, the corresponding arc value, $\frac{\Delta ltc}{\Delta x}$. The arc value lies between the two point values. If the lmc were linear over this segment, the arc value would be the mean of the two point values.

```
mc_arc:(ltclist[5]-ltclist[4])/(xlist[5]-xlist[4])$
matrix(["mc[4]", "mc_arc", "mc[5]"],
   [lmclist[4], mc_arc, lmclist[5]]);
```

$$(\%o131) \begin{bmatrix} \text{mc[4]} & \text{mc\_arc} & \text{mc[5]} \\ .091896 & .089416 & .088226 \end{bmatrix}$$

The first of the two graphs below plots the points in the table above and connects them with linear segments. It shows the general U shape of both curves. The bottom of the lmc curve occurs at the output rate at which increasing returns to scale give rise to decreasing returns to scale.[1] The minimum point on the long-run average cost curve is at the quantity at which $lac = lmc$.

```
wxdraw2d( ... Some commands omitted ...
 points(matrix(xlist, laclist)), ...
 points(matrix(xlist,lmclist)))$
```

[1]The first derivative of the total cost function, the marginal cost, is positive throughout the relevant output range. At first the second derivative of the total cost function is negative. At higher output rates the second derivative becomes positive. At the point of inflection, the second derivative equals zero.

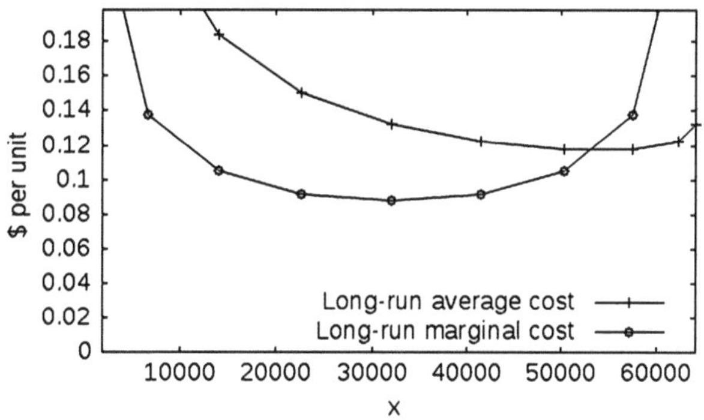

The graph below uses the `parametric` option to refine the graph that appears above. This option evaluates the parametric functions at 30 points (`nticks = 30` is the default value, but other values can be specified). It then interpolates between the points using a more sophisticated measure than the linear interpolation used above.

```
wxdraw2d( ... Some commands omitted ...points(
 matrix(xlist,laclist)),points(matrix(xlist,
 lmclist)), ... parametric(
 xaug(L,w0,r0,a0,b0,c0,e0),ltcaug(L,w0,r0,e0)/
 xaug(L, w0, r0, a0, b0, c0,e0), L,0.1*L2,0.99*L2),
... parametric( xaug(L, w0,r0,a0,b0,c0,e0),
  lmc(L,w0,r0,a0,b0,c0,e0), L, .1, .99*L2) )$
```

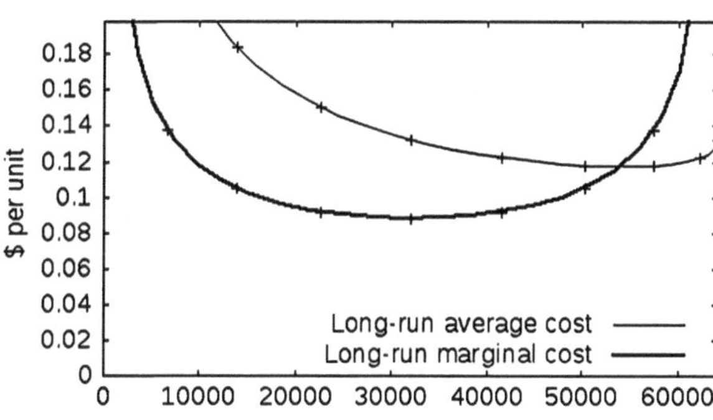

7.4 Short-Run Costs

We define short run as being so short that the employment level of K cannot be varied. For illustration, we bind the name `itemNo` to select a line from the table above. Here, we use the seventh row of the table, so that `KS0` $=$ 22.27 is the fixed employment level for K in the short run. The table below reproduces relevant information regarding output and cost in the long run given this level of K.

```
itemNo: 7$   /*Select line number from list*/
KS0:Klist[itemNo]$ LS0:Llist[itemNo]$
xS0:xlist[itemNo]$ tcS0:ltclist[itemNo]$
matrix(["K,SR","L,LR&SR", "x,LR&SR",
"ltc,LR&SR"],[KS0, LS0, xS0, tcS0]);
```

$$\begin{bmatrix} \text{K,SR} & \text{L,LR \& SR} & \text{x,LR \& SR} & \text{ltc,LR \& SR} \\ 22.27 & 83.141 & 50288. & 5938.7 \end{bmatrix}$$

The input group below defines the total cost curve, the average cost curve, and the marginal cost curve in the short run. The values of K are left variable here, but will be fixed in applications that follow.

```
stc(K,L,w,r):=r*K+w*L$ smc(K,L,a,b,c,e,w):=''(
  diff(stc(K,L,w,r),L)/diff(x(K,L,a,b,c,e),L))$
sac(K, L,a,b,c,e,w,r):=
     ''(stc(K,L,w,r)/x(K,L,a,b,c,e))$
```

The next table helps us compare short-run and long-run costs. The fixed value `KS0` appears as the value of K in all rows. The value of L varies as before from one-tenth of $L2$ to $L2$. The column `x` is derived by entering the items from `KS0list` and `Llist` (from above) into the production function to generate `x0list`.

The total cost list `stc0list` results from applying the total cost function to the fixed input `KS0` and the list of L values. Likewise, with the marginal cost list `smc0list`.

These lists are placed into a matrix, the transpose of which is added to the `names` matrix to generate the table of value. One value does not appear explicitly on the list: If $x = 0$, then $L = 0$ and $stc0 = r0 * KS0 = 1781.6$. This is the firm's per-period fixed cost, the part of total cost that does not vary with the firm's output level.

```
KS0list:makelist(KS0,i,1,10)$ x0list:x(KS0, Llist,
 a0,b0,c0,e0)$  stc0list:stc(KS0,Llist,w0,r0)$
sac0list:sac(KS0, Llist,a0,b0,c0,e0,w0,r0)$
smc0list:smc(KS0, Llist,a0,b0,c0,e0,w0)$  names:
 matrix(["KS0","L","x","stc0","sac0","smc0"])$
matrix(KS0list,Llist,x0list,stc0list,sac0list,
 smc0list)$   addrow(names, transpose(%));
```

KS0	L	x	stc0	sac0	smc0
22.27	11.877	5462.2	2375.5	.43489	.083517
22.27	23.755	13170.	2969.3	.22546	.073352
22.27	35.632	21383.	3563.2	.16664	.072112
22.27	47.509	29498.	4157.1	.14093	.074807
22.27	59.387	37168.	4750.9	.12782	.080663
22.27	71.264	44156.	5344.8	.12104	.090176
22.27	83.141	50288.	5938.7	.11809	0.105
22.27	95.019	55427.	6532.5	.11786	.12899
22.27	106.9	59462.	7126.4	.11985	.17209
22.27	118.77	62300.	7720.3	.12392	.26844

```
wxdraw2d( ... Some commands omitted ...points(
 [xS0,xS0],[0,tcS0]),... parametric(
 x(KS0,L,a0,b0,c0,e0),r0*KS0+w0*L,L,.1,1.2*L2),
 line_width=2,key="Long-run total cost",
 parametric(xaug(L,w0,r0,a0,b0,c0,e0),
 ltcaug(L,w0,r0,e0),L,.1,.99*L2))$
```

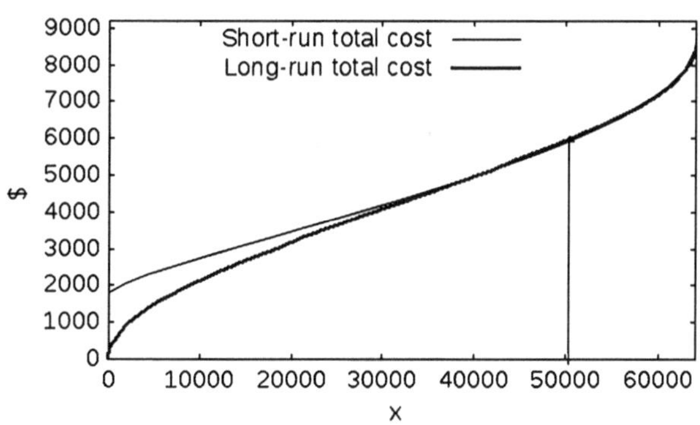

Each total cost curve above generates an average cost curve and a marginal cost curve. The short-run average cost curve is tangent to its long-run counterpart for the output level at which $K = KS0$ is the efficient employment level. The short-run curve lies above the long-run curve for all other output values.

Even though short-run total and average cost exceed their long-run counterparts for small output levels, the short-run marginal cost is less than the long-run marginal cost. This reflects the fact that only the cost of labor varies with output. Beyond the output level at which KS0 is the least-cost capital level, the diminishing marginal product of labor drives up marginal cost to an extent such that the short-run marginal cost exceeds the long-run marginal cost.[2]

[2]In the short run, $dx = \frac{dx}{dL} \cdot dL = mpl \cdot dL$ and $d(stc) = w \cdot dL$, so $smc = \frac{d(stc)}{dx} = \frac{w}{mpl}$.

```
wxdraw2d( ... Some commands omitted ...
 points([xS0,xS0],[0, tcS0/xS0]),...parametric(
 xaug(L,w0,r0,a0,b0,c0,e0),ltcaug(L,w0,r0,e0)/
 xaug(L,w0,r0,a0,b0,c0,e0),L,.1,.99*L2), ...
 parametric( x(KS0,L,a0,b0,c0,e0), sac(KS0,L,a0,
 b0,c0,e0,w0,r0),L,.1, 1.2*L2), ... parametric(
 xaug(L,w0,r0,a0,b0,c0,e0),lmc(L, w0,r0,a0,b0,
 c0,e0),L,.1,.99*L2), ... parametric(x(KS0,L,a0,
 b0,c0,e0),smc(KS0,L,a0,b0,c0,e0,w0),L,.1,L2))$
```

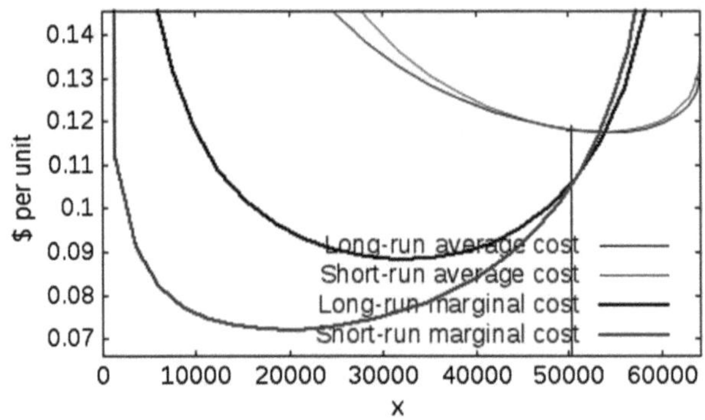

7.5 Long-Run and Short-Run Costs Using Isoquants

Using isoquants provides another view of the connection between the short run and the long run. The first input/output cell below defines the values required to draw three isoquants. It also provides the slope of each of these isoquants when K can be varied to minimize the cost of producing each of the quantities.

```
[L3,L4,L5]:[5,10,15]$  [K3,K4,K5]:
 float(m0*[L3,L4,L5])$ x3:x(m0*L3,L3,a0,b0,c0,e0)$
  x4:x(m0*L4,L4,a0,b0,c0,e0)$
x5:x(m0*L5,L5,a0,b0,c0,e0)$ slope:mrts(m0*L3,L3,e0)$
matrix(["slope =",slope]);   matrix(["L","K","x"],
 [L3, K3, x3], [L4, K4, x4], [L5, K5, x5] );
```

$$\begin{bmatrix} \text{slope} = -0.625 \end{bmatrix} \quad \begin{bmatrix} L & K & x \\ 5 & 1.3393 & 337.44 \\ 10 & 2.6786 & 1298.1 \\ 15 & 4.0179 & 2824.7 \end{bmatrix}$$

The graph below uses these values to produce three isoquants. The middle isoquant is for $x = 1298.1$ units. The line segment tangent to it is the isocost line that shows the minimum cost of producing 1298.1 units. The left-most isoquant depicts ways to produce $x = 337.44$ units. The least-cost way of doing this is to use the *K*, *L* pair indicated by the point of tangency with the lowest isoquant. In the short run, the firm employs too much capital. Cost minimization cannot occur on the isoquant at the right because the firm cannot get enough capital in the short run.

```
wxdraw2d(xlabel="L, when K = m0*L",ylabel="K",yrange=
  [0,(r0*K5+w0*L5)/r0], implicit(x(K,L,a0,b0,c0,e0)
  =x3,L,.1,1.2*L5,K,.1,1.5*K5),explicit((r0*K3+w0*L3)/
  r0+slope*L,L,0,(r0*K3+w0*L3)/w0),implicit(
  x(K,L,a0,b0,c0,e0)=x4,L,.1,1.2*L5,K,.1,1.5*K5),key=
   "",explicit((r0*K4+w0*L4)/r0+slope*L,L,0,(r0*K4
   +w0*L4)/w0), implicit(x(K,L,a0,b0,c0,e0)=x5,
   L,.1,1.5*L5,K,.1,2*K5),explicit((r0*K5+w0*L5)/
   r0+slope*L,L,0,
    (r0*K5 + w0*L5)/w0), explicit(K4,L,0,1.5*L5)    )$
```

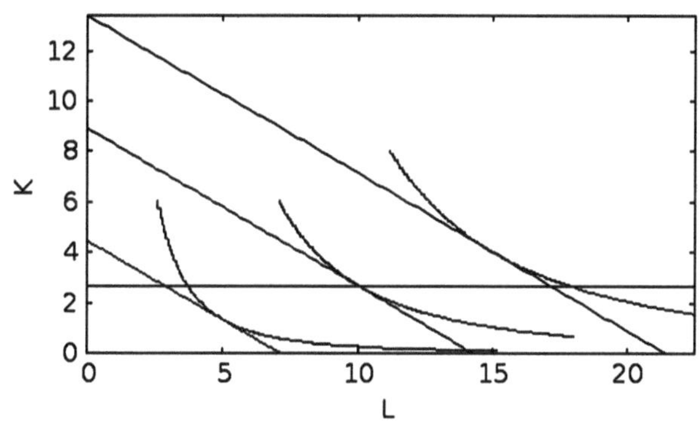

To determine the cost at the two points for which the isoquants are not tangent to the isocost lines, use *Maxima*'s `find_root` command. Set the production function, with *K* properly constrained, equal to the value of *x* along each of the two isoquants and evaluate that expression. The table contains the required employment levels for *L* and the implied short-run costs. The long-run values come from the earlier table of long-run values.

```
L3s:find_root(x(K4,L,a0,b0,c0,e0)=x3,L,0,(r0*K5
  +w0*L5)/w0)$L5s:find_root(x(K4,L,a0,b0,c0,e0)=
  x5,L,0,(r0*K5+w0*L5)/w0)$matrix(["x","K,LR","L,LR",
  "K,SR","L,SR","Cost,LR","Cost,SR"],[x3,K3,L3,K4,
  L3s,r0*K3+w0*L3,r0*K4+w0*L3s],
 [x4,K4,L4,K4,L4,r0*K4+w0*L4,r0*K4+w0*L4],
 [x5,K5,L5,K4,L5s,r0*K5+w0*L5,r0*K4+w0*L5s]);
```

x	K,LR	L,LR	K,SR	L,SR	Cost,LR	Cost,SR
337.44	1.3393	5	2.6786	3.715	357.14	400.04
1298.1	2.6786	10	2.6786	10	714.29	714.29
2824.7	4.0179	15	2.6786	17.847	1071.4	1106.6

7.6 Fixed and Variable Short-Run Costs

In the long run, a firm can vary the employment levels of all inputs, selecting the least-cost way of producing each input. When $x = 0$ the firm ceases operations, so that its total cost is zero. Thus, in the long run, all costs vary with the firm's output level.

In the short run, however, the firm must employ one or more inputs at some fixed level(s). This implies that the firm must incur some *fixed* costs for each period that it continues to operate.[3] In our illustration, the fixed cost equals $r0 \cdot$ KS0.

The two curves below show the total cost and the total *variable* cost. The vertical distance between the two, a constant, is the firm's total fixed cost.

```
wxdraw2d( ... Some commands omitted ... points(
 [xS0,xS0],[0, tcS0]), ... parametric(x(KS0,L,a0,b0,
 c0,e0),r0*KS0+w0*L,L,.1,1.2*L2), ... parametric(
 x(KS0,L,a0,b0,c0,e0),w0*L,L,.1,1.2*L2))$
```

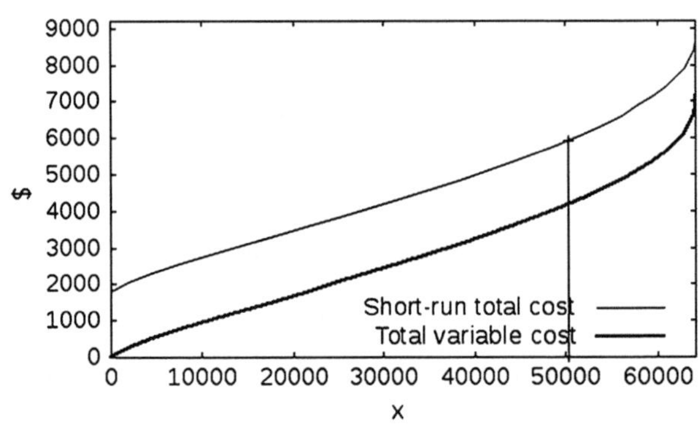

The data in the graph above can be stated in per-unit terms. The result is four curves.

- The short-run average cost curve shows short-run total cost divided by the output level.
- The average variable cost curve shows total variable cost divided by the output level.
- The short-run marginal cost curve shows the (common) slope of the short-run total cost curve and the total variable cost curve. The short-run marginal cost curve intersects each of these average cost curves at its minimum value.

[3]The formal analysis does not distinguish between fixed costs and sunk costs. They are not quite the same. Some fixed costs (i.e., costs that are the same for all positive output rates) might be avoided if the firm shuts down and sells off or sub-leases some of its capital. Those capital costs that cannot be avoided in this way are sunk costs, and they are part of the fixed cost.

- Finally, the average fixed cost curve is a constant, total fixed cost, divided by the output level. Analytically, this is the least important of the four curves and will generally be ignored henceforth. The height of this curve at any output level equals the vertical distance between the short-run average cost curve and the average variable cost curve, so it is redundant.

```
wxdraw2d( ... Some commands omitted ... points(
 [xS0,xS0],[0,tcS0/xS0]), ...
 parametric(x(KS0,L,a0,b0,c0,e0),sac(KS0,L,a0,
 b0,c0,e0,w0,r0)-r0*KS0/x(KS0,L,a0,b0,c0,e0),
 L,.1,L2), ... parametric(x(KS0,L,a0,b0,c0,e0),
 sac(KS0,L,a0,b0,c0,e0,w0,r0),L,.1,1.2*L2), ...
 parametric( xaug(L,w0,r0,a0,b0,c0,e0),smc(KS0,
 L,a0,b0,c0,e0,w0),L,.1,L2), ...
 explicit(r0*KS0/x, x, 1, x2) )$
```

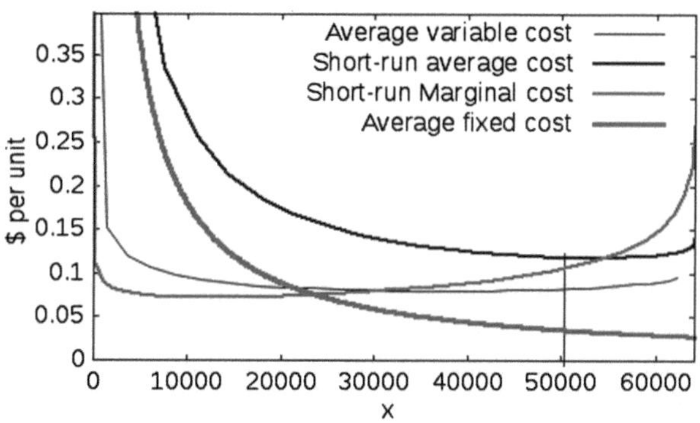

7.7 Long-Run Average Cost as a Planning Curve

At each moment, the firm is in some short run. Even so, the long-run average cost curve (or equivalently the long-run total cost curve) is pertinent as a planning curve. Suppose that the firm can choose between only two values of capital in the short run, $KS0 = 22.27$ and $KS1 = 9.5443$. Which to select depends on what output rate the firm expects to produce. The graph below shows that if the firm expects to produce $x \leq 3{,}000$, approximately, then it should employ the smaller amount of capital, keeping its per-unit cost at its minimum level. Likewise, larger expected output rates call for employing the larger amount of capital each time period.

If only these two options were available, the firm's long-run average cost curve would consist of the scalloped curve that consists of part of `sac, K = KS1` and then part of `sac, K = KS1`.

```
KS1: Klist[3]; wxdraw2d(xlabel="x",ylabel=
 "$ per unit",yrange=[.8*ac2,1.5*ac2],xrange=
 [0,1.1*xlist[10]],key="sac, K = KS0",
 parametric(x(KS0,L,a0,b0,c0,e0),sac(KS0,
 L,a0,b0,c0,e0,w0,r0),L,.1,1.2*L2),key="sac,
 K = KS1",color=gray,line_width=2,parametric(
 x(KS1,L,a0,b0,c0,e0),
   sac(KS1,L,a0,b0,c0,e0,w0,r0),L,.1,1.2*L2))$
```

9.5443

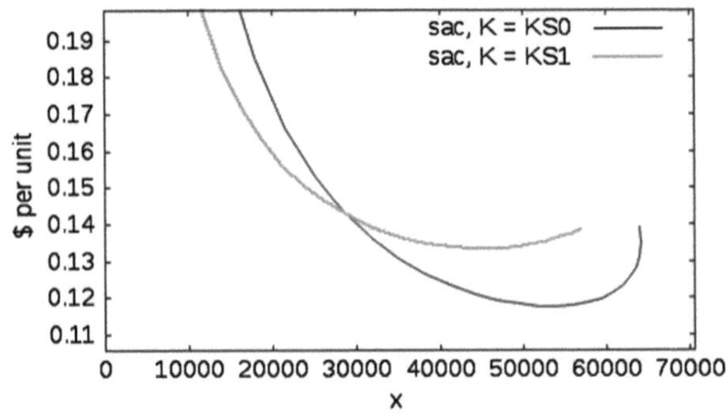

As the number of options increases, the scalloping becomes less pronounced, and the *envelope* of all of the relevant segments of various short-run average cost curves begins to look like the long-run average cost curve below. In the extreme, if capital employment were actually infinitesimally variable, then each point on the long-run curve would consist of a single point on the relevant short-run curve.

```
KS1:Klist[3];wxdraw2d( ... Some commands
 omitted ... points([[xlist[3],0],[xlist[3],
 laclist[3]]]),points([[xlist[7],0],[xlist[7],
 laclist[7]]]),key="sac, K = KS0",parametric(
 x(KS0,L,a0,b0,c0,e0),sac(KS0,L,a0,b0,c0,e0,w0,r0),
 L,.1,1.2*L2),key="sac, K = KS1", ...
 parametric(x(KS1,L,a0,b0,c0,e0),sac(KS1,L,a0,b0,
 c0,e0,w0,r0),L,.1,1.2*L2), ... key=
 "Long-run average cost",parametric( xaug(L,w0,r0,
 a0,b0,c0,e0),ltcaug(L,w0,r0,e0)/
   xaug(L,w0,r0,a0,b0,c0,e0),L,.1,.9*L2) )$
```

9.5443

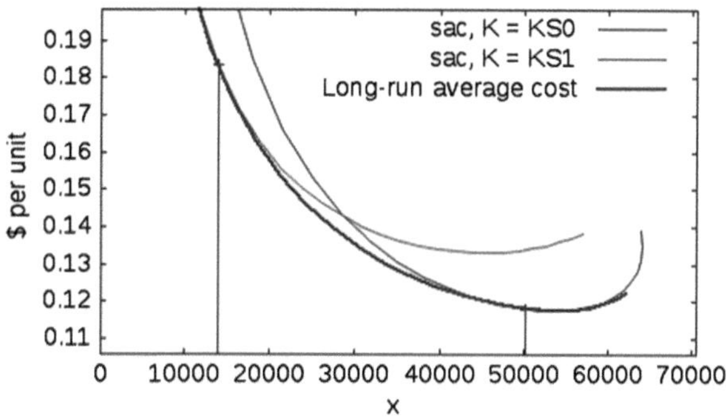
sac, K = KS0
sac, K = KS1
Long-run average cost
$ per unit
0.19
0.18
0.17
0.16
0.15
0.14
0.13
0.12
0.11
0
10000
20000
30000
40000
50000
60000
70000
x

Chapter 8
Firm and Industry Supply in the Short-Run

Abstract This chapter analyzes the output of a competitive industry in the short run. "Competitive" refers to the market structure in which each firm in the industry is a price taker. No single firm's output level appreciably affects the market price. The chapter begins by combining the implications of the firm's output selection for cost with the implications for revenue. We assume that the firm's goal is to earn maximum profit per time period, and we examine the conditions necessary for achieving this goal.

8.1 An Illustrative Cost Function

The function that represents the firm's cost resembles the one derived from the production function in the preceding chapter, but it has a simpler functional form and can be expressed explicitly.

Ideally, short-run cost curves should have two characteristics. First, the short-run average cost curve (`sac`) should be at least roughly U-shaped. At low levels of employment of the variable input(s), fixed cost is spread over a larger number of output units, so that component of average cost decreases as x increases. Also, over some range the marginal product(s) of the variable input(s) may be increasing, further driving down per-unit cost. As the output increases, however, the firm experiences diminishing marginal product(s) of the variable input(s), and this effect eventually dominates the effect of ever-decreasing average fixed cost.

The short-run marginal cost should also be U-shaped, because of initially increasing marginal product(s) of the variable input(s). When the employment level of a single input is variable, then $smc = \frac{w}{mpl}$ where we assume that labor is the variable input, w is the wage rate, and mpl is labor's marginal product. Later, we establish that the relationship continues to hold when more than one input's employment level can vary.

M.R. Hammock and J.W. Mixon, *Microeconomic Theory and Computation: Applying the Maxima Open-Source Computer Algebra System*, DOI 10.1007/978-1-4614-9417-1_8, © Springer Science+Business Media New York 2013

The simplest polynomial functional form that has these characteristics is cubic. We build the illustrative cubic function in a series of steps. First we define the average variable cost function as a function that has a minimum value a when $x = x0$. The values of both a and $x0$ can be user-selected. The first input/output cell contains expressions that are used to determine the value of the coefficient b that is consistent with a pair of selected values.

The expression for avc is as indicated above. The `expand` command instructs *Maxima* to expand the expression. The terms can be collected so that the expression is $(x0^2 - 2 \cdot x0) \cdot b \cdot x + b \cdot x^2 + a$. The expression for the short-run total cost is stated in order to produce its derivative, the marginal cost. Then the slope of the marginal cost is derived.

The marginal cost must decrease at first and then increase. The final expression provides the value of x at which smc reaches its minimum value: $x = \frac{2}{3} \cdot x0$.

```
avc:expand(a+b*(x-x0)^2)$ stc:expand(F+avc*x)$
smc:diff(stc,x)$       diff_smc:diff(stc, x, 2)$
matrix(["avc =",avc],["stc =",stc],["smc =",smc],
 ["diff_smc =",diff_smc]); solnmin:solve(diff_smc,x);
```

$$\begin{bmatrix} \text{avc} = & b\,x0^2 - 2\,b\,x\,x0 + b\,x^2 + a \\ \text{stc} = & F + b\,x\,x0^2 - 2\,b\,x^2\,x0 + b\,x^3 + a\,x \\ \text{smc} = & b\,x0^2 - 4\,b\,x\,x0 + 3\,b\,x^2 + a \\ \text{diff\_smc} = & 6\,b\,x - 4\,b\,x0 \end{bmatrix} \quad [x = \frac{2\,x0}{3}]$$

We solve for b in terms of $x0$ and a, removing b from the expression for the cost functions. First, `subst` is used to evaluate the smc expression when $x = \frac{2}{3} \cdot x0$. The `solve` command sets that value equal to one-fourth the value of a and solves it for b. Any proper fraction of a would work; the purpose is to ensure that smc remain positive for all values of x.

```
solnb: solve(subst(solnmin[1],smc)=a/4,b);
```

$$[b = \frac{9\,a}{4\,x0^2}]$$

Now we specify the functions that we use for the remainder of this section. We evaluate expressions for avc, stc, sac, and smc, given the expression for b that we have just derived. In each case, we phrase the statement as a function, with the variable x and the relevant parameters indicated within parentheses.

```
avc(x,x0,a)   := ''(subst(solnb[1],avc));
stc(x,x0,a,F) := ''(subst(solnb[1],stc));
sac(x,x0,a,F) := ''(expand(subst(solnb[1],stc/x)));
smc(x,x0,a)   := ''(expand(subst(solnb[1],smc)));
```

$$\text{avc}(x, x0, a) := -\frac{9\,a\,x}{2\,x0} + \frac{9\,a\,x^2}{4\,x0^2} + \frac{13\,a}{4}$$

$$\text{stc}(x, x0, a, F) := F - \frac{9\,a\,x^2}{2\,x0} + \frac{9\,a\,x^3}{4\,x0^2} + \frac{13\,a\,x}{4}$$

$$\text{sac}(x, x0, a, F) := \frac{F}{x} - \frac{9\,a\,x}{2\,x0} + \frac{9\,a\,x^2}{4\,x0^2} + \frac{13\,a}{4}$$

$$\text{smc}(x, x0, a) := -\frac{9\,a\,x}{x0} + \frac{27\,a\,x^2}{4\,x0^2} + \frac{13\,a}{4}$$

To evaluate the per-unit cost curves, we specify three parameters: the value of x at which avc reaches its minimum value, that minimum value, and total fixed cost, F. We arbitrarily set F equal to one half the total variable cost at the quantity for which avc reaches its minimum point. Because $avc = a+b\cdot(x-x0)^2$, $avc = a$, total variable cost equals $a \cdot x$ when $x = x0$. Therefore, $F0 = \frac{a0\cdot x0}{2}$ at $x = x0$. Another value could be used without affecting the analysis.

```
sac(x,x0,a0,F0); avc(x,x0,a0); smc(x,x0,a0);
```

$$\frac{F0}{x} - \frac{9\,a0\,x}{2\,x0} + \frac{9\,a0\,x^2}{4\,x0^2} + \frac{13\,a0}{4} \qquad -\frac{9\,a0\,x}{2\,x0} + \frac{9\,a0\,x^2}{4\,x0^2} + \frac{13\,a0}{4} \qquad -\frac{9\,a0\,x}{x0} + \frac{27\,a0\,x^2}{4\,x0^2} + \frac{13\,a0}{4}$$

```
[x0,a0]:[50,30]$ F0:a0*x0/2$ wxdraw2d( ... Some
commands omitted ... explicit(sac(x,x0,a0,F0),x,
0,2*x0), ... explicit(avc(x,x0,a0),x,0,2*x0),
... explicit(smc(x,x0,a0),x,0,2*x0),
... explicit(F0/x,x,0.1,2*x0))$
```

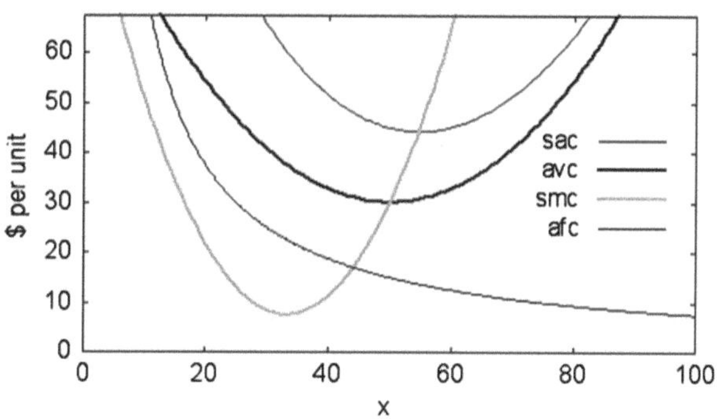

8.2 Selecting a Quantity

We assume that the firm maximizes the profit arising from the production and sale of a single product.[1] The definition of profit in this case is simple: $Profit = Revenue - Cost$.

[1]A more general model would let the firm produce a line of goods. Then, the firm would choose a vector of quantities that would maximize its profit. This selection might involve giving up profits from one good in the line in return for even more profits from other goods in the line. This extension would not change the nature of the analysis.

8.2.1 A General Statement

Maximizing profits requires finding the quantity at which the marginal profit equals zero. That is $\frac{dProfit}{dx} = 0$. Both revenue and cost are affected by the quantity sold, so we add the `depends` command to make these relationships explicit. The `depends` command tells *Maxima* that a term is a function of a variable, without specifying the functional form. Taking the derivative and setting it equal to zero confirms that the firm's profits are maximized if it produces the quantity for which marginal revenue equals marginal cost.

```
depends([Revenue,Cost],x)$
Profit:Revenue-Cost$
solve(diff(Profit,x)=0,x);
```

$[\frac{d}{dx} Revenue = \frac{d}{dx} Cost]$

For now, we add a second assumption that the firm charges the same price for each unit sold. Later chapters relax this assumption, and the next section sketches the model absent this assumption. Given this assumption, the profit function for a firm producing good x is $Profit = (p \cdot x) - Cost$.

8.2.2 The Price-Taking Firm

The cell below takes the derivative of profit with respect to x. Setting the result equal to zero and solving yields this rule: produce a quantity such that marginal cost = price, the proper response when the firm is a price taker—when the price for which it can sell its output is unrelated to the quantity that it sells.[2]

```
Profit:p*x-Cost$ solve(diff(Profit,x)=0,x);
```

$[\frac{d}{dx} Cost = p]$

8.2.3 The Price-Searching Firm

This section sketches the case in which a firm's quantity affects the price of its product.[3] (Chaps. 12–14 extend this analysis.) Price now depends on x, so the firm searches for the price that yields maximum profit. That price must be consistent with selling a quantity at which marginal cost equals $p + \frac{dp}{dx}$. For the price taker, the second term equals zero, so the condition collapses to the one in the previous cell. When $\frac{dp}{dx}$ is negative, however, marginal revenue is less than the price.

[2] *Maxima* assumes that p and x are unrelated, because we did not specify any relationship using `depends`.

[3] This section represents a slight digression. It can be omitted without loss of continuity.

Comparing this statement to the general statement above reveals that marginal revenue ($\frac{dProfit}{dx}$) equals $\frac{dp}{dx} \cdot x + p$. Manipulating this term reveals that the marginal revenue equals $p \cdot (1 + \frac{dp}{dx} \cdot \frac{x}{p})$, or $mr = p \cdot (1 + 1/Ep)$, where Ep is the own-price elasticity of demand for the good. In this introduction, we assume that the firm charges only one price for all units—it cannot engage in price discrimination.

8.2.3.1 Illustration

For simplicity, we use a linear demand curve for the firm's product. To establish a specific function for demand, we fix the maximum quantity (when $p = 0$) at $x = 3 \cdot x0$ (any multiple greater than 1.0 would work), and at $x0$ we set the height of the demand curve at 1.2 times (any multiple would work) the short-run average cost. The selected values ensure that the firm can earn a profit.

```
p(x,e,f) := e+f*x; solnp1: solve([p(3*x0,e,f)=0,
  p(x0,e,f)=1.2*sac(x0,x0,a0,F0)], [e,f])[1];
```

$$p(x,e,f) := e + f\,x \quad [e = 81,\ f = -\tfrac{27}{50}]$$

The input group below binds these parameter values to the names `e1` and `f1`.

```
[e1,f1]:[rhs(solnp1[1]),rhs(solnp1[2])]$
```

We next derive the marginal revenue and profit functions. The marginal revenue function that is generated by a linear demand curve[4] is also linear. The profit function is cubic.

```
mr(x,e,f):=''(diff(x*p(x,e,f),x));profit(x,e,f,a,F):=
   ''(expand(x*(p(x,e,f)-sac(x,x0,a,F))));
```

$$\mathrm{mr}(x,e,f) := 2\,f\,x + e$$
$$\mathrm{profit}(x,e,f,a,F) := -F - \frac{9\,a\,x^3}{10000} + f\,x^2 + \frac{9\,a\,x^2}{100} + e\,x - \frac{13\,a\,x}{4}$$

The table below summarizes the revenue information above for this example. It shows the expressions for the demand curve, the total revenue function, and the marginal revenue function.

```
expand(float([p(x,e1,f1),x*p(x,e1,f1),mr(x,e1,f1)]))$
matrix(["Demand","Total Revenue",
       "Marginal Revenue"],%);
```

$$\begin{bmatrix} \text{Demand} & \text{Total Revenue} & \text{Marginal Revenue} \\ 81.0 - 0.54\,x & 81.0\,x - 0.54\,x^2 & 81.0 - 1.08\,x \end{bmatrix}$$

Next, we determine the quantity that yields maximum profit. Setting marginal revenue equal to marginal cost and solving for x yields two values. At the first quantity, profit is minimized; at the second, profit is maximized.

[4]The demand curve is the average revenue function for a firm that charges the same price for all units.

```
solnq1:float(solve(mr(x,e1,f1)=smc(x,x0,a0),x));
```

$$[x = 4.141, \quad x = 49.192]$$

The first two items in the output list below show the profit levels associated with the profit-minimizing value of x and the profit-maximizing value of x. Only the latter value of x is of concern. It is assigned the name x1, the value of which appears as the third item in the list.

```
[profit(rhs(solnq1[1]),e1,f1,a0,F0), profit(rhs(
  solnq1[2]),e1,f1,a0,F0),x1:rhs(solnq1[2])]$
  matrix(["Min Profit","Max Profit",
      "Profit Maximizing x"],%);
```

Min Profit	Max Profit	Profit Maximizing x
−783.2	451.2	49.192

The table below shows relevant revenue, cost, and profit data for a list of output levels. The fifth row contains values for the profit-maximizing level of x. The values confirm that the profit level reaches its maximum value when marginal revenue equals marginal cost.

```
fpprintprec:3$ xlist:makelist(2*i*x1/10,i,1,8)$
plist:p(xlist,e1,f1)$     trlist:xlist*plist$
aclist:sac(xlist,x0,a0,F0)$ tclist:xlist*aclist$
mrlist:mr(xlist,e1,f1)$  proflist:trlist-tclist$
mclist:smc(xlist, x0, a0)$ names : matrix(["x",
 "p=ar","tr","sac","stc","profit","mr","smc"])$
numbers:transpose(matrix(xlist,plist,trlist,
  aclist,tclist,proflist,mrlist,mclist))$
addrow(names,numbers);    fpprintprec:5$
```

x	$p = ar$	tr	sac	stc	$profit$	mr	smc
9.84	75.7	745.	150.	1470.	−729.	70.4	52.2
19.7	70.4	1380.	92.9	1830.	−444.	59.7	22.6
29.5	65.1	1920.	66.7	1970.	−49.5	49.1	8.68
39.4	59.7	2350.	52.1	2050.	300.	38.5	10.4
49.2	54.4	2680.	45.3	2230.	451.	27.9	27.9
59.0	49.1	2900.	44.9	2650.	249.	17.2	61.0
68.9	43.8	3020.	50.5	3480.	−461.	6.62	110.
78.7	38.5	3030.	61.8	4860.	−1830.	−4.0	174.

The graph below confirms the results above. The firm can earn profits for any value of x for which the `sac` curve lies below the demand (average revenue) curve. Maximum profits occur when the marginal cost curve cuts the marginal revenue curve from below.[5]

```
p(x,e1,f1); x1;
```

$81 - \frac{27x}{50}$ 49.192

```
wxdraw2d(... Some commands omitted ... explicit(
 p(x,e1,f1),x,0,2*x1), ... explicit(mr (x,e1,f1),
 x,0,2*x1), ... explicit(avc(x,x0,a0),x,0,2*x1),
 ... explicit(sac(x,x0,a0,F0),x,0,2*x1)
 ...explicit(smc(x,x0,a0),x,0,2*x1))$
```

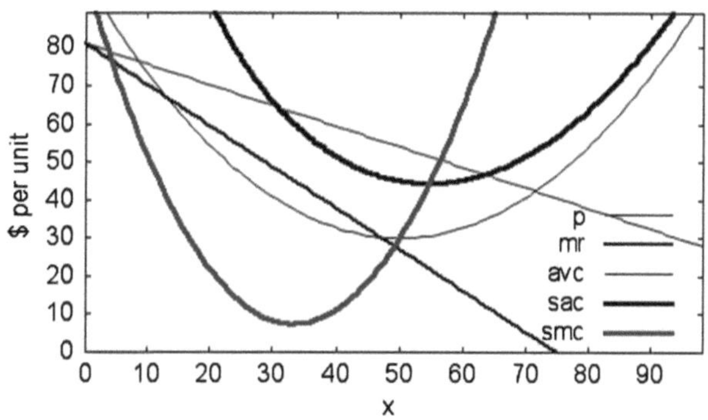

A firm is not assured a profit in the short run. The example below shows the same firm facing a reduced demand curve. The curve is shifted so that, at $x = x0$, its height is halfway between *avc* and *sac*.

```
solnp2:solve([p(4*x0,e,f)=0,p(x0,e,f)=
 (sac(x0,x0,a0,F0)+avc(x0,x0,a0))/2],[e,f])$
[e2,f2]:[rhs(solnp2[1][1]),rhs(solnp2[1][2])]$
matrix(["e1","e2"],[float(e1),float(e2)]);
```

$$\begin{bmatrix} e1 & e2 \\ 81.0 & 50.0 \end{bmatrix}$$

The graph shows that the demand curve lies below *sac* for all x. Therefore, the firm cannot earn a profit in the short run. It can, however, cover its variable cost, partially offsetting the per-period fixed cost.

[5] We leave it as an exercise to confirm that, at the profit-minimizing x, the marginal cost curve cuts marginal revenue from above. To do this copy the input group and paste it into a new cell. Change the maximum value of x values in all `explicit` commands from `x1` to 0.1, and change the `yrange` to `[97, 98]`.

```
wxdraw2d( ... Some commands omitted
  ... explicit(p(x,e2,f2),x,0,2*x1),
  ... explicit(mr(x,e2,f2),x,0,2*x1),
  ... explicit(avc(x,x0,a0),x,0,2*x1),
  ... explicit(sac(x,x0,a0,F0),x,0,2*x1),
  ... explicit(smc(x,x0,a0),x,0,2*x1))$
```

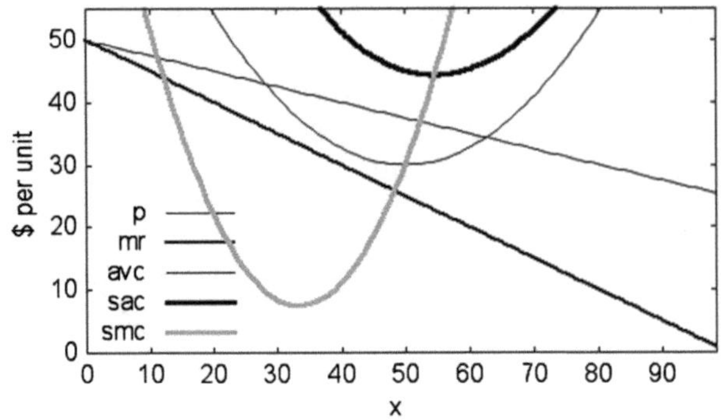

Again, marginal revenue equals marginal cost at two values of x. The second of these, $x = 49.4$ leads to a maximum profit level. The first value of x yields minimum profits (maximum losses) for any positive value of x. The second output line shows the loss if the firm were to produce $x = 10.4$ units. This loss is less than the firm would incur at $x = 0$, which is $F = 750$.

```
solnprof2:float(solve(mr(x,e2,f2)=
smc(x,x0,a0),x))$
x2 : rhs(solnprof2[2])$
matrix(["x2","profit"],[x2,
profit(x2,e2,f2,a0,F0)]);
```

$$\begin{bmatrix} \text{x2} & \text{profit} \\ 48.37 & -370.98 \end{bmatrix}$$

The table below shows that a loss results from any output levels, but that the loss is minimized when $x = 49.4$ units. As above, this is the loss-minimizing value of x for which marginal revenue equals marginal cost.

```
fpprintprec:3$ xlist2:makelist(2*i*x2/10,i,1,8)$
plist2:p(xlist2,e2,f2)$  trlist2:xlist2*plist2$
aclist2:sac(xlist2,x0,a0,F0)$ tclist2:xlist2*aclist2$
mrlist2:mr(xlist2,e2,f2)$ proflist2:trlist2-tclist2$
mclist2:smc(xlist2,x0,a0)$ names:matrix(
   ["x","p","tr","sac","stc","profits","mr","smc"])$
numbers:transpose(matrix(xlist2,plist2,trlist2,
  aclist2,tclist2,proflist2,mrlist2,mclist2))$
addrow(names,numbers); fpprintprec:5$
```

x	p	tr	sac	stc	profits	mr	smc
9.67	47.6	460.	151.	1460.	−1000.	45.2	52.8
19.3	45.2	874.	94.1	1820.	−947.	40.3	23.3
29.0	42.7	1240.	67.7	1970.	−725.	35.5	9.01
38.7	40.3	1560.	52.8	2040.	−484.	30.7	9.83
48.4	37.9	1830.	45.6	2200.	−371.	25.8	25.8
58.0	35.5	2060.	44.7	2590.	−533.	21.0	57.0
67.7	33.1	2240.	49.6	3360.	−1120.	16.1	103.
77.4	30.7	2370.	60.0	4640.	−2270.	11.3	165.

8.2.4 Section Summary

Before returning to the important special case of the price-taking firm, we review the points raised so far. A firm that maximizes its profits produces its selected output level in a least-cost fashion (per-unit cost is not above the height of the average cost curve for that quantity). Furthermore, it produces the level at which marginal revenue equals marginal cost.

These conditions can be thought of in either of two ways. First, they might be taken as *normative*: They define what the firm must do if it is to achieve maximum profits. In doing so, they define a set of questions, or an analytical framework within which the firm determines whether it can increase profits by changing either its input mix or its output level. This use of the analysis forms the basis for *managerial economics*.

The alternative view, and the one taken here, is that these conditions constitute a set of *positive* propositions. The analysis proceeds *as if* firms do satisfy the stated criteria. Such positive assertions are invariably incorrect if used as an exact description of reality. This does not imply, however, that they should not be made. Even though firms may never actually achieve the goal of profit maximization, and this might not even be the firm's goal, models based on these assertions can be useful. Furthermore, we should expect firms in competitive markets that persistently stray from these propositions to eventually be displaced by firms that act as though they *were* trying to follow these propositions. See [1].

As long as profit is an important and consistent part of what firms set out to achieve and as long as they seek ways to move toward higher profits, the model of profit maximization will be generally consistent with observed behavior.

8.3 The Price-Taking Firm in the Short Run

The price for which the price-taking firm sells its product is determined by the market and does not depend on how much the firm produces. Before examining the implications of this market structure for the firm's output decision, we consider why this model might apply beyond the markets in which it precisely describes the firm's situation.

8.3.1 Market Elasticity, Firm Size, and the Firm's Marginal Revenue

Consider a simple model in which a firm is embedded in a market with other similar firms. The firm is large enough that its output affects the market price. Indeed, we assume that the firm sets a price and that the other firms react to this price. The demand for the firm's product is this: $x = X - xo$ where x is the quantity demanded of the firm's product, X is the market quantity demanded, and xo (not $x0$) is the quantity supplied by other firms.

Take the derivative of this expression with respect to p:

$$\frac{dx}{dp} = \frac{dX}{dp} - \frac{dxo}{dp}.$$

That is, the change in the quantity demand of this firm's product per-unit price change is determined by two responses: the response of the overall market demand to this price change, and the supply response by other firms. Convert this expression to elasticities:

$$\left(\frac{dp}{dx} \cdot \frac{x}{p}\right) = \left(\frac{dX}{dp} \cdot \frac{X}{p} \cdot \frac{x}{X}\right) - \left(\frac{dxo}{dp} \cdot \frac{xo}{p} \cdot \frac{x}{xo}\right).$$

Define this firm's share of the market as $s = x/X$ and note that $xo = X - x$. With this information and a few manipulations, the equation above can be rephrased as follows:

$$Ef = Em/s - Eo \cdot \frac{1-s}{s}.$$

Here, Ef, Em, and Eo are the elasticity of demand for the firm's product, elasticity of demand for the product at the market level, and the elasticity of *supply* from the other firms in the market.

```
Ef(Em,Eo,s):=Em/s-Eo*(1-s)/s$
```

The table below shows a set of elasticities for a list of s values. In the first case, `"Less Elastic,"` $Em = -0.5$ and $Eo = 0.25$. In the second case, `"More Elastic,"` $Em = -1.5$ and $Eo = 0.75$. The implied values of Ef appear in the second and fourth columns. In all cases, the demand for the firm's product is quite elastic.

The third and fifth columns show the ratio of marginal revenue to price. Even for the less elastic case, if the firm's share is as low as 0.05, then the marginal revenue exceeds 90% of the price. In the case where the elasticities are somewhat higher, marginal revenue exceeds 90% of price for all values of s less than 0.20. With marginal revenue so close to price, the distinction between selecting the quantity at which $p = mc$ and the quantity at which $mr = mc$ becomes small, especially given that firms do not know their demand curves but must estimate them.

```
slist:makelist(i*.05,i,1,6)$ Ef1list:
  Ef(-0.5,0.25,slist)$  mr1list:1 + 1/Ef1list$
Ef2list:Ef(-1.5,0.75,slist)$ mr2list:1+1/Ef2list$
numbers:transpose(matrix(slist,Ef1list,mr1list,
  Ef2list,mr2list))$     names : matrix(["s",
  "Less Elastic","mr/p,low","More Elastic","mr/p,
  high"])$    addrow(names,numbers);
```

s	$LessElastic$	$mr/p, low$	$MoreElastic$	$mr/p, high$
0.05	−14.75	0.932	−44.25	0.977
0.1	−7.25	0.862	−21.75	0.954
0.15	−4.75	0.789	−14.25	0.93
0.2	−3.5	0.714	−10.5	0.905
0.25	−2.75	0.636	−8.25	0.879
0.3	−2.25	0.556	−6.75	0.852

8.3.2 Determining Quantity

The input group below determines the value of x for a price-taking firm with the cost function used above. The first output item is a list with two expressions of x as a function of p, the results of setting price equal to marginal cost. Neither expression in this list yields a real number unless $p \geq \frac{15}{2}$. Only the second expression yields positive x values whenever $p \geq \frac{15}{2}$, so this is the expression for the supply curve.[6]

---

[6]The second output item shows where this critical value of p comes from. It is the minimum value of the marginal cost curve. The value $x = x0 \cdot \frac{2}{3}$ is, as we saw earlier, the value of x at which marginal cost reaches it minimum value. If price is less than this value, then the solutions would involve imaginary numbers.

```
[solnpt:solve(p-smc(x,x0,a0),x),   smc(x0*2/3,x0,a0)];
```

$$[[x = -\frac{25^{\frac{3}{2}}\sqrt{2p-15}-300}{9},\ x = \frac{25^{\frac{3}{2}}\sqrt{2p-15}+300}{9}],\ \frac{15}{2}]$$

Next, we define a profit function and a supply curve for this price-taking firm, `ptprofit`. Contrast this price taker's supply curve with the analysis for the price-searching firm. The price-searching firm must determine both price and quantity together. Such firms cannot have supply curves, because a supply curve expresses quantity in terms of an externally determined price.

```
ptprofit(p):=''(ratsimp(subst(solnpt[2],p*x-
stc(x,x0,a0,F0))));      x(p):=''(rhs(solnpt[2]));
```

$$\text{ptprofit}(p) := \frac{\sqrt{2p-15}\left(4\,5^{\frac{3}{2}}\,p-6\,5^{\frac{5}{2}}\right)+900\,p-54000}{27}$$

$$\text{x}(p) := \frac{25^{\frac{3}{2}}\sqrt{2p-15}+300}{9}$$

The table below shows a range of prices, stated as multiples of $a0$, the minimum average variable cost. The `Quantity` column shows the quantity of x that the function above implies. Notice that for all values of x below 30, the resulting loss exceeds fixed cost, 750. Thus, at all prices below 30, $x = 0$ is the proper selection. At $x = 30$ the firm can produce either $x = 0$ or $x = 50$, but no intermediate value, and lose 750 per time period. As the price rises, so does the profit level. Also, as the price rises, the output level is determined by the function developed above.

```
plist:float(makelist(i*a0/4,i,1,10))$ xlist:float(
 x(plist))$ ptprofitlist:float(ptprofit(plist))$
names:matrix(["Price","Quantity","Profit"])$
numbers:transpose(matrix(plist,xlist,
ptprofitlist))$ addrow(names,numbers);
```

Price	Quantity	Profit
7.5	33.333	−1750.0
15.0	42.956	−1451.9
22.5	46.942	−1113.9
30.0	50.0	−750.0
37.5	52.578	−365.1
45.0	54.85	37.914
52.5	56.904	457.11
60.0	58.792	891.06
67.5	60.55	1338.7
75.0	62.201	1799.0

The graph below shows the firm's supply curve in the short run. It is the segment of the firm's short-run marginal cost curve that lies above the average variable cost curve. The short-run average cost curve, while shown here, is irrelevant to the firm's output decision. It does, however, determine whether the firm is maximizing its profit or minimizing its loss.

```
wxdraw2d( ... Some commands omitted
  ... explicit(avc(x0,x0,a0),x,0,x0),
  ... explicit(avc(x,x0,a0),x,0,2*x1),
  ... explicit(sac(x,x0,a0,F0),x,0,2*x1),
  ... explicit(smc(x,x0,a0),x,0,2*x1),
  ... explicit(smc(x,x0,a0),x,x0,2*x1))$
```

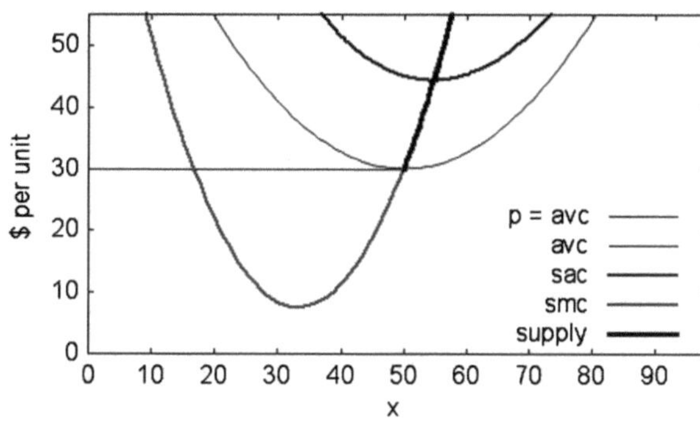

8.3.3 *The Firm and the Industry*

In the short-run the number of firms is constant. Profits do not induce new firms to enter and, as long as variable costs are covered, incumbent firms do not leave the industry. Therefore, the logic of the relationship between firms' supply curves and the industry supply curve is simple: at each price the quantity supplied in the market is the sum of the quantities produced by the individual firms. Another way of stating this relationship—one that relates to the firm supply curves like the one depicted above—is to say that the industry supply curve is the result of a horizontal summation of the firms' supply curves.

The firms need not be identical, but to keep the analysis compact we assume that they are. In that case, horizontal summation of n identical curves is equivalent to multiplication, as indicated in the commands below, which define the industry supply curve, $X(p,n)$.

```
X(p, n) := ''(n*x(p));
```

$$X(p,n) := \frac{n\left(25^{\frac{3}{2}}\sqrt{2p-15}+300\right)}{9}$$

The firm and industry supply curves appear below. In this simplified case in which all firms have identical supply curves, the two supply curves shown are identical except for scale.

```
n0:100$ firmsupply:gr2d(title="Firm Supply",yrange=
[0,2*a0],xlabel="x",ylabel="$ per unit",implicit(
 x =x(p), x, x0/2, 1.5*x0, p, a0, 3*a0))$
industrysupply:gr2d(title="Industry Supply",yrange=
 [0,2*a0],xlabel="X",ylabel="$ per unit",xtics=2000,
 line_width=2,implicit(X =X(p,n0),X,100*x0/2,150*x0,
    p,a0,3*a0))$
wxdraw(firmsupply,industrysupply,columns=2)$
```

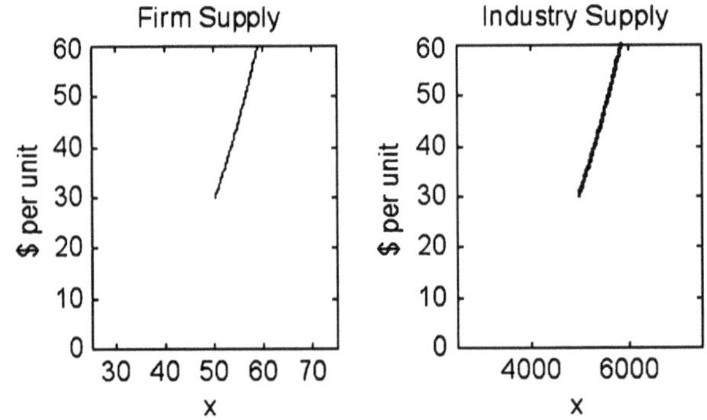

Adding a market demand curve completes the model. We use a constant-elasticity demand curve $X = A \cdot P^{Em}$, where Em is the price elasticity of market demand for this good. We solve for the case in which $p = 1.5 \cdot a0$ at $X = 5{,}000$ (arbitrarily selected) in order to determine the value of the parameter A. The initial functional expression $X = A \cdot p^{Em}$ is replaced in *Maxima*'s memory by the expression $X = A0 \cdot p^{Em0}$.

```
Em0:-1.5$    Xd(p,Em):=A*p^Em;   solve(Xd(1.5*a0,Em0)=
  5000,A)$   A0:rhs(%[1])$  Xd(p):=A0*p^Em0;
```

$$\mathrm{Xd}\,(p, Em) := A\, p^{Em} \qquad \mathrm{Xd}\,(p) := A0\, p^{Em0}$$

We now determine the equilibrium price and quantity by setting supply equal to demand: $X(p, n0) = Xd(p)$. Use of the constant-elasticity demand curve results in an expression that solve cannot evaluate, so we use `find_root`.

We let `find_root` search over the range $p = a0$, the smallest possible value of p if firms are to exist, to three times that value. Any large value could be used.[7]

[7]Exercise: replace `3*a0` with `300*x0`—or 1,000,000, or some other large number—and confirm that the solution remains the same.

```
peq0:find_root(Xd(p)-X(p,n0),p,a0,3*a0)$ Xeq0:float(
   X(peq0,n0))$   matrix(["Equilibrium Price =",peq0],
  ["Equilibrium Quantity =",Xeq0]);
```

$$\begin{bmatrix} \text{Equilibrium Price} = & 42.661 \\ \text{Equilibrium Quantity} = & 5416.8 \end{bmatrix}$$

The graphs below replicate the ones above, but add three items: the demand curve, horizontal lines to indicate the equilibrium price, and a vertical line to indicate the equilibrium quantity. The horizontal line in the left-hand panel corresponds to the demand curve for the firm's product.

```
firmx: gr2d(title="Firm Supply and Demand",
  yrange=[0,2*a0],xlabel="x",ylabel="$ per unit",
  implicit(x=x(p), x, x0/2, 1.5*x0, p, a0, 3*a0),
  explicit(peq0,x,x0/2,1.5*x0) )$
industryX : gr2d(title="Market Supply and Demand",
 yrange=[0, 2*a0],xlabel="x",ylabel="$ per unit",
 xtics=2000,line_type=dots,points_joined=true,
 points([ [Xeq0,0],[Xeq0,peq0]]),
 points([ [n0*x0/2, peq0], [Xeq0, peq0] ]),
 line_type=solid,line_width=2,implicit(X=X(p,n0),
   X, 100*x0/2, 150*x0,p,a0,3*a0),color=gray,
 implicit(Xd(p)=X,X,100*x0/2,1.5*n0*x0,p,a0,3*a0))$
wxdraw(firmx, industryX, columns=2)$
```

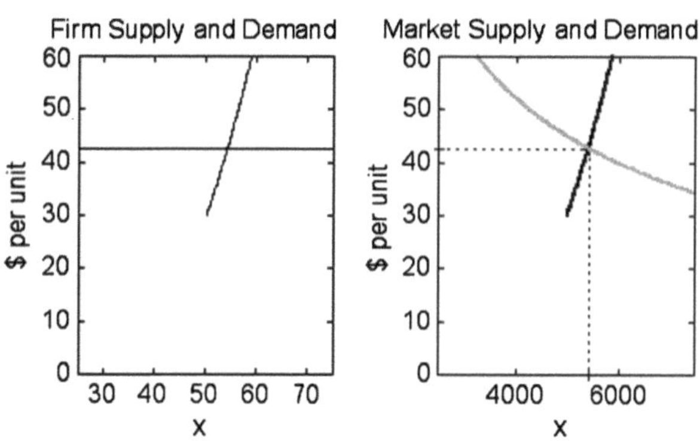

Reference

1. Alchian A (1950) Uncertainty, evolution, and economic theory. J Polit Econ 58:211–221

Chapter 9
Long-Run Supply

Abstract This chapter develops the long-run supply curve for a competitive industry. It begins with the simplest case, in which the industry consists of identical firms whose cost curves are not affected by the number of firms. Next it allows for these identical firms' costs to be influenced by the number of firms. Then it analyzes the case in which each firm has unique cost curves. Finally, it addresses the implication of allocative efficiency.

9.1 Introduction

Chapter 8 shows that in the short run the supply curve for an industry of price-taking firms is the horizontal sum of the firms' supply curves. For each price the quantity supplied to the market equals the sum of the firms' quantities supplied, and the number of firms is fixed (i.e., independent of the price).

In contrast, the number of firms can vary with the price in the long run. Indeed, part of the derivation of the long-run supply involves determining the number of firms. This chapter illustrates this process.[1]

This chapter proceeds in a series of steps. First, it considers the case in which two conditions are met: all firms are identical (have the same cost curves) and the firms' common cost curves do not depend on the number of firms in the industry. Next, it relaxes the second condition and allows for the still-common cost curves to be affected by the number of firms in the industry. Finally, it restores the assumption

[1]The economist does not "determine" the number of firms in any industry, of course. Rather, economic analysis analyzes the forces that determine the number of firms. The material below shows the number of firms given the information contained in the illustrative functions. No analyst could ever have this much information.

M.R. Hammock and J.W. Mixon, *Microeconomic Theory and Computation: Applying the Maxima Open-Source Computer Algebra System*,
DOI 10.1007/978-1-4614-9417-1_9,

that firms' cost curves are not affected by the number of firms in the industry but allows each firm to have a unique set of cost curves.[2]

We also discuss the question of least-cost production in an industry that consists of many independently operated firms. Our analysis shows that an industry of uncoordinated price-taking firms distributes production among those firms and establishes the number of firms, in a way that yields least-cost production.

9.2 A Constant-Cost Industry

This section develops the supply curve for a "constant-cost" industry that consists of identical firms. The term "constant-cost" does not imply that any firm's costs are not affected by the quantities that it produces. Rather, it means that no firm's cost curve is affected by the number or size of other firms in the industry. An implication of this characteristic is that the long-run supply curve is horizontal.[3]

The first cell establishes the parameters of the long-run cost curve for each firm in this industry. This function, $ltc = a + b \cdot x^2$ should be treated as *illustrative* over a relevant range of x values. It does not describe the firm's cost function over all values of x because it contains a constant term, and in the long run all costs are variable.[4]

We specify a minimum long-run average cost, `lac0`, and a quantity, `x0`, at which that average cost is realized.

```
[ltc:a+b*x^2, lmc:diff(ltc,x)];
[lac0,x0]:[30,50];
```

$$[b\,x^2 + a, 2\,b\,x] \quad [30, 50]$$

For this simple function, the marginal cost function is $\frac{d\,ltc}{dx} = 2{\cdot}b{\cdot}x$ and it equals lac at the minimum lac. Thus, $b0 = \frac{lac0}{2{\cdot}x0}$, where $lac0$ and $x0$ are the values defined above. Once $b0$ is determined, $a0$ can be calculated from the fact that $ltc = x \cdot lac$. The values are used throughout the chapter.

```
solve([(a+b*x0^2)/x0=lac0,2*b*x0=lac0],
[a,b])$  [a0,b0]:[rhs(%[1][1]),rhs(%[1]
[2])]$ matrix(["a0 =",a0],["b0 =",b0]);
```

$$\begin{bmatrix} \text{a0} = & 750 \\ \text{b0} = & \frac{3}{10} \end{bmatrix}$$

---

[2]A fourth case, in which each firm has a unique cost structure and that structure depends on the number of firms, is not illustrated. The illustration would be more complex than that of the third case, but it would yield qualitatively similar results.

[3]The number of firms does not affect the cost curves of member firms if changes in the number of firms do not affect input prices *and* if the number of firms does not directly affect member firms' production functions (no technical externalities).

[4]The constant term could be interpreted as follows. Suppose that if the firm produces even the smallest positive amount, it must incur a cost of a, but if it shuts down entirely ($x = 0$), this term vanishes, as does the firm itself. With this interpretation, $a + b * x^2$ is the cost curve for all but the smallest positive values, and $ltc = 0$ for $x = 0$.

Part of the development of the equilibrium condition is an illustration of how an industry moves from one long-run equilibrium value to another when a change occurs. To illustrate this process requires that the short run be taken into account. We use a simpler version of the short-run total cost than the one used in the preceding chapter. The function used here does not allow for an increasing marginal product of the variable input(s). Thus, the short-run marginal cost curve begins at the origin and increases in a monotonic fashion.[5]

```
[stc:d+e*x^3, smc:diff(stc,x)];
```

$$[e\,x^3+d, 3\,e\,x^2]$$

When an industry of identical firms is at equilibrium, each firm operates at the minimum point on its lac. Therefore, the relevant short-run average cost curve is tangent to the long-run average cost curve at the minimum value of the latter. Also, at this point short-run marginal cost, $3 \cdot e \cdot x^2$ equals $lac0$. The next cell provides values for the parameters of the short-run average and marginal cost curves, by solving for the $e0$ that makes $smc0$ equal to $lac0$, and the $d0$ that makes ltc equal to stc.

```
[e0 :lac0/(3*x0^2), d0:lac0*x0-e0*x0^3]$
matrix(["d0","e0"],[d0,e0]);
```

$$\begin{bmatrix} \mathrm{d0} & \mathrm{e0} \\ 1000 & \frac{1}{250} \end{bmatrix}$$

The next cell states the four relevant per-unit cost functions in terms of output and the parameters defined above.

```
[lac(x,a0,b0):=''(a0/x+b0*x),lmc(x,b0):=
  ''(2*b0*x)];[sac(x,d0,e0):=''(d0/x+e0*x^2),
smc(x,e0):=''(3*e0*x^2)];
```

$$[\mathrm{lac}\,(x, a0, b0) := \tfrac{3\,x}{10} + \tfrac{750}{x}, \mathrm{lmc}\,(x, b0) := \tfrac{3\,x}{5}]$$

$$[\mathrm{sac}\,(x, d0, e0) := \tfrac{x^2}{250} + \tfrac{1000}{x}, \mathrm{smc}\,(x, e0) := \tfrac{3\,x^2}{250}]$$

9.2.1 *Firm Supply*

We review the firm's short-run and long-run supply curves. For the price-taking firm, price equals marginal revenue. The profit-maximizing quantity is the one at which marginal revenue equals marginal cost. Therefore, the inverse of the marginal cost curve (the value of x at each level of cost) corresponds to the supply curve.

A qualifier must be added to the preceding statement in the long run: price must also equal or exceed long run average cost.[6] A lower price would force the firm to

---

[5]This is less restrictive than it appears. In an industry of price takers, production always occurs in a range such that the marginal cost curve is increasing, so having it increase from the beginning does not constitute a serious restriction.

[6]Analogously, in the short run, price must be at least as high as the minimum average variable cost.

leave the industry. Therefore, the expression that characterizes the firm's long-run supply curve below contains an `if ...then` statement.[7]

The cell below generates a list of prices beginning at 0 and increasing in increments of 10 to a maximum value of 80.[8] The list of short-run quantities is created by inserting `plist` into the function for the short-run supply curve, resulting in `xslist`. Generating the list of long-run values, `xllist`, requires using map.[9]

```
xl(p):=''(if p < lac0 then 0 else p/(2*b0))$
xs(p):='' (sqrt(p/(3*e0)))$
plist:makelist(10*i,i,0,8)$   xslist:
float(xs(plist))$ xllist:float(map(xl,plist))$
names:matrix(["p, dollars/unit)","x,short run",
"x, long run"])$   numbers:transpose(matrix(
plist,xslist,xllist))$    addrow(names,numbers);
```

p, dollars/unit)	x,short run	x, long run
0	0.0	0.0
10	28.868	0.0
20	40.825	0.0
30	50.0	50.0
40	57.735	66.667
50	64.55	83.333
60	70.711	100.0
70	76.376	116.67
80	81.65	133.33

The graph below generalizes the information in the table. The short-run supply curve is the gray curve that extends from the origin. The long-run supply curve consists of a line that corresponds to part of the long-run marginal cost curve for which $p \geq lac0$. When $p = lac0$, the firm can produce either $x =$ `x0` or $x = 0$, earning no economic profits either way. That is, when $p = lac0$, the firm's owner would be indifferent between remaining in this industry or leaving it. It will not produce a quantity other than one of the two indicated above, so the interior points on the dashed segment in the graph are not part of the firm's long-run supply curve.

[7] Part of that statement involves the quote-quote (″) operator, which activates the expression in parentheses for use in function calls.

[8] The `makelist` command now contains four arguments: `10*i` multiplies the value of `i` by 10, `i` identifies the variable, and the final two terms define the endpoints.

[9] The arguments of the function are not included in the `map` command. The expression must be a function of a single variable so that *Maxima* treats the values in `plist` as the values for the function's argument.

The analysis that follows reveals that the curves that matter are the long-run average cost and the short-run marginal cost. The lac curve defines the condition for entry and exit and determines the quantity produced at long-run equilibrium. The smc determines quantity changes in the short run, between long-run equilibria, in response to price changes.

```
wxdraw2d( ... Some commands omitted ...
 key="sac",explicit(sac(x,d0,e0),x,0,2*x0),
 line_width=2,key="lac",explicit(lac(x,a0,b0),x,
 0,2*x0), ... key="smc",explicit( smc(x,e0),
 x,0,2*x0), ... key="lmc",explicit(lmc(x,b0),
 x,x0,2*x0), ... points([[0,lac0],[x0,lac0]]))$
```

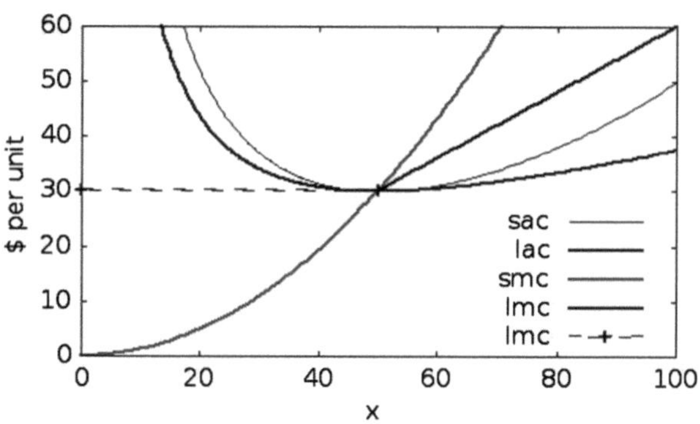

9.2.2 Long-Run Equilibrium

The graph below combines the preceding graph with a representation of the industry's long-run supply curve. A constant-cost industry's long-run supply curve is a horizontal line, the height of which is $lac0$. Any higher price would generate profits and induce entry by new firms. Any lower price would generate losses and force exit of some of the incumbent firms.

```
firmsupply:gr2d(user_preamble="set key bottom",
... Some commands omitted ...
 key="smc",explicit(smc(x, e0), x, 0, 2*x0),color=
 black,line_width=3, key= "relevant lmc",explicit(
 lmc(x,b0),x,x0,2*x0),line_type=dots,key="",
 points_joined=true,points([[0,lac0 ],[x0,lac0]]))$
industrysupply : gr2d(xlabel="Industry Quantity", ...
 explicit(lac0, x, 0, 100*x0) )$ wxdraw(firmsupply,
 industrysupply,columns=2,dimensions=[540,400])$
```

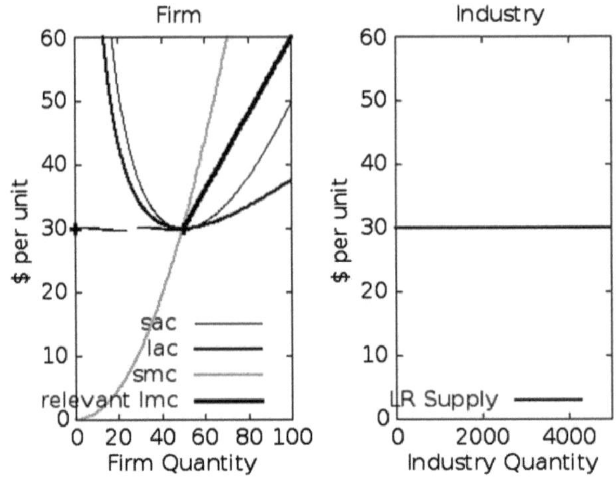

9.2.3 Long-Run and Short-Run Equilibria

The key to long-run equilibrium and, hence, to the nature of the long-run supply curve is the number of firms in the industry. If incumbent firms earn profits in the short run, new firms enter; losses force exit. This section describes how an industry moves from one long-run equilibrium to another, via a short-run equilibrium.

As Chap. 8 notes, this long-run, short-run dichotomy is stylized. In fact, entry occurs over time, as do the adjustments that incumbent firms make. Therefore, a change is likely to result in a relatively large short-run price shock and then the price moves over time to a new equilibrium value. We offer three snapshots of that process: the initial equilibrium, a short-run equilibrium in which the firms remain on their short-run supply curves and the number of firms is fixed, and the new long-run equilibrium in which the number of firms has fully adjusted. That is, our illustration is limited to *comparative statics*, which considers the nature of equilibrium positions and not the nature of the path from one such equilibrium to another.[10]

The cell below produces a table of values that trace changes due to a demand shift. The illustration begins with a specified number of firms, $n0 = 200$, and the industry output given that each firm produces at the minimum point on its lac curve, $200 \cdot 50 = 10000$. We establish this as an equilibrium by defining an initial demand curve that has a height equal to $lac0$ at $X = 10{,}000$ (X denotes industry quantity; x denotes firm quantity).

[10]Some authors refer to the move from one long-run equilibrium to another as dynamic. In some sense this is correct, but we avoid this usage. Speaking of dynamics suggests a time path than can be discovered. Such a discernible path is not likely to exist, however, for if it did traders would exploit it and in the process change it.

In the constant-cost industry, the long-run supply curve is a horizontal line with a height of $lac0$. We arbitrarily define the demand curve so that its value when $p = 0$ is 5000 greater than its value when $p = lac0$. The term Xeq is the equilibrium value of X, the quantity at which the demand curve intersects the long-run supply curve, and the term T is a demand shift term, initially set at $T = 5000$.

The number of firms $n0$ is specified as the first entry in the cell. Then the number of firms is later calculated by setting the quantity demanded equal to the quantity supplied. This computation serves as a check; if the specification is correct, the computed value equals the specified value.

```
n0:200$   Xeq0:n0*x0$   Xs(p,n):=''(n*xs(p)); Xl(p,n):=
 ''(n*xl(p));    Xd(p,Xeq,T):=Xeq+T-(5000/lac0)*p;
solve(Xd(lac0,Xeq0,5000)-Xs(lac0,n),n)$
n0calc:float(rhs(%[1]))$
matrix(["Check: n0calc = ",float(n0calc)]);
```

$$\mathrm{Xs}\,(p,n) := \frac{5\sqrt{10}n\sqrt{p}}{\sqrt{3}}$$

$$\mathrm{Xl}\,(p,n) := n\left(\text{if } p < 30 \text{then } 0 \text{else } \tfrac{5p}{3}\right)$$

$$\mathrm{Xd}\,(p, Xeq, T) := Xeq + T + \left(-\tfrac{5000}{lac0}\right)\, p$$

$$200.0 \qquad \left[\text{Check: n0calc} = \ 200.0\right]$$

Next, we shift the demand curve, changing T from 5000 to 15000, and trace the effects of this 10,000 unit demand increase. First, we determine the new price, $p1$ in the short run. The short-run supply curve is defined with $n = 200$. The price rises to 63.031, and the quantity increases from 10,000 to 14,495 units. This is a short-run equilibrium because firms cannot enter.

To find the new long-run equilibrium number of firms, we set demand equal to the new short-run supply with the price constrained to $lac0$, but the number of firms unknown (as more firms will enter due to the higher short-run price). Solving this equation for n yields $n1 = 400$, so the number of firms doubles. The new long-run equilibrium quantity is 20,000 units, so quantity increases by an amount equal to the rightward shift in the demand curve.

In this special case, *and only in this special case*, the long-run supply curve determines the price, and the demand curve determines the quantity.

```
p1:find_root(Xd(p,Xeq0,15000)-Xs(p,n0),p,lac0,
 10*lac0)$  solve(Xd(lac0,Xeq0,15000)-Xs(lac0,n),n)$
n1:float(rhs(%[1]))$          matrix(
["No. Firms","Price","Xs","Xd"],
[n0,lac0,float(Xs(lac0, n0)),Xd(lac0,Xeq0,5000)],
[n0,p1,float(Xs(p1,n0)),Xd(p1,Xeq0,15000)],
[n1,lac0,float(Xs(lac0,n1)),Xd(lac0,Xeq0,15000)]);
```

$$\begin{bmatrix} \text{No. Firms} & \text{Price} & Xs & Xd \\ 200 & 30 & 10000. & 10000 \\ 200 & 63.031 & 14495. & 14495. \\ 400.0 & 30 & 20000. & 20000 \end{bmatrix}$$

To draw a graph of the demand and supply curves requires that we express the demand price (the height of the demand curve) as a function of X. The next cell produces the requisite function.

```
solve(Xd(p,Xeq,T)-X,p)$ p(X,Xeq,T):=''(rhs(%[1]));
```

$$\mathrm{p}(X, Xeq, T) := -\frac{3\,X - 3\,T - 3\,Xeq}{500}$$

The next graph generalizes the information in the table. The initial demand curve passes through the long-run supply curve at $X = 10000$, as does the initial short-run supply curve. The demand shift raises the price and quantity to the values shown above. At a price of \$63.031, each of 200 firms produces $14495/200 = 72.475$ units.

Entry of 200 additional firms in response to profits causes an increase in quantity to 20,000 units and in a shift in the short-run supply curve as depicted in the graph. Once again each firm produces $x = 50$.[11] The values in the `points` commands are from the table above.

```
wxdraw2d( ... Some commands omitted ... key=
 "Short-run supply curves",explicit(smc(
 X/n0,e0),X,0,25000), ... explicit(smc(
 X/n1,e0),X,0,25000),...key="Demand curves",
 explicit(p(X, Xeq0,5000),X,0,25000),key="",
 explicit(p(X, Xeq0,15000),X,0,25000), ...
 key="Long-run supply curve",explicit(lac0,
 x,0,25000),   ... points([Xeq0,Xeq0],
 [0,lac0]),points([0,Xs(p1,n0)],[p1,p1]),
 points([Xs(p1,n0),Xs(p1,n0)],[0,p1]),
 points([Xs(lac0,n1),Xs(lac0,n1)],[0,lac0]))$
```

[11] The increased demand eventually causes a shift in the short-run supply curve, but a movement along the (relevant) long-run supply curve. In the short-run, the increased demand causes a movement along the (relevant) short run supply curve.

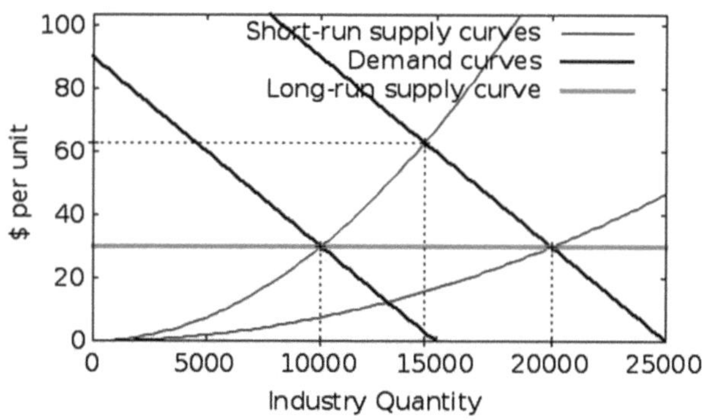

9.3 A Variable Cost Industry

Next, let the number of firms in an industry affect the cost levels of the firms that constitute the industry. This effect may be either positive or negative.

In *increasing-cost industries*, the entry of new firms bids up the price(s) of any resource(s) for which the industry demand is large enough to affect the resource market(s). Most industries probably are of this type, though the effect may be small enough for many that the constant-cost model provides a close approximation.

In *decreasing-cost industries*, the entry of new firms reduces incumbents' costs. This condition might apply in frontier settings, where more firms in an industry allow the spreading of some common costs among more firms.[12] We limit our analysis to the case of the increasing-cost industry.

We begin by deriving the illustrative long-run average and marginal cost functions for the firms in the industry. Both are derived from this long-run total cost function:

$$ltc2 = n^{\alpha} \cdot (a + b \cdot x^2),$$

where n is the number of firms in the industry and α is the elasticity of change in total cost per 1 % increase in the number of firms.[13]

The new cost functions $lac2$ and $lmc2$ keep the values $a0$ and $b0$ from the previous section and use the value $\alpha = 0.1$. Setting $lac = lmc$ and solving for

---

[12] According to [1], the market for dynamic random access memory (DRAM) chips is an example of this type of industry, where cost-saving discoveries made by any one firm are quickly adopted by other firms.

[13] A more flexible functional form would allow a and b to be multiplied by different functions of n. This generalization would increase computational complexity with little gain.

x confirms that the minimum value of lac occurs at a value that depends only on a and b and not on n or α.

The two commands in the last input line confirm that, when $a = a0$ and $b = b0$, the firm's lac curves attain a minimum value at $x = 50$ and that when $n = 1$, that minimum value is $lac = 30$.

```
[lmc2(x,n,%alpha,b):=2*b*n^%alpha*x,
lac2(x,n,%alpha,a,b):=a*n^%alpha/x + b*n^%alpha*x];
solve(lmc2(x,n,%alpha,b)=lac2(x,n,%alpha,a,b),x);
matrix( [float(subst([a=a0,b=b0],%[2])),"lac =",
 lac2(x0,1,.1,a0,b0)]);
```

$$[\mathrm{lmc2}\,(x, n, \alpha, b) := 2\,b\,n^{\alpha}\,x, \mathrm{lac2}\,(x, n, \alpha, a, b) := \frac{a\,n^{\alpha}}{x} + b\,n^{\alpha}\,x]$$
$$[x = -\sqrt{\tfrac{a}{b}}, x = \sqrt{\tfrac{a}{b}}] \quad \begin{bmatrix} x = 50.0 & lac = & 30.0 \end{bmatrix}$$

9.3.1 The Value of α

The value selected for α, 0.1, might appear low. Consider how such a value is determined. Suppose that R is the quantity of some resource employed in this industry. For this resource, $R = Rt - Ro$, where R is the industry quantity of the resource, Rt is the total quantity, and Ro is the quantity employed in other industries (not other firms in this industry).

Reasoning that we developed previously regarding the price elasticity of demand for a single firm's product can be applied to convert this equation to a counterpart stated as elasticities: $E = \frac{Et}{s} - Eo \cdot \frac{1-s}{s}$, where E is the elasticity of supply to the industry, Et is the elasticity of supply to the entire resource market, Eo is the elasticity of demand from firms in other industries, and s is this industry's share of the total resource market.

Suppose, as an extreme case, that $Et = 0$. That is, the supply curve in the resource market is vertical. Then $E = -Eo \cdot \frac{1-s}{s}$. Let $Eo = -0.5$ and $s = 1/10$. Then $E = 0.5 \cdot 9 = 4.5$. Even with these conservative values, the supply to this industry is quite elastic. In this case, a 4.5 % increase in the quantity of the resource employed in this industry would result in a mere 1 % increase in the resource price.

Even if this were the only resource employed by firms in the industry, so that firms' costs would increase in the same proportion as resource cost, the implied value of α is small: α is $1/4.5 = 0.22\ldots$.

For two reasons, α will be lower than this value. First, no resource makes up the entire cost of producing a good. Even if no substitution were possible, the percentage impact on the firms' costs would be less than the value above. Furthermore, firms can substitute away from resources whose prices increase, further moderating the cost impact. Therefore, we consider the value $\alpha = 0.1$ to be reasonable.

9.3.2 Industry Supply

We begin to examine the nature of supply in this type of industry by determining the relationship between price and the number of firms in the industry. We set $lmc = p$ at $x = x0$. We have established that $x0$ is the quantity at which firms in this industry achieve least-cost production.

We solve this equation twice. The first solution defines the number of firms as a function of price and the parameter α. The second solution defines the inverse of the first one, stating price as a function of the number of firms.

```
assume(p>0)$  declare(%alpha,noninteger)$
solnn:solve(lmc2(x0,n,%alpha,b0)-p,n)$
solnp:solve(lmc2(x0,n,%alpha,b0)-p,p)$
matrix([n(p, %alpha)  := ''(rhs(solnn[1])),"  or",
p(n, %alpha)  := ''(rhs(solnp[1]))]);
```

$$\left[\mathrm{n}\,(p,\alpha) := \frac{p^{\frac{1}{\alpha}}}{30^{\frac{1}{\alpha}}} \;\; or \;\; \mathrm{p}\,(n,\alpha) := 30\,n^{\alpha} \right]$$

As with the constant-cost industry, we examine the adjustment process by including the short-run effects. Therefore, we define the short-run marginal cost curves for firms in this industry. We specify the short-run marginal cost curve as its long-run counterpart, using n^{α} as the impact of the number of firms.

```
smc2(x,n,%alpha,e):=
3*e*n^%alpha*x^2;
```

$$\mathrm{smc2}\,(x,n,\alpha,e) := 3\,e\,n^{\alpha}\,x^2$$

The next cell produces points on the long-run supply curve for this industry. It lists five values of n, starting with 100, ending at 500, and at increments of 100. For each number for firms, the second command determines the lac at its minimum point, where $x = x0$. These lac values must equal price for equilibrium to be established. The quantities supplied are computed by multiplying the number of firms by $x0$.

```
nlist:makelist(n,n,100,500,100)$   plist:
 lac2(x0,nlist,0.1,a0,b0)$  Xlist:x0*nlist$ names:
matrix(["No. of Firms","Price","Quantity"])$ numbers:
matrix(nlist,plist,Xlist)$addrow(names,numbers);
```

No. of Firms	Price	Quantity
100	47.547	5000
200	50.959	10000
300	53.068	15000
400	54.617	20000
500	55.849	25000

The graph below generalizes the results from the table, showing the relationship between X and p for a range of X values. It also shows the short-run supply curves when the number of firms is set equal to the second item in `nlist` and again for the fifth item in that list.

The values of `Xmax` and `pmax` are computed for use in defining the ranges for the graph. The value of n in the function $p(n, \alpha)$ is entered as $X/x0$, which converts it to a function of X for graphing purposes.

```
Xmax : 1.25*Xlist[5]$ pmax:1.5*p(Xlist[5]/x0,0.1)$
wxdraw2d( ... Some commands omitted ... key=
  "Long-run Supply",explicit(p(X/x0,0.1),X,0,Xmax),
  line_width=2,key=concat("Short-run Supply, n = ",
  string(nlist[2])),explicit(smc2(X/nlist[2],
  nlist[2],.1,e0),X,0,Xmax), ...
  explicit(smc2(X/nlist[5],nlist[5],
  0.1,e0),X,0,Xmax ),... key="New Price",
  points([[0,plist[5]], [Xmax,plist[5]]])  )$
```

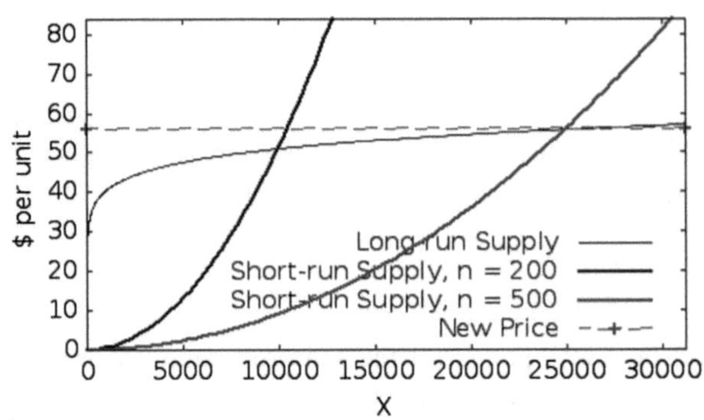

No demand curve appears in the graph above. None is needed in order to analyze the effect of a price change. To see this, suppose we begin with the point at which $X = 10000$ and $p = 50.959$. Raise the price to 55.849 (the fifth price in the list above). Examination of the graph shows that the quantity will increase only slightly above 10,000 in the short run, while the number of firms remains at $n = 200$. In the long run, however, the number of firms increases to $n = 500$ and the quantity increases proportionately to $X = 25000$.

We can dispense with the insertion of a demand curve into the analysis of the supply curve because in this case each price yields a determinate number of firms, and hence quantity, so the demand curve is not required to determine quantity responses. Thus, the model could now be applied to cases in which a price is set by fiat, rather than by market forces.

The presentation above determines the supply price for each quantity, providing a conventional graph with price on the vertical axis. It is a straightforward matter to specify price and to determine the number of firms that will inhabit the industry given long-run equilibrium, and the quantity that they will produce. The table below shows an example.

For each of the two prices listed, the number of firms is fractional. We could remove the fractional firm by rounding down. This would change the analysis only quite slightly, reducing output by a fraction of $x0$ and resulting in very small profits for firms in the industry. The analysis in the next section rules out a fractional number of firms.

```
p1 : 51$ n1 : n(p1, 0.1)$    X1 : n1*x0$
p2:  55$  n2 : n(p2, 0.1)$   X2 : n2*x0$
matrix(["Price","No. of Firms","Quantity"],
[p1, n1, X1], [p2, n2, X2]);
```

$$\begin{bmatrix} \text{Price} & \text{No. of Firms} & \text{Quantity} \\ 51 & 201.6 & 10080. \\ 55 & 428.96 & 21448. \end{bmatrix}$$

9.4 An Industry with Non-identical Firms

So far, we have assumed that all firms in an industry have identical cost curves. This simplification helps clarify the connection between firms' supply curves in the short run and the connection between firms' average cost and the industry supply curve in the long run. In doing so, it provides an important insight into the relationship between industry adjustment in the short run and its adjustment in the long run:

- The short-run price change is large relative to that in the long run. Indeed, in a constant-cost industry, changes in demand result in no price change in the long run, while short-run prices may change dramatically.
- The short-run quantity change in response to a price change is small compared to its long-run counterpart. Of course, in the constant-cost industry, the quantity change in response to a price change is "infinite" so demand curves must be added to determine industry quantities.

This section replaces the assumption that all firms have the same cost curve with its polar opposite, that no two firms have the same cost curve. The function below defines the long-run average cost for the ith firm. Firms are ranked with the lowest-cost firm ($i = 1$) having a long-run cost curve $lac = a0 + b0 \cdot x$.

The index i is mathematically similar to the number of firms n above, and all firms achieve minimum average cost at $x = x0$.[14] The commands in the cell below impose that condition and derive a function that relates the maximum number of

[14]The cost curves could be constructed so that the lowest average cost occurs at different output levels for different firms. This change would have no effect of consequence. In a market of price-taking firms, entry will occur until the marginal firm earns an economic profit of zero.

firms in the industry ($imax$) to the industry price. As before, the value of α is a parameter in the function. We constrain the number of firms to whole numbers, using the `floor` command.

We do not incorporate the possibility that the number of firms would affect each firm's cost. Doing so would require a bit more mathematics but would change the presentation only slightly.

```
a0*i^%alpha/x0+b0*i^%alpha*x0;solve(%=p,i)$
imax(p,%alpha):=''(floor(rhs(%[1])));
```

$$30\,i^{\alpha} \quad \text{imax}\,(p,\alpha) := \text{floor}\left(\frac{p^{\frac{1}{\alpha}}}{30^{\frac{1}{\alpha}}}\right)$$

The cell below sets a price and determines the number of firms given that price, with $\alpha = 0.25$. These values are used throughout the remainder of this illustration. When firms' technologies differ, we have less reason to expect very low values of α, which is now an indicator of cost differences and does not reflect any effect of resource cost changes (which are assumed not to exist).

```
p0:100$    imax0: imax(p0, 0.25);
```

123

Marginal cost of the ith firm is set equal to p. We solve for the ith firm's output as a function of p, i, and α, and limit analysis to the long run. Each firm has a different cost function, so each firm's long-run output differs from those of the other firms. The next cell establishes the long-run supply curve for each firm.[15]

```
x_i(p,i,%alpha):= ''(rhs(
solve(2*b0*i^%alpha*x-p,x)[1]));
```

$$\text{x\_i}\,(p,i,\alpha) := \frac{5\,p}{3\,i^{\alpha}}$$

The cell below applies the function that is developed above to determine how much the least-cost firm ($i = 1$) and the marginal firm ($i = 428$) would produce.

```
[x_1:x_i(p0,1,0.25),
x_imax0:x_i(p0,imax0,0.25)];
```

[166.67, 50.046]

The graph shows the same results as above. The long-run average cost curves (bold black lines) for the least-cost firm and for the highest-cost firm appear. Also, the firms' long-run marginal cost curves (bold gray lines) appear, along with the price (horizontal black line).

Each firm produces a quantity such that $p = lmc$. The lowest-cost firm produces 91.667 units per time period, and the highest-cost firm produces 50.046 units per time period. The lowest-cost firm earns a profit. The highest-cost firm's profit is negligible.

[15] We do not develop the firms' short-run cost curves. We could do so, following the approach developed in the preceding section. The result would be similar to that above: The long-run industry supply curve would be more elastic than the short-run industry supply curves.

Each firm's supply curve consists of the segment of its marginal cost curve that lies above the firm-specific minimum long-run average cost. These marginal cost curves are linear and pass through the origin for this specification. Therefore, each firm's supply curve exhibits unitary elasticity.

```
wxdraw2d(user_preamble="set key top center",
yrange = [0,  2*b0*imax0^0.25*(3*x0)],
xlabel = "Output Levels", ylabel = "$ per unit",
key="p0", explicit(p0, x, 0, 3*x0),
line_width = 2, key = "Average Cost Curves",
explicit(a0/x + b0*x, x, 0, 3*x0),key="",
explicit(a0*imax0^0.25/x+b0*imax0^0.25*x,x,0,3*x0),
color="dark-gray", key = "Marginal Cost Curves",
explicit(2*b0*x, x, 0, 3*x0), key = "",
explicit(2*b0*imax0^0.25*x, x, 0, 3*x0) )$
```

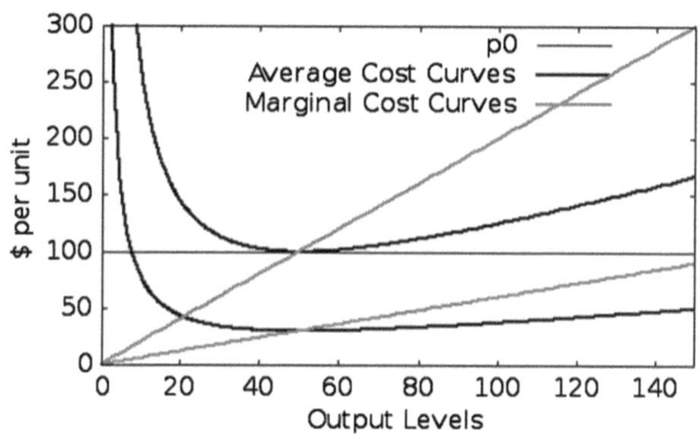

9.4.1 *Profits, Rents, and Shifting Cost Curves*

The cost curves above may be thought of as being derived directly from a set of production functions, in a manner developed in preceding chapters. The firm with the index $i = 1$ has a production function that yields more output for a given bundle of inputs than other firms.

All firms except the lowest-cost firm produce at a point to the right of the minimum point on the firm's lac. This implies that all firms except the marginal firm operate in a range of *decreasing returns to scale.*

The favorable cost conditions must reflect some real differences among the firms. Location is one possibility. Firms may have different costs because some are farther from supplies or from markets. In such a case, the proximity would become a priced feature of production. That is, firms nearer to supplies or markets would pay more *rent* for their location. Once this rental gradient is factored in, all firms would tend to have lac's with a minimum equal to that of the marginal firm.

The fact that the rent-inclusive lac tends to equalize across all firms does not change one important implication of the analysis above: All firms except the marginal one produce quantities such that they experience decreasing returns to scale. The rents simply weaken the link between average cost and economies of scale. We return to this point when we consider the efficiency of an industry of price-taking firms.

9.4.2 Industry Supply

We derive the industry supply curve by applying the relationships above to a given set of prices. The first list below creates a list of prices that are expressed as a percentage of the minimum lac of the least-cost firm. The second list applies the `imax` equation to the list of prices, determining the maximum number of firms for each price.

Given that $\alpha = 0.25$, increasing the price brings firms into the industry at an increasing rate. For this specification, doubling the price causes the number of firms to rise from 1 to 16. The quantity supplied at the industry level increases dramatically as well, mostly because of the entry of new firms.

Comparing $p = 30$ and $p = 33$ shows one reason for increased X: each firm in the industry produces more when p increases. (Likewise, for the effects of $p = 36$ and $p = 39$). The other reason, of course, is entry.

```
pList:makelist(lac0*(1+(j-1)/10),j,1,11)$ imaxList:
  imax(pList,0.25)$ X_list: makelist(sum(
  x_i(pList[j],i,0.25),i,1,imaxList[j]),j,1,11)$
  perfirmxList: X_list/imaxList$ numbers:transpose(
  matrix(pList,imaxList,X_list,perfirmxList))$
  names:matrix(["Price","No. of Firms",
 "Industry X","Avg. Firm x"])$ addrow(names,numbers);
```

Price	No. of Firms	Industry X	Avg. Firm x
30	1	50.0	50.0
33	1	55.0	55.0
36	2	110.45	55.227
39	2	119.66	59.829
42	3	182.05	60.684
45	5	298.24	59.649
48	6	369.24	61.54
51	8	495.12	61.89
54	10	626.81	62.681
57	13	814.87	62.683
60	16	1010.3	63.142

Earlier, we observed that each firm's supply curve exhibits unitary elasticity. Therefore, doubling the price should double the lowest-cost firm's output, from $x = 50$ to $x = 100$. The result below shows that it does. It also shows the output for Firm 8 and for Firm 16, the highest-cost firm.

```
matrix(["Firm1's x", "Firm8's x", "Firm16's x"],
[x_i(pList[11],1,0.25),x_i(pList[11],8,0.25),
x_i(pList[11],16,0.25)]);
```

Firm1's x	Firm8's x	Firm16's x
100.0	59.46	50.0

The graph below shows the 11 points on the industry supply curve that appear in the table above. It is apparent that, even though each firm's supply curve has an elasticity of 1.0, the industry supply curve becomes quite elastic as price increases. The step at X just over 100 occurs when the price increase is not large enough to induce an added firm but does lead to increased output per firm (the third and fourth lines in the table above).

```
wxdraw2d(user_preamble="set key bottom",yrange=
 [0,1.5*pList[11]],ylabel="$ per unit",xlabel=
 "Industry Quantity, X",line_width=2,
 points_joined=true,key= "Industry Supply Curve",
 points(X_list,pList))$
```

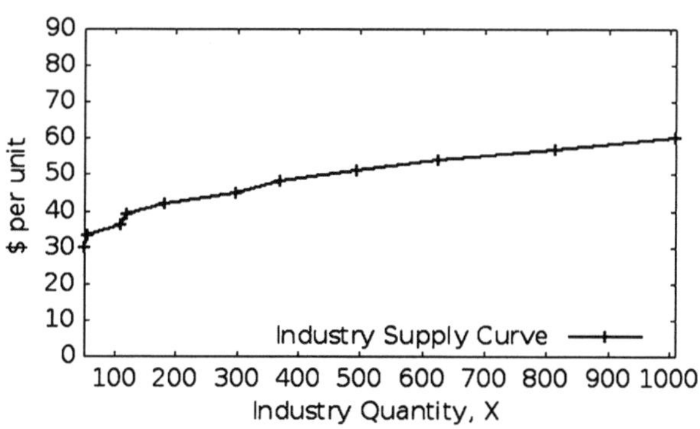

9.4.3 *Least-Cost Production in a Competitive Industry*

A theme that pervades economic analysis is that of efficiency, which entails the production of the highest-valued set of goods and services given a finite supply of resources. One aspect of overall efficiency is *productive efficiency*. This type of efficiency is achieved when a specified quantity of a good or service is produced at the lowest possible cost. An important criterion to consider when evaluating a market structure (or any other institution within which production occurs) is whether it tends to deliver productive efficiency.

This section addresses two questions. The first is the condition(s) necessary for the achievement of productive efficiency. That is, it deduces the implications of the normative assertion that efficiency is desirable. The second question is whether a competitive industry tends to satisfy these condition(s). This is positive analysis; it addresses the questions of what markets tend to do and not how to evaluate what they tend to do.

To anticipate the results of the analysis, the answers are these: (1) The quantity produced at the industry level must be produced by individual firms in quantities such that all firms have the same marginal cost and that the average cost of the marginal (high-cost) firm must equal the marginal costs of all firms, and (2) Yes, competitive industries do tend to satisfy those conditions.

A common way of "demonstrating" the productive efficiency of perfect competition is to refer to a graph like the one below (copied from earlier in the chapter) that shows a firm and to point out that all firms are forced to operate at the minimum point on their lac curves when all are price takers.

This approach does not work, however, when cost curves differ across firms, because every firm except the marginal one produces more than the minimum-lac quantity. Firm 1 above produces 1,000 units, not the 50 that would result in minimum lac, when $p = 60$. Given this result, efficiency must be sought somewhere other than at the minimum point on firms' lac curves.

```
firmsupply:gr2d(user_preamble="set key bottom right",
 yrange = [0, 2*lac0], xlabel="Firm Quantity",
 ylabel = "$ per unit", key = "sac",title=
 "Firm",explicit(sac(x, d0, e0),x,0,2*x0),
 line_width=2, key="lac",explicit(lac(x,a0,b0),x,
 0,2*x0),color=gray,key="smc",explicit(smc(x,e0),x,
 0,2*x0),color=black,key="lmc",explicit(lmc(x,b0),
 x,x0,2*x0),line_type=dots,points_joined=true,
 points([[0,lac0],[x0,lac0]]))$
industrysupply:gr2d(xlabel = "Industry Quantity",
yrange=[0,2*lac0],ylabel="$ per unit",title=
"Industry",key="LR Supply",xtics={100*x0/2,100*x0},
line_width = 2, explicit(lac0, x, 0, 100*x0) )$
wxdraw(firmsupply, industrysupply, columns = 2)$
```

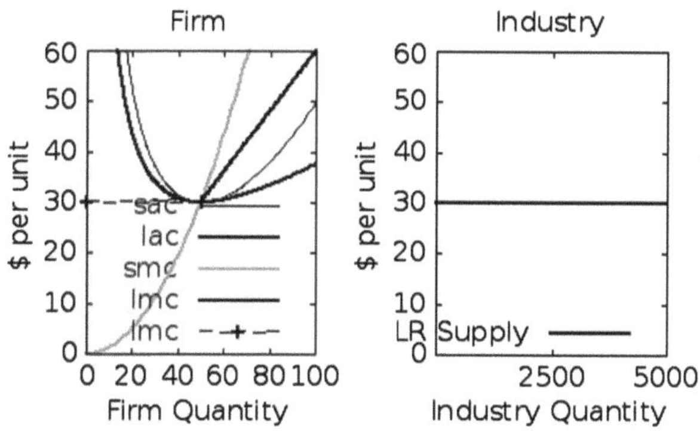

The conditions for efficiency have been stated. Now we elaborate.

- Each firm in the industry must achieve minimum-cost production. That is, each firm must incur no more cost at any output level than its lac mandates.
- Firms in the industry must produce quantities such that every firm has the same marginal cost. The easiest way to see why this must be true is to reason by contradiction. Suppose that the current industry output is forthcoming in quantities such that Firm 49 has a marginal cost of \$55 while Firm 67 has a marginal cost of \$62. Then simply allocating a single unit of Firm 67's output to Firm 49 would reduce industry cost by \$7. This reasoning is general, applying for any set of firms and set of marginal costs.
- The marginal (highest-cost) firm's lac must equal every firm's lmc, including its own. Suppose that each firm in the industry currently is producing a quantity such that their marginal costs are all \$68 (any number would serve in illustrating this point) but that an additional firm could enter the industry with an average cost of \$52. Suppose further that this potential firm's lac achieves this minimum value at $x = 50$ units of output. For each unit of output that is diverted from the other firms to this one, a saving of about $\$16 (= \$68 - \$52)$ occurs. Thus the total saving by having this firm enter would be about $\$\ 16 \cdot 50$ or \$800. ("About" rather than exactly because marginal costs will change slightly as firms make discrete changes in their output rates.)

Thus, we have established the (normative) conditions that must hold if the industry's cost is to be minimized. Now we establish that a competitive industry tends to achieve this goal. Keep in mind that no one is in charge of an industry. Industry supply is the result of the uncoordinated supply decisions of a large number of relatively small firms. Consider each of the requirements above in turn.

- That the first of these criteria is met is a matter of assumption. That is, a behavioral assumption is that the price-taking firms in this industry maximize profits. If costs are higher than the need be (that is, if the firm is incurring a per-unit cost above its lac to produce a given quantity), then the firm is not maximizing profits.

- The second condition follows from the fact that the firms are price takers. Each firm independently selects a quantity such that its $lmc = p$. Because all firms face the same price, the transitive property of equality ("things equal to the same thing equal each other") implies that all firms have the same marginal cost.
- The third condition is met if entry and exit are not blocked. As long as $p > lac$ for potential entrants, new firms will enter and whenever $p < lac$ for any firms in the industry, some firms will leave. Thus $p = lac$ for the marginal is an equilibrium (positive) condition. The marginal firm operates on the minimum part of its lac, at which $lac_m = lmc_m = p$, where m indicates marginal firm. And all other firms face the same price. Accordingly, the marginal firm's lac does equal its own lmc and the lmc of every other firm in the industry.

These conditions are met whether or not firms in the industry are identical. When firms all have the same cost curves, then equilibrium occurs when every firm operates at the point of minimum lac (every firm is the marginal firm). In this unlikely case, and only in this case, least-cost production does coincide with the condition that all firms are forced to produce quantities that minimize per-unit costs.[16]

9.4.4 The Fourth Case

An earlier note indicated why the fourth case—many different firms whose cost curves are affected by industry size—is not illustrated formally. The note asserts that the results do not differ qualitatively from the third case. We support that assertion here.

Suppose that an indefinite number of firms exist. Each firm has a unique U-shaped long-run average cost. Each cost curve can be affected by the number of firms in the industry. The firms' cost curves can be ranked by the height of the minimum points on their long-run average cost curves. This relationship is ordinal; no explicit functional form need describe the ranking relationship.

Determining the quantity at each price in this case is simple: The minimum point of the highest-cost firm must equal the price. Meeting this condition implies a number of firms. Once the number of firms is established, then the long-run quantity supplied is also determined: It is the sum of the quantities that the firms produce at the specified price.

A price change has three effects. It changes the number of firms. This change shifts the inframarginal firms' cost curves, in particular their long-run marginal cost curves. It also causes a movement along each firm's new marginal cost curves.

---

[16]Actually, it is not quite necessary for all firms to have identical costs for this result to hold. It will hold if all firms have the same minimum lac even if the common value does not occur at the same quantity for every firm.

At each price, firms produce quantities such that the long-run marginal cost of each firm equals that of every other firm (because marginal cost = price). Also, the marginal firm's average cost equals price.

Reference

1. McAfee RP (2006) Introduction to economic analysis. Available at www.introecon.com

Chapter 10
Competitive Markets: Extensions and Application

Abstract This chapter extends the analysis in the preceding two chapters in two directions. First, it focuses on shifts in demand and supply. Then it addresses allocative efficiency which, combined with the productive efficiency the previous chapters have introduced, generates economic efficiency. Economic efficiency is realized when resources are used in a way that results in the highest-valued set of goods and services being produced. This chapter also addresses the distorting effects of taxes placed on the exchange of specific goods or services.

10.1 Demand and Supply Shifts

This section provides an alternative, summary phrasing of material from the preceding development of demand and of supply. The analysis of consumer behavior shows that demand curves can shift if any of the following changes: tastes (the utility function), prices of related goods, or money income. Likewise, supply curves shift in response to changes in technology (production functions), resource prices, expectations, and so on.

We can phrase these shifts in terms of the effects that two parameters have on the equilibrium quantity and the equilibrium price.[1] The quantity demanded can be written as a function of the price of that good and of one or more shift parameters. We illustrate with a single shift function. $Xd = f(p, \beta)$, where β is the shift parameter. Likewise, the supply curve can be phrased as $Xs = g(p, \alpha)$. Equilibrium requires that $Xd = Xs$.

[1]See [3], Chap. 11, and [2], Chap. 10.

M.R. Hammock and J.W. Mixon, *Microeconomic Theory and Computation: Applying the Maxima Open-Source Computer Algebra System*,
DOI 10.1007/978-1-4614-9417-1_10,

To analyze the comparative statics of this system of two equations, take the total differential of the Xd and Xs functions:

$$dXd = \frac{\partial Xd}{\partial p} dp + \frac{\partial Xd}{\partial \beta} d\beta,$$

and

$$dXs = \frac{\partial Xs}{\partial p} dp + \frac{\partial Xs}{\partial \alpha} d\alpha.$$

To restore the market to equilibrium after any change in either shift parameter requires that $dXd = dXs$, i.e., the change in quantity demanded equals the change in quantity supplied. To examine the effects of individual parameters, we alternatively set $d\alpha = 0$ and $d\beta = 0$. This restriction is not necessary, but it facilitates presentation. First, shift the demand curve only, by setting $d\alpha = 0$. Then

$$\frac{\partial Xd}{\partial p} dp + \frac{\partial Xd}{\partial \beta} d\beta = \frac{\partial Xs}{\partial p} dp.$$

Thus, the change in the equilibrium price per one-unit shift in the demand curve can be expressed as

$$\frac{dp}{d\beta} = \frac{\frac{\partial Xd}{\partial \beta}}{\frac{\partial Xs}{\partial p} - \frac{\partial Xd}{\partial p}}.$$

Multiplying both sides of this equation by $\frac{\beta}{p}$ and then dividing both numerator and denominator of the right-hand side by Xd (which equals Xs) converts this expression to one involving three elasticities:

$$Epb = \frac{Edb}{Esp - Edp},$$

where `Epd` is the percentage change in *price* per 1% change in β, `Edb` is the elasticity of demand with respect to β, and `Esp` and `Edp` are the price elasticities of supply and demand. All elasticities on the right-hand side measure quantity responses; the elasticity on the left-hand side measures a price response.

Supply shifts affect price in an analogous fashion:

$$Epa = \frac{Esa}{Esp - Edp},$$

where `Epa` is the elasticity of price with respect to a change in α, and `Esa` is the percentage shift in the supply curve per 1% change in α.

The three elasticities on the right of either of these expressions determine price level changes. Suppose, for example, that `Edb` = 2.0, `Edp` = −1.25, and `Esp` = 5.0. Then `Epb` = 2.0/6.25 = 0.32. A 1% change in β shifts the demand curve rightward by 2%. The large elasticity of supply coupled with a slightly elastic demand curve dampens the effect, so that price rises by only 0.32%.[2]

10.2 Examples

This section demonstrates that the approximation above works quite well for at least two classes of demand and supply curves, those that exhibit constant elasticity and those that exhibit constant slope.

The analysis proceeds as follows. First, we establish a base quantity and price and determine the demand and supply functions that have the elasticities indicated above at the initial values. Then we shift the demand or supply curve by 10% and determine the new values. In particular, we evaluate the percentage change in price in each case.

We consider a system of two curves, the general nature of which is this:

- Demand: $Xd = f(p, I)$
- Supply : $Xs = g(p, w)$.

Here Xd is the quantity demanded, Xs is the quantity supplied, p is price, I is income, and w is the wage rate in the industry. We assume that changes in I and w are not related, though this assumption could be relaxed with little effect on the analysis.

10.2.1 Base Values

We begin the analysis with a set of elasticities and of variable values that define the parameters of the demand and supply curves used in this illustration. The four elasticity values are applied directly to the constant-elasticity equations; they are used to determine the slope coefficients in the linear equations.

```
[Esp0,Edp0,Esa0,Edb0]:[5.0, -1.25, -0.5,2.0]$
[X0,p0,I0,w0]:[5000.0, 7.0, 40.0, 25.0]$
```

[2]Previous chapters indicate that long-run supply curves are likely to be quite elastic. In contrast, short-run supply curves can be inelastic. Thus, in the short run price is affected more by the shift than in the short run, as we have already seen.

10.2.2 Constant-Elasticity Curves

First, consider the case in which both the demand curve and the supply curve exhibit constant elasticities as regards price and the shift variable. The demand curve is $Xd = B \cdot p^{Edp} \cdot I^{Edb}$, and the supply curve is $Xs = A \cdot p^{Esp} \cdot w^{Esa}$.

Values of A and B must be determined such that the curves pass through the designated values of p and X for the designated values of I and w and of the elasticities. The cell below defines the demand function.

The first command determines the value of B, given the values that we specified above. The second command substitutes the value of B that the `solve` command produces and places it in the equation that defines the demand curve. The third command generates a demand curve for the specified income.

```
solve(X0 = B*p0^Edp0*I0^Edb0, B);
[Xd(p,I,Edp,Edb):='' (float(
 subst(%,B*p^Edp*I^Edb))),
Xd(p,I0,Edp0,Edb0)];
```

$[B = \frac{1371293000}{38539643}]$

(%o4) $[\mathrm{Xd}(p, I, Edp, Edb) := 35.581\, p^{Edp}\, I^{Edb}, \frac{56930.}{p^{1.25}}]$

The supply function is solved in the same fashion.

```
solve(X0 = A*p0^Esp0*w0^Esa0, A);
[Xs(p,w,Esp,Esa):='' (
float(subst(%,A*p^Esp*w^Esa))),
Xs(p,w0,Esp0,Esa0)];
```

$[A = \frac{25000}{16807}]$

(%o6) $[\mathrm{Xs}(p, w, Esp, Esa) := 1.4875\, p^{Esp}\, w^{Esa}, 0.297\, p^{5.0}]$

We confirm that the price at which the quantity demanded equals the quantity supplied is $7 per unit, and that the quantity is 5,000 units.

```
[solnp: find_root(Xd(p,I0,Edp0,Edb0)-
   Xs(p,w0,Esp0,Esa0),p,0.1,500),
   Xs(solnp,w0,Esp0,Esa0)];
```

[7.0, 5000.0]

We also determine the elasticity of price with respect to a change in I (0.32) and with respect to a change in w (−0.08).

```
[Epb0:Edb0/(Esp0-Edp0),
   Epa0:Esa0/(Esp0-Edp0)];
```

[0.32, − 0.08]

We shift the demand curve with a 10% increase in I. Given that Edb = 2, the demand curve shifts rightward by 20%. Based on the relationships developed earlier, the price elasticity with respect to the income change should be 0.32 (the calculation appears below).

As a result of the shift, the price rises from 7.0 to 7.2168. The midpoint formula for the price elasticity used below is consistent with the value from the formula based on elasticity values, at least to the second decimal place.

```
[Epb0:Edb0/(Esp0-Edp0), I1:1.1*I0]$  matrix(
 ["New price = ",p1:find_root(Xd(p,I1,Edp0,Edb0)-Xs(
 p,w0,Esp0,Esa0),p,p0,2*p0),"and Arc Elasticity =",
 ((p1-p0)/(p1+p0))/((I1-I0)/(I1+I0))]);
```

[New price = 7.2168 and Arc Elasticity = 0.32]

We also determine the new equilibrium quantity. This value is bound to the name `X1` for use in the graph below. It is produced twice, once using the demand function and again using the supply function, as a check.

```
[X1:Xd(p1,I1,Edp0,Edb0),
 Xs(p1,w0,Esp0,Esa0)];
```

[5823.7, 5823.7]

The graph shows the two demand curves, the supply curve, and the two points that correspond to equilibria.

```
wxdraw2d(... Some commands omitted ...
 points([[0, p0],[X0, p0]]), points([[X0,0],
 [X0,p0]]),points([[0,p1],[X1,p1]]),points([[X1,0],
 [X1, p1]]),line_type=solid,key="Initial Demand",
 implicit(Xd(p,I0,Edp0,Edb0)-X,X,X0/10,2*X0,p,p0/10,
 2*p0),line_width=2,key="New Demand",implicit(...) )$
```

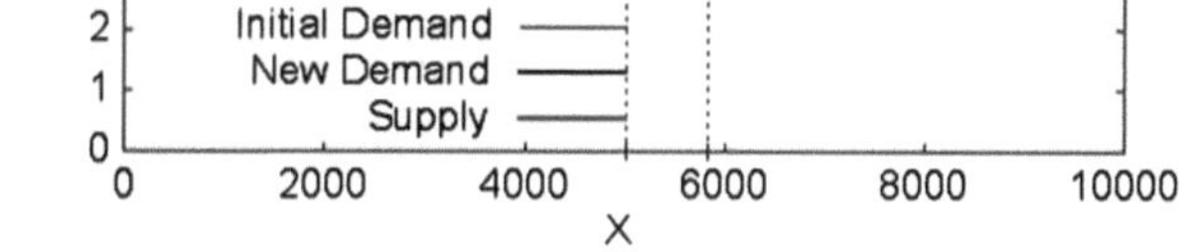

Exercise: Analyze the effect of a 10% supply shift, holding demand constant

10.2.3 Linear Curves

Next, we fit linear demand and supply through an initial point and shift the demand curve as before. The cell below contains the commands needed to fit the curves and to define the functions. The functional names are changed to XsL and XdL to indicate the linear nature of the functions.

The first command below determines the value of B that is consistent with the parameters and variable values specified above. The second command defines the resulting demand function. The third and fourth commands do the same for the supply function.

```
solve(X0=B+p0*Edp0*X0/p0+I0*Edb0*X0/I0,B);XdL(
 p,I,Edp,Edb):=''(float(rhs(%[1]))+p*Edp0*X0/p0+
 I*Edb0*X0/I0);   solve(X0=A+p0*Esp0*X0/p0+
 w0*Esa0*X0/w0,A);XsL(p,w,Esp,Esa):=''(float(
 rhs(%[1]))+p*Esp0*X0/p0 + w*Esa0*X0/w0);
```

$$[B = 1250] \qquad \mathrm{XdL}\,(p, I, Edp, Edb) := 250.0\, I - 892.86\, p + 1250.0$$

$$[A = -17500] \quad \mathrm{XsL}\,(p, w, Esp, Esa) := -100.0\, w + 3571.4\, p - 17500.0$$

The cell below contains the calculation of the new price and the new quantity (again, quantity is calculated twice as a check), given increased income. These values differ slightly from those generated when the functions exhibit constant elasticities.

The output also shows the arc elasticity of price with respect to the demand shift parameter using the midpoint formula. The calculated value differs from the value computed above (0.32) slightly. The reason for this small difference is that the elasticities change slightly as the values of X and p change, holding the slopes constant.

```
p2:find_root(XdL(p,I1,Edp0,Edb0)-
   XsL(p,w0,Esp0,Esa0),p,0,2*p0); matrix(
  ["Xsupply =",X2:XsL(p2,w0,Esp0,Esa0),"Xdemand =
   ",XdL(p2,I1, Edp0, Edb0 ),"and Arc Elasticity =",
  ((p2-p0)/(p2+p0))/((I1-I0)/(I1+I0))]);
```

$$7.224 \quad \left[\mathrm{Xsupply} = 5800.0\ \mathrm{Xdemand} = 5800.0\ ;\ \mathrm{Arc\ Elasticity} = 0.331\right]$$

As before, the graph displays the demand shift and shows the implied price and quantity changes.

```
wxdraw2d(ip_grid=[100,100], ... Some commands omitted
 ... points([[0,p0],[X0, p0]]),points([[X0, 0],
 [X0,p0]]),points([[0, p2],[X2,p2]]),points([[X2,0],
 [X2,p2]]), ... key="Initial Demand",
 implicit(XdL(p,I0,Edp0,Edb0)-X,X,X0/10,2*X0,p,p0/10,
 2*p0),line_width=2,key="New Demand",implicit(...)$
```

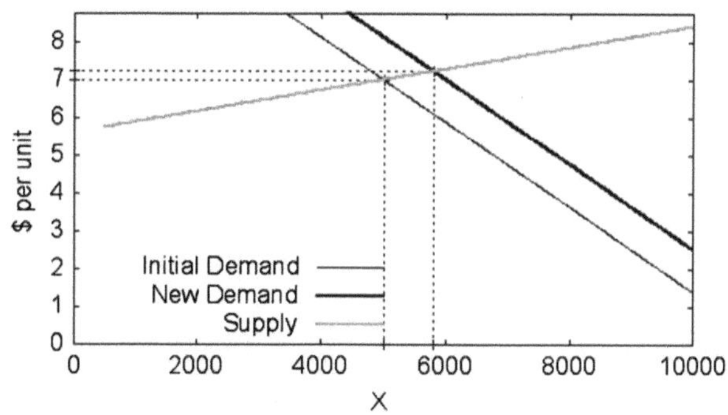

10.3 Allocative Efficiency

The preceding chapter shows that in an industry of profit-maximizing firms that take price as given, the total industry output is produced at the lowest possible cost. This section extends the question of efficiency to include the determination of the quantity that yields maximum value and a comparison with that quantity to the equilibrium quantity.

We use constant-elasticity demand and supply curves, the equations for which appear below.

```
[Xs(p,A,Esp):=A*p^Esp,Xd(p,B,Edp):=B*p^Edp];
```

$$[\mathrm{Xs}(p, A, Esp) := A\, p^{Esp}, \mathrm{Xd}(p, B, Edp) := B\, p^{Edp}]$$

For purposes of analysis, it is useful to define the inverse demand and supply functions. The following cell accomplishes this task. The `assume` and `declare` commands are entered in order to avoid a dialog with *Maxima* while carrying out this restatement of the supply and demand functions.

```
assume(A>0,B>0,X>0)$declare([Esp,Edp],noninteger)$
[ps(X,A,Esp) := ''(rhs(solve(Xs(p,A,Esp)-X,p)[1])),
pd(X,B,Edp):=''(rhs(solve(Xd(p,B,Edp)-X,p)[1]))];
```

$$[\mathrm{ps}(X, A, Esp) := \frac{X^{\frac{1}{Esp}}}{A^{\frac{1}{Esp}}}, \quad \mathrm{pd}(X, B, Edp) := \frac{X^{\frac{1}{Edp}}}{B^{\frac{1}{Edp}}}]$$

We begin by specifying the equilibrium quantity, as expressed in terms of the four parameters of the system defined above. As above, we set values for the equilibrium quantity and price, and for the elasticity values, and we determine the values of A and B that are consistent with these equilibrium values.

```
[X0,p0,Esp0,Edp0]:[2000,50,2.5,-1.5]$  A0:rhs(float(
 solve(Xs(p0,A,Esp0)=X0,A))[1])$   B0:rhs(float(
 solve(Xd(p0,B,Edp0)=X0,B))[1])$ matrix(["X0","p0",
 "A0","Esp0","B0","Edp0"],[X0,p0,A0,Esp0,B0,Edp0]);
```

$$\begin{bmatrix} X0 & p0 & A0 & Esp0 & B0 & Edp0 \\ 2000 & 50 & 0.113 & 2.5 & 707110. & -1.5 \end{bmatrix}$$

A graph based on the material above facilitates the analysis of the conditions required for efficiency. The graph below consists of two curves, labeled `Marginal Value` and `Marginal Opportunity Cost`.

```
wxdraw2d( ... Some commands omitted ...
 explicit(pd(X,B0,Edp0),X,0,1.5*X0),key=
 "Marginal Opportunity Cost",color=gray40,
 explicit(ps(X,A0,Esp0),X,0,1.5*X0),key="",
 ... points([[0,p0],[X0,p0],[X0,0]]) )$
```

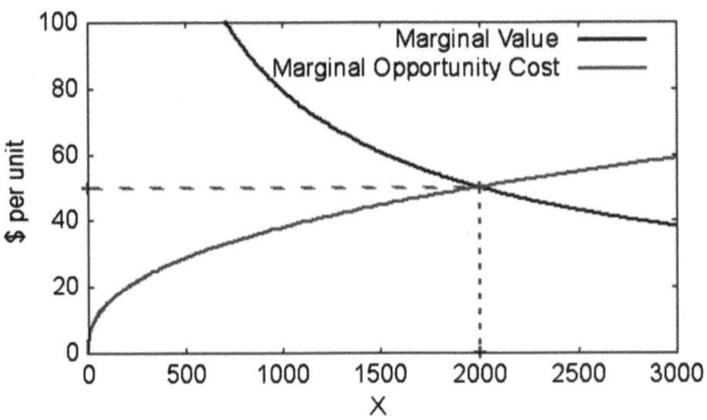

The graph above shows that $X = 2{,}000$ results in an efficient allocation of resources to this industry. Mathematically, the two value curves (cost is the value of resources in their next best use) coincide with the inverse demand and supply curves defined above. Whether this coincidence is a correct representation depends on whether the Marginal Value curve actually coincides with the demand curve and the Marginal Opportunity Cost curve coincides with the supply curve. If they do not, an externality may be present.

10.3.1 Consumer Surplus and Producer Surplus

In the analysis of efficiency that follows, we assume that the two conditions are met. Before turning to that analysis, we consider the circumstances under which this assumption is warranted. In doing so, we revisit the concept of Consumer Surplus and introduce its counterpart, Producer Surplus.

10.3.1.1 Consumer Surplus Revisited

Chapter 5 developed the concept of Consumer Surplus for an individual. It showed that the area defined by the price of the good, the vertical axis, and the height of the uncompensated demand curve provides a close approximation to the surplus that an individual enjoys from being able to buy the optimal quantity at a given price.

For each price, the quantity on the market demand curve is the sum of individual quantities demanded at that price. Therefore, the price is the marginal willingness to pay. Therefore, we can think of the demand curve as a marginal willingness-to-pay curve. For the marginal willingness to pay to coincide with marginal value, these conditions must be met.

- The income distribution must be "acceptable." Each income distribution will result in a different demand (marginal willingness to pay) curve. We could drop this assumption and refer to *an* efficient outcome rather than to *the* efficient outcome.
- Buyers of the product must receive all of the benefits from consuming the product and bear all costs of doing so.
- Buyers must be accepted as being competent to judge the value of the good for themselves (i.e., no paternalist concerns).

10.3.1.2 Producer Surplus

We have established that the height of the supply curve at each quantity corresponds to the marginal cost of producing that quantity. Therefore, the area between the supply curve and the price line in the graph above corresponds to Producer Surplus.

Producer Surplus does not have the complicating factor of an income effect that we encountered in the analysis of Consumer Surplus. It does, however, have its own subtleties as regards who receives the Producer Surplus. First, recall the first and simplest type of industry, the constant-cost industry. In this industry all firms are identical and their costs do not depend on the number of firms in the industry. In this case, the long-run supply curve is horizontal, so no Producer Surplus is generated.

In the second case, all firms are identical but the cost of each firm depends on the number of firms in the industry. For now, assume that this dependency is due to the effect of the number of competing firms on some resource price(s). In this case, the long-run equilibrium occurs when firms earn zero economic profits. This result implies that all Producer Surplus accrues to the owners of resources that are used in the industry, not to the firms' owners.

Finally, when the cost structure differs from firm to firm, low-cost firms earn profits. If the costs of these firms are independent of the number of firms, then Producer Surplus accrues to the owners of the inframarginal firms (those with costs lower than the cost of the highest-cost, marginal firm), or to owners of peculiar resources that account for the low-cost status. If the number of the firms affects the prices of resources used by firms in the industry, as in the preceding paragraph, then

Producer Surplus is shared with owners of resources whose prices vary with the number of firms in the industry.

A separate question is whether the industry's supply curve does correspond to the Marginal Opportunity Cost Curve. For this correspondence to occur, these conditions must be met.

- The firms' production costs must be borne entirely by the firms in the industry. That is, "externalities" are not generated.[3] Also, their production cannot generate benefits such as technological discoveries that can accrue to others elsewhere in the economy.
- Firms in the industry must not impose either benefits or costs on each other, except via the effects on resource prices mentioned above. Strictly, this is a subset of the previous condition, but it merits attention. Suppose that the firms affect each other via congestion, perhaps of roads to the market. In that case, the supply curve would lie below the Marginal Opportunity Cost curve, and Producer Surplus would overstate the gains from production.

  In the opposite direction, suppose that as new firms enter an industry they generate external benefits on incumbents and on each other. Then the supply curve would lie above the Marginal Opportunity Cost curve, and Producer Surplus would understate the gains from production. To see this, suppose that these external benefits are large enough to generate a downward sloping supply curve. Then the area showing the sum of the differences between the price and the marginal (private) cost would be negative, implying that firms lose from being members of this industry even though all earn at least zero economic profits.

If the conditions above are met, then the efficient production level corresponds to the equilibrium production level. At that level, the sum of Consumer Surplus and Producer Surplus reaches its maximum value.[4]

10.3.2 Efficiency and Equilibrium

Henceforth, we assume that the conditions above are met, so that maximizing the sum of Producer Surplus and Consumer Surplus corresponds to achieving allocative efficiency in this industry.

We now determine the efficient value of X, as defined above. The measure of gain from producing a specified quantity of this good is the integral of the difference

[3]We do not address the important question raised by Coase [1] regarding the institutional settings within which externalities are likely to be an issue.

[4]Note, however, that neither of the surpluses individually reaches its maximum value at this quantity. This fact can be the basis of pressures for public policies to favor one group over the other. It can also be the basis for attempts to coordinate actions among either sellers or buyers to affect the market price.

between the height of the demand curve (marginal willingness to pay) and the height of the supply curve (marginal opportunity cost). As above, the `assume` command is used to avoid dialog. The function named `surpluses` expresses the net gain from producing various levels of X as a function of the system's parameters.

```
assume(Edp<0, Esp>0, notequal(Edp+1,0))$
integrate(pd(X,B,Edp)-ps(X,A,Esp),X);
surpluses(X, A, B, Esp, Edp) := ''%$
```

$$\frac{X^{\frac{1}{Edp}+1}}{\left(\frac{1}{Edp}+1\right) B^{\frac{1}{Edp}}} - \frac{X^{\frac{1}{Esp}+1}}{\left(\frac{1}{Esp}+1\right) A^{\frac{1}{Esp}}}$$

The first command below evaluates the `surpluses` function given the values above and takes the first derivative of that function. The second command solves for X given that the derivative evaluates to zero, necessary for establishing the efficient (surplus-maximizing) quantity. The third command shows the level of combined surpluses.

```
diff(surpluses(X,A0,B0,Esp0,Edp0),X);   find_root(
    %,X,.1,2*X0); S0:surpluses(%,A0,B0,Esp0,Edp0);
```

$$\frac{7937.0}{X^{0.667}} - 2.3909\,X^{0.4} \quad 2000.0 \quad 228570.$$

The next cell confirms that the equilibrium condition returns the initial values for quantity and price and that these correspond to the quantities that yield maximum total surpluses (consumer surplus *plus* producer surplus).

```
Xeq0:find_root(ps(X,A0,Esp0)-pd(X,B0,Edp0),X,.1,
 2*X0)$    pseq0:ps(X0,A0,Esp0)$ pdeq0:pd(X0,B0,Edp0)$
matrix(["Equilibrium Quantity", "Supply Price",
"Demand Price"],  [Xeq0, pseq0, pdeq0]);
```

$$\begin{bmatrix} \text{Equilibrium Quantity} & \text{Supply Price} & \text{Demand Price} \\ 2000.0 & 50.0 & 50.0 \end{bmatrix}$$

10.3.3 An Excise Tax

One source of deviation from this efficient outcome is the imposition of a tax on the sale or purchase of this good. Chapter 2 shows that whether the tax is nominally imposed on the sales or on the purchase does not matter. Therefore, we treat only one of the two cases, imposing the tax on the sellers. We consider an excise tax and leave analysis of an ad valorem tax as an exercise.

Impose an excise tax, $T = 10$, on the sale of X. Suppose that the tax is nominally paid by the sellers. The output does not provide an expression that allows an explicit expression for X in terms of the system's parameters and the value of T, so, we use the numerical method `find_root` to determine the new equilibrium quantity.

The price received by the sellers and the price paid by the buyers are calculated separately. The difference is the value of T.

```
T0:10$  X1: find_root(pd(X, B0, Edp0) -
   (ps(X, A0, Esp0) + T0), X, .1, 2*X0)$
ps1:ps(X1,A0,Esp0)$  pd1:pd(X1,B0,Edp0)$
Revenue1: (pd1-ps1)*X1$   matrix(
 ["Quantity","Net Price","Price w/Tax",
 "Tax Revenue"], [X1,ps1,pd1,Revenue1]);
```

$$\begin{bmatrix} \text{Quantity} & \text{Net Price} & \text{Price w/Tax} & \text{Tax Revenue} \\ 1666.0 & 46.476 & 56.476 & 16660. \end{bmatrix}$$

The cell below shows the surplus without and with the tax and the change in total surpluses due to the tax. The tax has two effects on the combined surplus. It reduces the total quantity traded, and it extracts revenues from buyers and sellers. Evaluating the `surpluses` function accounts for the first effect, and subtracting `Revenue1` accounts for the second.

```
S1: surpluses(X1,A0,B0,Esp0,Edp0)-Revenue1$
change:surpluses(X1,A0,B0,Esp0,Edp0)-surpluses(
 X0,A0,B0,Esp0,Edp0)-Revenue1$    matrix(
 ["Surplus w/o Tax","Surplus w/Tax","Change",
 "Net Change"],[S0,S1,change,change+Revenue1]);
```

$$\begin{bmatrix} \text{Surplus w/o Tax} & \text{Surplus w/Tax} & \text{Change} & \text{Net Change} \\ 228570. & 210310. & -18265. & -1604.2 \end{bmatrix}$$

The graphs below show the effects of the tax on equilibrium values and on the total surpluses (labeled `Net Gains`).

```
market: gr2d(... Some commands omitted ... key=
 "Supply w/o Tax",explicit(ps(X,A0,Esp0),X,0,
 1.5*max(X0,X1)),...  key="Supply w/Tax",
 explicit(ps(X,A0,Esp0)+T0,X,0,1.5*max(X0,X1)),
 ... key ="Demand",explicit(pd(X, B0, Edp0),X,
 0,1.5*max(X0,X1)),...points([[X1,0],[X1,pd1]]))$
gains:gr2d( xlabel="X",ylabel="$",key="Net Gains",
 yrange = [0.8*S1, 1.1*S0],explicit(surpluses(
   X,A0,B0,Esp0,Edp0),X,.1,1.5*max(X0,X1)),key="",
line_type=dots,points_joined=true,points(
  [[0,S1],[X1,S1]]),points([[0,S0],[X0,S0]]) )$
wxdraw(market, gains,dimensions=[480,460])$
```

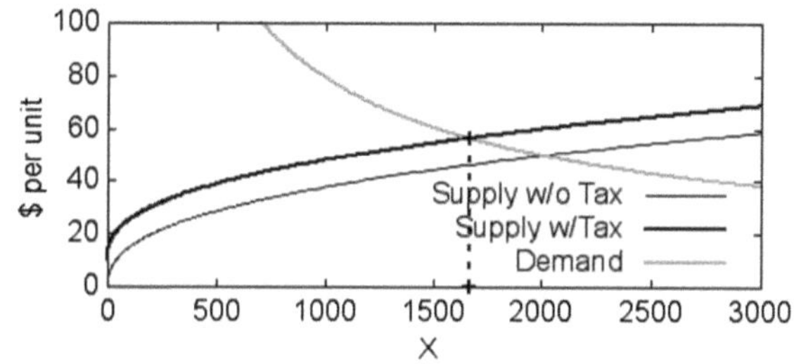

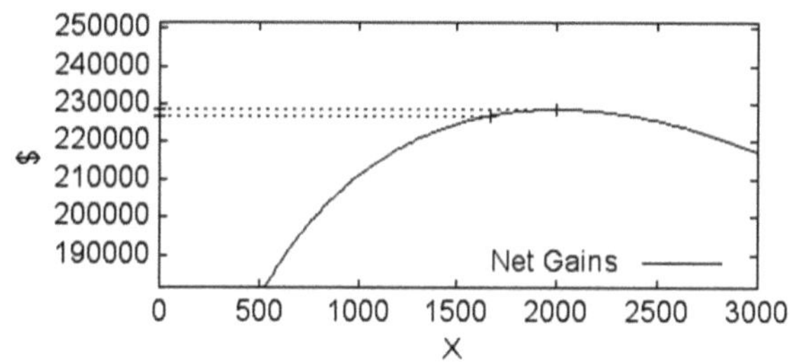

10.3.4 Demand, Supply, and Value

To conclude, we review the question of whether the demand and supply curves should be treated as value curves. Given the conditions stated earlier, they should, and the reasoning behind the calculations above apply exactly. Otherwise, then the reasoning above leads to answers that are more or less appropriate, depending on the size of the deviations from the stated conditions.

We can say a bit more. If the equilibrium quantity produced exceeds the value at which Marginal Value equals Marginal Opportunity Cost, then the reasoning above results in an overstatement of the cost of taxes. In this case (the case of a negative externality), taxes up to some rate can enhance efficiency. Such corrective taxes are referred to as *Pigouvian taxes*.

If the equilibrium quantity is less than the efficient quantity, in contrast, then imposing a tax results in an efficiency loss that exceeds the loss indicated by the reasoning above. Chapter 12 treats price-searcher markets, in which the equilibrium output is less than the efficient level, and shows the impact of imposing taxes under this condition.

References

1. Coase RH (1960) The problem of social cost. J Law Econ 3:1–44
2. Mas-Colell A, Whinston MD, Green JR (2005) Microeconomic theory. Oxford University Press, New York
3. Nicholson W, Snyder C (2008) Microeconomic theory: basic principles and extensions, 11th edn. South-Western, Mason

Chapter 11
General Equilibrium

Abstract This chapter develops a model of a perfectly competitive economy. It does so from a viewpoint of *general equilibrium*, rather than examining subsets of the economy. The chapter begins with production and then addresses consumption. General equilibrium values for quantities produced, product prices, and input prices are determined.

11.1 The Model's Assumptions

The assumptions that provide the foundation for the model developed below are these:

- No single buyer or seller can affect any market's demand or supply enough to have a discernible effect on price. Also, all potential buyers and sellers know the market prices and product characteristics.
- Consumers buy bundles that maximize their utility, and producers sell quantities that maximize their profits.
- Consumer preferences and production technologies are given. They do not change as a result of any changes that the model is being used to analyze. They may change for reasons that are outside the model ("exogenous" changes).
- Both buyers and sellers may freely enter or leave any market.

In addition to these assumptions, which underlie both partial-equilibrium analysis and general-equilibrium analysis, our simple general-equilibrium model requires one additional assumption. This assumption is that the quantities of the resources used in production are given. In terms of partial-equilibrium analysis, this means that the supply curves of resources are vertical (have zero price elasticities). This assumption can be relaxed, but doing so increases the model's complexity.

M.R. Hammock and J.W. Mixon, *Microeconomic Theory and Computation: Applying the Maxima Open-Source Computer Algebra System*,
DOI 10.1007/978-1-4614-9417-1_11,

The formal development of this model is limited to the competitive case. We discuss at two points during the development how non-competitive aspects of an economy can change the model's results. These discussions are informal, however, and provide only qualitative indications of how results change.

11.2 Production

The cell below defines two of the equations that are components of the system to be examined, namely the production functions.[1] Production has the following characteristics:

- The total amount of good X produced in the economy is XT. Production is governed by the Cobb–Douglas production function

$$XT = A \cdot LX^a \cdot KX^{1-a}.$$

- The amount of labor used to produce X is LX, and the amount of capital used to produce X is KX. The total supplies of labor and capital, LT and KT, respectively, are fixed and are fully employed. Thus, $LY = LT - LX$, and $KY = KT - KX$.
- The coefficient A shifts the production function, reflecting technology and the exponent a reflects the relative input intensities of labor and capital in producing good X.
- The terms YT, LY, KY, B, and b have corresponding interpretations regarding good Y.
- The production functions exhibit constant returns to scale at all production rates, so the number of firms is indeterminate and is immaterial to the analysis.

11.2.1 The Edgeworth Production Box

The Edgeworth Production Box provides a direct way to represent the allocation of fixed quantities of two resources between two competing uses. Generalization to more than two uses or more than two resources would not allow the relatively simple graphic representation used here. Fortunately, most of the lessons learned from the simplest cases can be generalized.

The figure below shows an Edgeworth Box for two resources, labor and capital, and two goods, X and Y. For now ignore the box itself and focus on the horizontal and vertical axes at the left and at the bottom. These two axes relate capital used in

[1] The rest of the system consists of utility functions of the consumers, who are also resource owners, and a specification about how much of each resource each consumer owns.

X production (KX) to labor used in X production (LX). Thus, this part of the graph is the same as that used in earlier chapters to analyze production.

The Edgeworth Box allows representation of a second good's production to be superimposed on the same graph. The upper right-hand corner of the graph represents a second origin—the point at which the amount of labor and capital employed in producing good Y are both zero.

The first input line in the next cell defines the Edgeworth box's dimensions, `LT0` and `KT0` and the initial allocation of inputs. The values `LX0` and `KX0` specify the amounts of inputs assigned to producing good X; the remainder are used to produce Y.[2] The expression `EBox` is a logical statement that returns a value of $KT0$ or zero, depending on LX's value. Finally, we plot two relationships, the Edgeworth box and the point that represents resource allocation.[3]

```
[LT0,KT0,LX0,KX0]:[700,50,350,40]$
EBox : if LX < LT0 then KT0 else 0$     wxdraw2d(
 xlabel="Labor, LX",ylabel="Capital,KX",yrange=
 [0,1.2*KT0],user_preamble="set key top left",
 xrange=[0,1.2*LT0],point_type=circle,points(
 [LX0],[KX0]),key="Edgeworth Box",line_width=2,
 explicit(EBox,LX,0,LT0),label(["Good Y origin",
 LT0,1.05*KT0])    )$
```

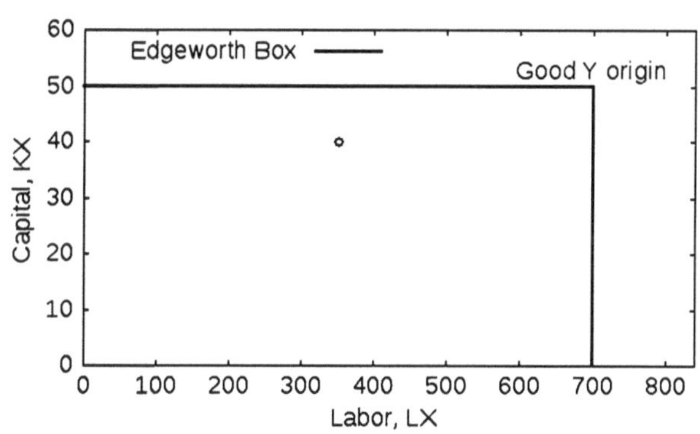

11.2.2 Initial Production Levels and Isoquants

The next cell provides a set of values for the parameters of the two production functions, $XT = A \cdot LX^a \cdot KX^{1-a}$ and $YT = B \cdot LX^b \cdot KY^{1-b}$. It also specifies

[2]Most of the algebra that follows is phrased in terms of LX and KX. The analyst must keep track of the corresponding values of LY and KY.

[3]Strictly speaking, the Edgeworth box, named after the nineteenth-century economist who first popularized its use, is defined by the bold line in the graph and segments of the x and y axes.

full employment. Given these relationships, *Maxima* determines the initial quantity produced of each good, XT = `XT0` and YT = `YT0`.

Next, expressions for isoquants are defined. Defining the `XT0` isoquant is direct: Set XT = `XT0` and solve the production function for KX as a function of LX. Logically, the `YT0` isoquant is solved the same way, but *Maxima* graphs the function in terms of LX, recognizing that LY = `LT0` $- LX$. It does not, however, know to treat the top and right sides of the Edgeworth box as axes. To make the graph treat KY as a vertical distance between `KT0` and KY requires subtracting the latter from the former, as is done in defining the expression `IsoY0`.

The graph is created by copying and pasting from above and then adding two more explicit expressions. Each explicit expression is a function of LX and the range is from $0.01LX$ to $0.99LX$. This range is used to avoid division by zero.

The output shows the values of `XT0` and `YT0` in a list, along with a text entry to indicate the meaning of the numbers. The output also shows the expressions for the `XT0` and `YT0` isoquants. The expression `IsoX0` shows the height of the `XT0` isoquant. That is

$$KX = \left(\frac{\texttt{XT0}}{A\ LX^{a}}\right)^{\frac{1}{1-a}}.$$

Likewise,

$$KY = \left(\frac{\texttt{YT0}}{B\ LY^{b}}\right)^{\frac{1}{1-b}},$$

but the expression in the cell below subtracts this value from `KT0` because the values refer to the axes for good Y.

```
[A,a]:[10,0.75]$ XT : A*LX^a*KX^(1-a)$
[B,b]:[5,0.25]$ YT:B*(LT0-LX)^b*(KT0-KX)^(1-b)$
[XT0, YT0] : [XT, YT], LX = LX0, KX = KX0$
 matrix(["XT0", "YT0"], [XT0, YT0],["- - - - -",
 "- - - - -"],["IsoX0 expression","IsoY0 expression"
 ], [IsoX0 : (XT0/(A*LX^a))^(1/(1 - a)),
IsoY0 : KT0-(YT0/(B*(LT0-LX)^b))^(1/(1-b))] );
```

$$\begin{bmatrix} \text{XT0} & \text{YT0} \\ 2035.0 & 121.61 \\ ----- & ----- \\ \text{IsoX0 expression} & \text{IsoY0 expression} \\ \frac{1.715\,10^{9}}{\left(LX^{0.75}\right)^{4.0}} & 50 - \frac{70.473}{\left((700-LX)^{0.25}\right)^{1.3333}} \end{bmatrix}$$

```
wxdraw2d ( xlabel="Labor, LX",ylabel="Capital, KX",
 xrange = [0, 1.25*LT0], yrange=[0, 1.2*KT0],
 point_type=circle,point_size=2,points([[LX0,KX0]]),
 color =gray40,line_width=2,explicit(EBox,LX,0,
 1.01*LT0), key="XT0 isoquant, refer to X origin",
 explicit(IsoX0, LX, 0.01*LT0, LT0),line_width=3,
 key = "YT0 isoquant, refer to Y origin",
 explicit(IsoY0, LX, 0.01*LT0, .99*LT0) )$
```

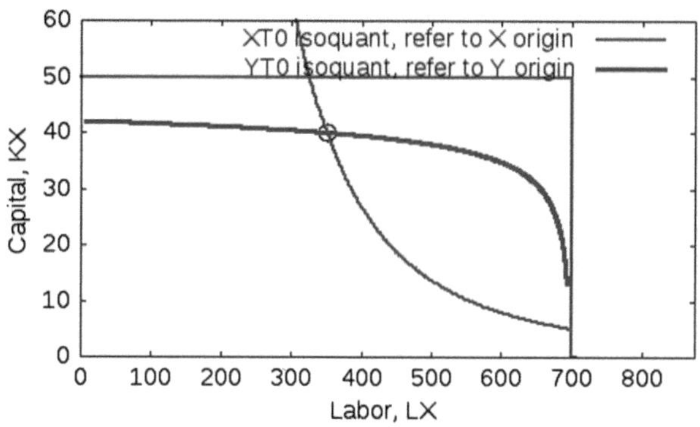

11.2.3 A More Efficient Allocation

The two isoquants intersect and define a lens-shaped area. This lens-shaped area consists of a set of alternative allocations of labor and capital between the two goods' production such that more of both goods can be produced.[4]

At the initial resource allocation, the Y isoquant is flatter than the X isoquant. The slope of an isoquant is the ratio of the marginal product of labor to the marginal product of capital. The first two commands below instruct *Maxima* to take the relevant derivatives and compute the slopes and functions of LX. The functions for the slopes appear in the first output line, with the text entry to indicate that these are functions for slopes.

The second output item shows values of the slope functions at the point shown on the graph. The slope value for the X isoquant is 0.343, and the slope value for the Y isoquant is 0.00952 (signs are ignored), so increasing LX by one unit would allow KX to decrease by approximately 0.343 units without changing XT. The reduction of LY by one unit can be offset by increasing KY by only 0.00952 units. Therefore, if the entire 0.343 units of capital were allocated to Y production, then YT would

[4]If the edge of the lens is included, then it is possible to increase the output of one good without decreasing the production of the other good.

increase. If any number of units between 0.00952 and 0.343 units of capital were diverted, then both XT and YT would increase.

Generally, movements into the lens-shaped area, with LX increasing and KX decreasing, increase the output of both goods. We arbitrarily select the new allocation $LX = 500$ and $KX = 30$. The third output line shows the resulting values, when $XT =$ XT1 and $YT =$ YT1, and the fourth line shows the output increase.

The last two input lines define two new expressions. These are used subsequently to graph the isoquants that correspond to these two output rates.

```
MPLX:diff(XT,LX)$  MPKX:diff(XT, KX)$
MPLY : -diff(YT, LX)$ MPKY : -diff(YT, KX)$
[LX1,KX1]:[500,30]$   matrix( ["SlopeFunctions:",
 Xslope: MPLX/MPKX, Yslope:MPLY/MPKY],
["SlopeValues:",subst([LX=LX0,KX=KX0],Xslope),
 subst([LX=LX0,KX=KX0],Yslope)],
["New X and Y Values:",  XT1: subst([LX=LX1,KX=
 KX1],XT),YT1: subst([LX=LX1,KX=KX1],YT)],
["X and Y Output Change: ",XT1-XT0,YT1-YT0] );
IsoX1 : (XT1/(A*LX^a))^(1/(1 - a))$
IsoY1 : KT0 - (YT1/(B*(LT0-LX)^b))^(1/(1-b))$
```

$$\begin{bmatrix} \text{SlopeFunctions:} & \frac{3.0\,KX}{LX^{1.0}} & \frac{.33333\,(50-KX)}{(700-LX)^{1.0}} \\ \text{SlopeValues:} & .34286 & .0095238 \\ \text{New X and Y Values:} & 2474.6 & 177.83 \\ \text{X and Y Output Change:} & 439.61 & 56.213 \end{bmatrix}$$

11.2.4 The Production Contract Curve

As noted, whenever the slopes of the X and Y isoquants differ, resource reallocation can increase the output of at least one good, and perhaps both. Thus, efficient resource allocation requires equality of the two slopes. We impose that condition below, using `solve`.

The first output line shows the result. The solution is assigned the name $KXCC$. $KXCC$ is this expression:

$$KXCC = \frac{25LX}{3150 - 4LX}.$$

The `CC` stands for "contract curve." This expression shows how much capital must be allocated to X production, given LX, for efficient resource allocation. (Values of LY, KY, XT, and YT are implied.)

The resulting graph is much like the one above. Now it contains two more isoquants. Also, it shows the contract curve.[5]

```
solve(Xslope = Yslope, KX); KXCC : rhs(%[1])$
wxdraw2d( user_preamble = "set key bottom left",
 xlabel="Labor, LX",ylabel="Capital, KX",xrange=
 [0, 1.2*LT0],yrange=[0,1.2*KT0],point_type=circle,
 point_size=2, points([[LX0,KX0]]),points(
  [[LX1,KX1]]),line_width=2,explicit(
  EBox,LX,0,1.01*LT0),key="XT isoquants",color=
  gray40,explicit(IsoX0,LX,0.01*LT0,LT0),key = "",
 explicit(IsoX1,LX,0.01*LT0,LT0),line_width=3,
 key="YT isoquants",explicit(IsoY0,LX,0.01*LT0,
 .99*LT0), key = "", explicit(IsoY1, LX, 0.01*LT0,
 .99*LT0), color=black,line_width=1,key=
 "Contract Curve", explicit(KXCC,LX,0,LT0))$
```

$$[KX = -\frac{25\,LX}{4\,LX-3150}]$$

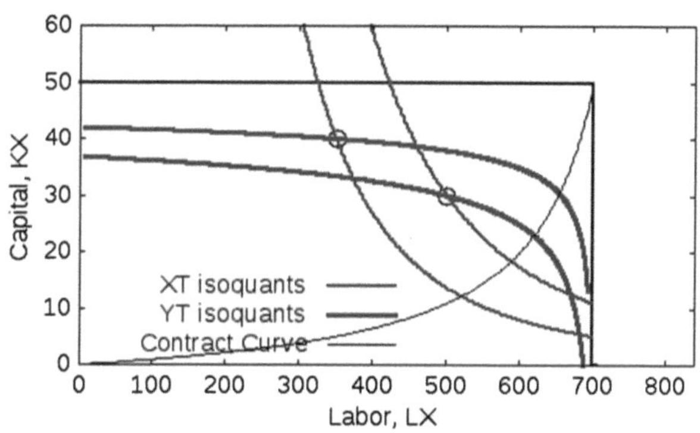

Two points on the contract curve merit attention, the points at which the curve intersects the initial X and Y isoquants. A value of LX below the smaller of these two would reduce X production; a value above the larger of the two would reduce Y production. We execute `find_root` twice to determine LX at each point.

```
LowerLX: find_root(KXCC - IsoX0, LX, 400, 700)$
UpperLX: find_root(KXCC - IsoY0, LX, 500, 699)$
matrix(["Lower LX value", "Upper LX value"],
[LowerLX, UpperLX ]);   kill(LowerLX, UpperLX)$
```

$$\begin{bmatrix} LowerLXvalue & UpperLXvalue \\ 520.34 & 653.31 \end{bmatrix}$$

[5]For the functions used in this illustration, this is a curve with a constantly increasing positive slope. Contract curves can be wavy. The curve lies below the diagonal of the Edgeworth box because X production is relatively labor intensive.

11.2.5 Contract Curve and Production Possibilities

Any point on the contract curve corresponds to an efficient resource allocation. Accordingly, we can use the contract curve to generate a Production Possibilities Curve (PPC).

Specify an LX value, named LXCC0, where CC indicates that the point is on the contract curve. The next two commands create two new variables: XTS is the amount of X produced (i.e., supplied) given LX and the amount of KX implied by the expression for $KXCC$. YTS is the amount of Y determined likewise. Both XTS and YTS are expressions. To determine specific values, evaluate these expression for LXCC0. The resulting values are named XTS0 and YTS0.

The last command in this group defines maxXT0, which is the amount of X that would result if all resources were allocated to its production, and its Y counterpart, maxYT0. These values are generated for use in graphs that follow but are not reported here.

```
LXCC0:580$  KXCC0:float(subst(LX=LXCC0,KXCC))$
XTS: A*LX^a*KXCC^(1-a)$
YTS:B*(LT0-LX)^b*(KT0-KXCC)^(1-b)$
[XTS0,YTS0]: subst(LX=LXCC0,
  [XTS,YTS:B*(LT0-LX)^b*(KT0-KXCC)^(1-b)])$
matrix(["LXCC0:", LXCC0], ["KXCC0:",KXCC0],
 ["XTS Expression:",XTS],["YTS Expression:",YTS],
 ["XTS value:",XTS0],["YTS value:",YTS0]);
IsoXCC0 : (XTS0/(A*LX^a))^(1/(1 - a))$
IsoYCC0:KT0-(YTS0/(B*(LT0-LX)^b))^(1/(1-b))$
maxXT0:subst([LX=LT0, KX = KT0],XT)$
maxYT0:subst([LX=0,KX =0],YT)$
```

$$\begin{bmatrix} \text{LXCC0:} & 580 \\ \text{KXCC0:} & 17.47 \\ \text{XTS Expression:} & 22.361\, LX^{0.75} \left(-\frac{LX}{4\, LX-3150}\right)^{0.25} \\ \text{YTS Expression:} & 5\,(700-LX)^{0.25} \left(\frac{25\, LX}{4\, LX-3150}+50\right)^{0.75} \\ \text{XTS value:} & 2416.3 \\ \text{YTS value:} & 225.41 \end{bmatrix}$$

Finally, we graph the contract curve and the isoquants associated with output rates XTS0 and YTS0 and show how the information in this graph, where inputs appear explicitly, translates into the PPC, where outputs appear explicitly.

The first of the two graphs below shows the results of the initial selection of values for LX and KX and the results when a point on the contract curve is used. Both XT and YT increase as the result of the efficiency gained by this reallocation of labor and capital.

```
wxdraw2d( ... Some commands omitted ...
 explicit(EBox,LX,0,1.01*LT0),key="XT isoquants",
 ... explicit(IsoX0, LX, 0.01*LT0, LT0),
 key="", ... explicit(IsoXCC0,LX,0.01*LT0,LT0),
 key="YT isoquants", ... explicit(IsoY0,LX,
 0.01*LT0,.99*LT0),key="",explicit(IsoYCC0,LX,
 0.01*LT0,0.99*LT0),... key="Contract Curve",
 explicit(KXCC,LX,0,LT0))$
```

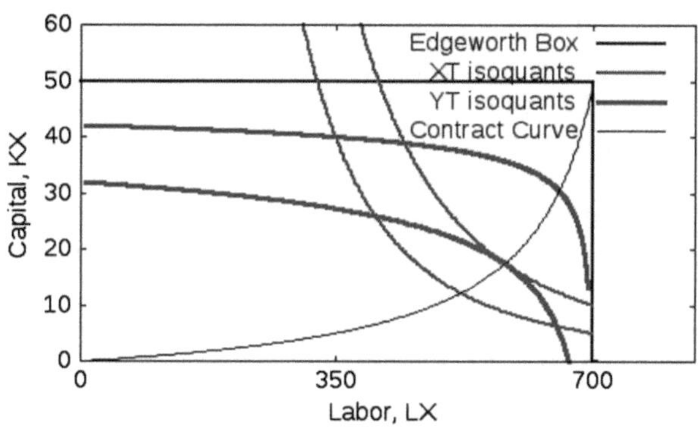

The second graph shows the values of XT and YT in output space, because they cannot be seen in input space. We use `parametric` to draw the PPC. Both XTS and YTS are expressions stated in terms of LX, so *Maxima* treats LX as a common parameter. The command `parametric` tells *Maxima* to draw the graph by changing value of the parameter `LX`, and then plotting the resulting values of XTS and YTS.

```
wxdraw2d( ... Some commands omitted ...  key=
 "Initial Output",point_type=circle,point_size=2,
 points([[XT0, YT0]]), key = "Efficient Output",
 point_type=diamant,points([[XTS0,YTS0 ]]),
  line_width=2,key="Production Possibilities",
  parametric(XTS,YTS,LX,0, LT0)  )$
```

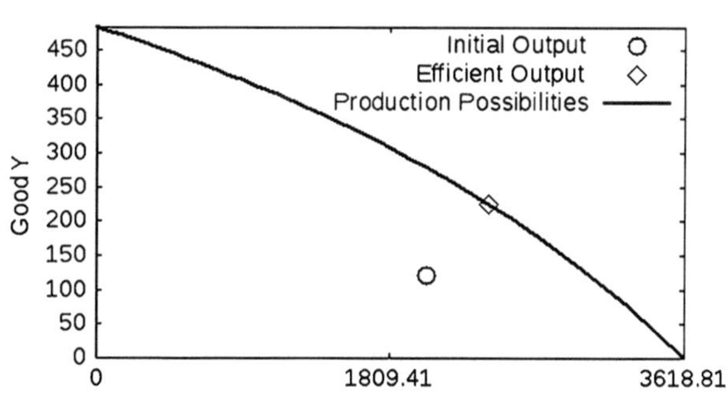

As indicated earlier, each point on the contract curve in the first graph above corresponds to a point on the PPC in the second graph. Likewise, the point off the contract curve at which the `XT0` isoquant intersects its `YT0` counterpart corresponds to a point inside the PPC.

11.2.6 Efficiency and Resource Prices

So far, the discussion of efficiency has been technical, specifying the condition that must be met if efficiency is to be achieved. The technical condition has implications for an important part of an economy, the relative cost of resources (or, from another perspective, the relative income of resource owners). Competitive firms achieve minimum cost production when $MPL/MPK = w/r$, where MPL and MPK are the marginal products of labor and capital in the production of the firm's output and w/r is the ratio of the wage rate to the per-unit cost of employing capital.

In a competitive economy all firms face the same values for w and r, so all firms have the same MPL/MPK. They will not generally have the same value of either MPL or MPK; it is the ratio that will be common to all firms.

Consider the case above. `LXCC0` is specified and the amounts of KX, LY, and KY are determined, consistent with the ratio of marginal products being the same in both sectors. The cell below shows the value of w/r consistent with the values above. First, compute `KXCC0`, the amount of capital used to produce X. Once that is determined, compute the slopes of the two isoquants. Both slopes equal the same value, 0.0904. Finally, assign one of these two values the name `wrRatio0`, which will be used subsequently.

```
[Xslope0,Yslope0]:subst([LX=LXCC0,KX=KXCC0],[Xslope,
 Yslope])$ wrRatio0:Xslope0$  matrix(
 ["X Isoquant Slope","Y Isoquant Slope"],
 [Xslope0, Yslope0]);
```

$$\begin{bmatrix} \text{X Isoquant Slope} & \text{Y Isoquant Slope} \\ .090361 & .090361 \end{bmatrix}$$

The graph below incorporates `wrRatio0` into the analysis. For firms that produce good X the line is the familiar isocost line, viewed from the X-origin. The same is true for firms that produce good Y, except they view the isocost line "upside down" from the Y-origin (the upper right-hand corner of the Edgeworth box). Only a segment of the X producers' isocost line appears on the graph. For Y producers the segment inside the Edgeworth Box is their isocost line; negative capital use makes no economic sense. An exercise: Confirm that the equation for the isocost line facing Y producers is $KY = 40 - 0.1 \cdot LY$.

```
wxdraw2d( ... Some commands omitted ...  key=
 "XTS0 isoquant", ... explicit(IsoXCC0,LX,
 .01*LT0,LT0), ... key="YTS0 isoquant",explicit(
 IsoYCC0,LX,0.01*LT0,.99*LT0),key="IsoCost Line",
 ... explicit(wrRatio0*LXCC0+KXCC0-wrRatio0*LX,
 LX,0,1.5*LT0),key ="Contract Curve",
 explicit(KXCC,LX,0,LT0))$
```

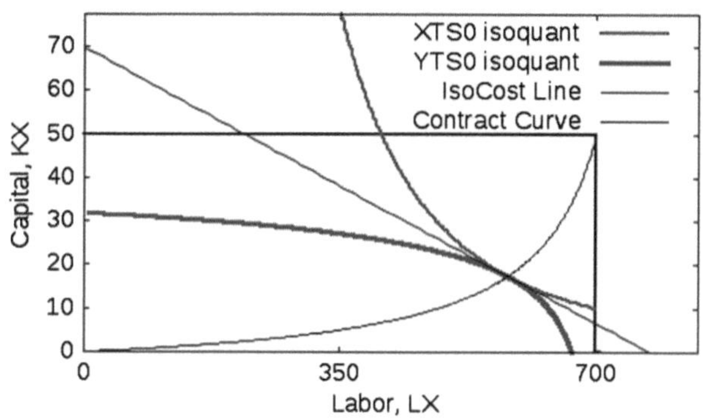

11.3 Relative Product Prices

The preceding section shows that the ratio of input prices ($\frac{w}{r}$) equals the ratio of the marginal products of labor and capital in *either* sector. This section shows that the relative price of the two goods, PX/PY is determined by the relative marginal products of *either* labor or capital in the two sectors. We demonstrate this in terms of labor, but capital could have been used.

The change in the amount of X produced when LX changes, holding KX constant, is $\Delta XT = MPLX \cdot \Delta LX$. This follows from the definition of marginal product. Likewise, $\Delta YT = MPLY \cdot \Delta LY$. The ratio of these two relations is as follows: $\frac{\Delta YT}{\Delta XT} = \frac{MPLY \cdot \Delta LY}{MPLX \cdot \Delta LX}$. Because $\Delta LY = -\Delta LX$ in our application, $\frac{\Delta Y}{\Delta X}$, the slope of the PPC, equals $-\frac{MPLY}{MPLX}$.

The added cost of producing a (small) additional unit of X is $w\Delta LX$. This, along with the marginal product of L in X implies that the marginal cost of X is calculated as follows: $MCX = \frac{\Delta TCX}{\Delta X} = \frac{w}{MPLX}$. Likewise, $MCY = \frac{\Delta TCY}{\Delta Y} = \frac{w}{MPLY}$. Firms in the two industries pay the same wage rate, so $\frac{MCX}{MCY} = \frac{MPLY}{MPLX}$. Given that this is a competitive economy, $\frac{PX}{PY} = \frac{MCX}{MCY}$.

At this point, we see how a competitive economy produces two price *ratios*, but not how any single price is determined. This reflects the important economic reality that relative prices, not absolute prices, matter in an economic system that is in equilibrium.

Nevertheless, we require a system of prices in order to determine nominal total spending/income. To state individual prices, we require a *numeraire*. We could define the units of either of the two inputs or either of the two goods such that the unit price is \$1. We arbitrarily select good Y, so that $PY \equiv 1$. Given this, we can determine the other three prices: $PX = MCX/MCY$, $w = MPLY$ (or $w = PX \cdot MPLX$), $r = MPKY$ (or $r = PX \cdot MPLX$). These values, given `LXCC0`, appear below.

```
KXCC0 : subst([LX = LXCC0], KXCC)$
[MPLXCC0, MPKXCC0, MPLYCC0,MPKYCC0 ]:
 subst( [LX=LXCC0,KX=KXCC0],[MPLX,MPKX,MPLY,MPKY])$
PX0: MPLYCC0/MPLXCC0$    PY:1.0$     w0: MPLYCC0$
r0:MPKYCC0$  matrix(["MPLX","MPLY","MPKX","MPKY"],
   [MPLXCC0,MPLYCC0,MPKXCC0,MPKYCC0] );
matrix(["PX0","PY","w0","r0"],
  [PX0,PY,PX0*MPLXCC0,PX0*MPKXCC0]);
```

$$\begin{bmatrix} MPLX & MPLY & MPKX & MPKY \\ 3.1245 & .46961 & 34.577 & 5.197 \end{bmatrix} \quad \begin{bmatrix} \mathtt{PX0} & PY & \mathtt{w0} & \mathtt{r0} \\ 0.1503 & 1.0 & .46961 & 5.197 \end{bmatrix}$$

11.4 Product Value and Resource Cost

We now determine the economy's Gross Domestic Product (GDP) or, equivalently, its Gross Domestic Income (GDI). The GDP is the market value of final goods and services produced, in this case the value of XT plus the value of YT (or $PX \cdot XT + 1 \cdot YT$). The GDI is the total income paid to resource owners, $(w \cdot \mathtt{LT0}) + (r \cdot \mathtt{KT0})$. Given the parameter values above and the selected value of LX, the GDP equals $0.15 \cdot 2416.3 + 1 \cdot 225.41 = 588.58$ and the GDI equals $(0.47 \cdot 700) + (5.197 \cdot 50) = 588.58$ (all in dollar units).

That the market value equals the cost of producing each of the two goods follows from the fact that production in both sectors is defined by production functions that exhibit constant returns to scale.

```
[XTS0,YTS0]  : subst([LX=LXCC0,KX =KXCC0],[XT,YT])$
GDP0: PX0*XTS0 + YTS0$       GDI0: w0*LT0+r0*KT0$
matrix(["X Level","Y Level","GDP0","GDI0"],
    [ XTS0, YTS0, GDP0, GDI0 ]);
matrix(["Xvalue", "Xcost", "Yvalue", "Ycost" ],
[PX0*XTS0, w0*LXCC0+r0*KXCC0, YTS0,
    r0*(KT0 - KXCC0) + w0*(LT0 - LXCC0) ] );
```

$$\begin{bmatrix} \text{X Level} & \text{Y Level} & \text{GDP0} & \text{GDI0} \\ 2416.3 & 225.41 & 588.58 & 588.58 \end{bmatrix} \quad \begin{bmatrix} \text{Xvalue} & \text{Xcost} & \text{Yvalue} & \text{Ycost} \\ 363.17 & 363.17 & 225.41 & 225.41 \end{bmatrix}$$

```
wxdraw2d( ... Some commands omitted ...
 key="Production Possibilities",parametric(
 XTS,YTS,LX,0,LT0),
 key="IsoValue",line_width=2,explicit(
 GDP0-PX0*X,X,0,1.1*maxXT0),line_type=dots,
 key="",points_joined=true,line_width=1,
 points([[XTS0,0],[XTS0,YTS0]]),
 points([[0,YTS0],[XTS0,YTS0]]))$
```

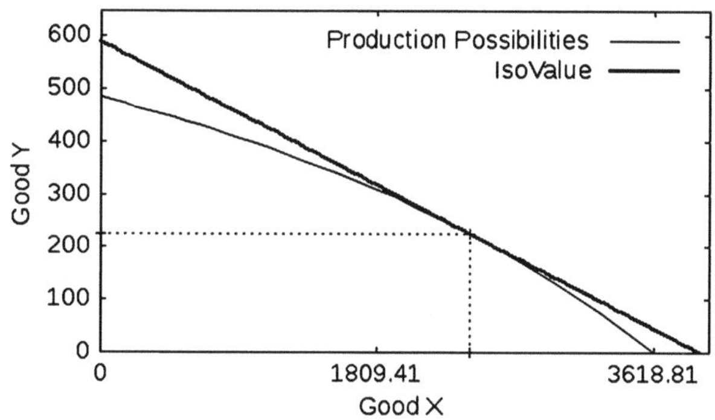

11.4.1 The Supply Curve for Good X

The preceding section calculated a specific price given the value of LX. We now generalize this, making the price of X a function of LX, given that for each value of LX the associated value of KX is determined by the equation for the contract curve. This function, named $PXCC$, is the price that must be paid to have the indicated amount of LX. This function provides the basis for the supply curve for good X, which relates the price of X to the quantity of X.

```
[MPLXCC,MPLYCC]: subst(
KX=KXCC, [MPLX,MPLY])$
PXCC : MPLYCC/MPLXCC;
```

$$\frac{.074536\, LX^{0.25} \left(\frac{25\, LX}{4\, LX-3150}+50\right)^{0.75}}{(700-LX)^{0.75} \left(-\frac{LX}{4\, LX-3150}\right)^{0.25}}$$

The graph below shows the supply curve. Again we use `parametric`. The logical function above the `wxdraw2d` command allows an `explicit` command that shows the result of $LX = LX0$.

```
PX_QX:if XTsup < XTS0 then PX0 else 0$   wxdraw2d(
 xlabel="Good X (XT)",ylabel="$ per unit",key=
 "Supply Curve",parametric(XTS,PXCC,LX,.01*LT0,
 .99*LT0),key="Point on Supply Curve",line_type=
 dots,explicit(PX_QX, XTsup, 0, XTS0) )$
```

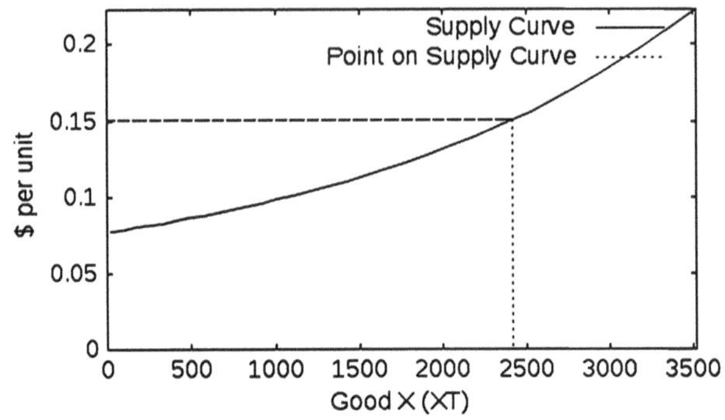

11.4.2 Finding LX, Given PX

We can reverse the process. Suppose that PX is determined outside the system (as it would be if this economy were small enough that world markets set the prices). We can enter the price below and determine the quantity of LX that would yield that price.

The process is broken into two cells. The first cell draws a graph that guides the use of `find_root`. The graph shown below is *not* a supply curve. It relates LX and PX, not XT and PX. The `xaxis` option is set to true, so the point at which $LX = 0$ becomes easier to discern.

```
/* Select a price */ PXset:0.125$
wxdraw2d(xaxis=true,xlabel="LX",ylabel = concat(
 "PX - ",string(PXset)),explicit(PXCC-PXset,LX,
 0,LT0), title="Not a Supply Curve--relates
 PX to LX. Used as a guide")$
```

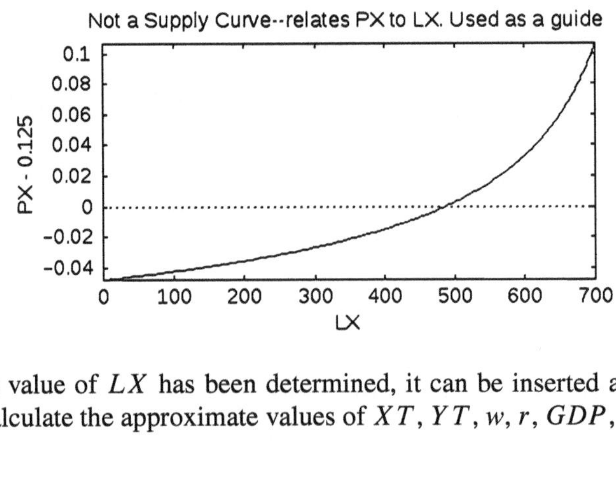

Once this value of LX has been determined, it can be inserted as the value of `LXCC0` to calculate the approximate values of XT, YT, w, r, GDP, GDI.

```
LXset:find_root(PXCC-PXset,LX,400,600)$
matrix(["PXset","LXset"],[PXset,LXset]);
```

$$\begin{bmatrix} \text{PXset} & \text{LXset} \\ 0.125 & 487.5 \end{bmatrix}$$

11.5 Distributing Output: Consumption

The graph below shows the PPC once more. This time, however, it contains a second Edgeworth box. In this case the $XTandYT$ values that are implied by the selection of $LXCC0$ define the sides of the box. In this example, $LXCC0 = 580$ implies that $XTS0 = 2416.3$ and $YTS0 = 225.41$. We create a new variable XTS which is used to define a logical function to create a line that draws the top and right sides of the Consumption Edgeworth Box.[6]

```
wxdraw2d( ... Some commands omitted ...
 parametric(XTS,YTS,LX,0,LT0),key=
 "Consumption Edgeworth Box", ... explicit(
 if XTsup<XTS0 then YTS0 else 0, XTsup,0,XTS0))$
```

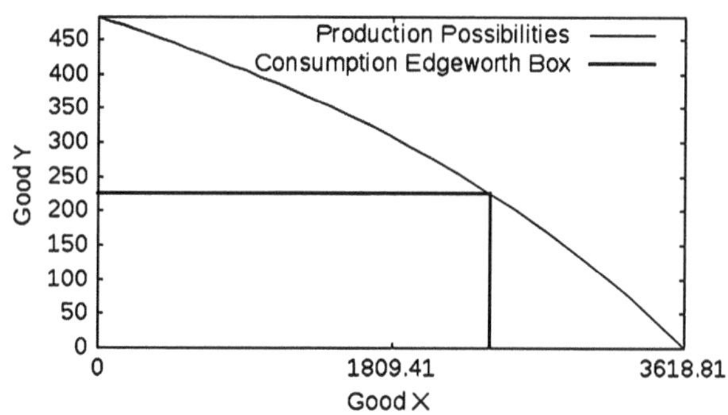

11.5.1 Distributing Output

The next two input cells define an arbitrary distribution of the two goods to our consumers, named Anderson and Brooks for ease of reference.

```
[XA0,YA0]:[1500,75]$  [XB0,YB0]:[(XTS0-XA0),
 (YTS0-YA0)]$  matrix( ["XT","YT","XA","YA",
 "XB","YB"],[XTS0,YTS0,XA0,YA0,XB0,YB0]);
```

$$\begin{bmatrix} XT & YT & XA & YA & XB & YB \\ 2416.3 & 225.41 & 1500 & 75 & 916.26 & 150.41 \end{bmatrix}$$

[6]The line is drawn by placing the logical expression inside the draw command, in contrast to the approach used when graphing the supply curve above. *Maxima* treats the two approaches as identical, so the choice depends on user preference.

The next cell depicts this arbitrary distribution of the two goods between the two consumers.

```
wxdraw2d( ...Some commands omitted ...   points(
  [XA0],[YA0]),label(["Brooks origin",XTS0,
  1.05*YTS0]),key="Consumption Edgeworth Box",
  line_width=2,explicit(
  if XTsup < XTS0 then YTS0 else 0,XTsup,0,XTS0))$
```

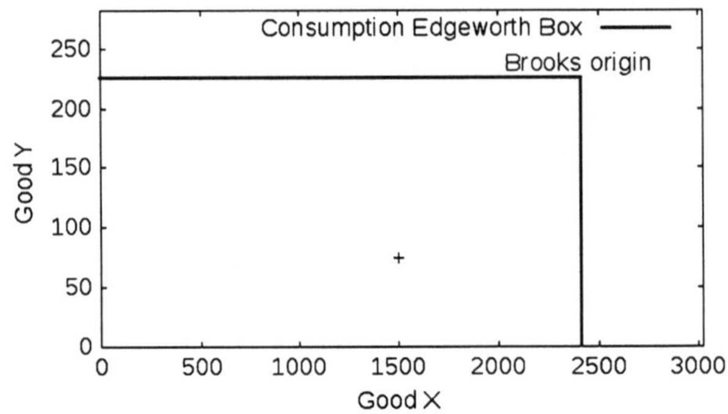

11.5.2 Indifference Curves

The next cell defines the utility functions of these two consumers. Each utility function is a function of the consumer's quantities of X and Y, and it has a single parameter, the exponent of good X. The same type of function as was used to illustrate production, the Cobb–Douglas function, is used here for utility functions. The coefficients are set at 1.

Unlike production, utility units do not matter. Any order-preserving transformation of this function would work equally well. For example, the coefficients could be multiplied by any positive number, numbers (positive or negative) could be added to the function, and the exponents could be increased in any proportion.

We cannot compare utility levels. That Brooks's utility is a larger number than Anderson's, given this distribution of products, has no economic meaning.

```
[c,d]:[0.3,0.7]$ [UA,UB]:[XA^c*YA^(1-c),(XB)^d*(YB)^
 (1-d)]$    matrix( ["UA","UB","UA0","UB0"],
  [UA,UB,UA0:XA0^c*YA0^(1-c), UB0:XB0^d*YB0^(1-d)]);
```

$$\begin{bmatrix} \text{UA} & \text{UB} & \text{UA0} & \text{UB0} \\ XA^{0.3}\,YA^{0.7} & XB^{0.7}\,YB^{0.3} & 184.23 & 532.84 \end{bmatrix}$$

Given the utility functions above, defining equations for indifference curves is a direct matter. Consider Anderson first. Anderson's utility function is $UA = XA^d YA^{1-d}$. Given the value above, $XA^d YA^{1-d} = 186.99$, so

$$YA^{1-d} = \frac{186.99}{XA^d}.$$

This implies that

$$YA = \left(\frac{186.99}{XA^d}\right)^{\frac{1}{1-d}}.$$

We bind the expression on the right-hand side to the name `ICA0`.

The expression for the initial Brooks indifference curve is defined likewise, but with one important difference. The values of YB, Brooks's consumption of good Y, must be the vertical distance from the upper axis of the Edgeworth box, but must be stated in terms of XA for graphing. Therefore, an expression like the one above, for YB, is subtracted from `YTS0`. The two expressions, given the values above, appear below.

```
/*Indifference curves */  [ICA0:(UA0/XA^c)^(1/(1-c)),
ICB0 : YTS0-(UB0/(XTS0-XA)^d)^(1/(1-d))];
```

$$[\frac{1722.8}{(XA^{0.3})^{1.4286}}, \; 225.41 - \frac{1.22648\,10^9}{((2416.3-XA)^{0.7})^{3.3333}}]$$

The graph below shows the two indifference curves. They intersect at the initial distribution point, where they have different slopes. The two curves form a lens-shaped area. Like its counterpart in the discussion of production, this area consists of points that are consistent with increased utility for either Anderson or Brooks, or both.

```
wxdraw2d( ... Some commands omitted ... explicit(
 if XTsup<XTS0 then YTS0 else 0,XTsup,0,XTS0),
 ... key ="ICA0, refer to SW origin",
 explicit(ICA0,XA,.01*XTS0,.99*XTS0),...
 key="ICB0, refer to NE origin",
 explicit(ICB0,XA, .01*XTS0, 0.99*XTS0)       )$
```

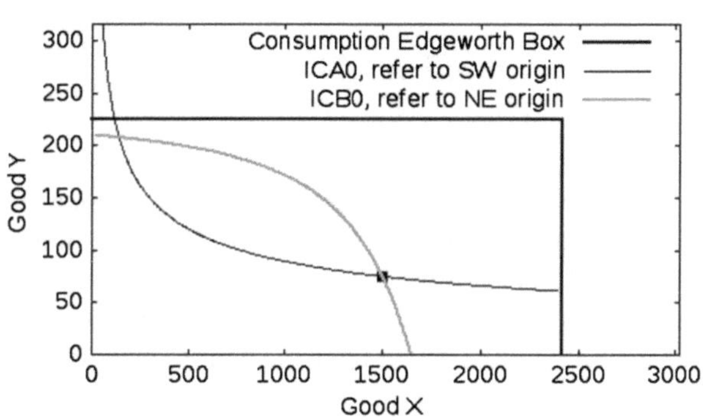

11.5.3 The Consumption Contract Curve

The next input group contains commands that create expressions for the slopes of Anderson's and Brooks's indifference curves. Then these two expressions are set equal to each other. The resulting equation is an expression for YA in terms of XA that defines the Consumption Contract Curve.

```
MUXA:diff(UA,XA)$   MUYA:diff(UA,YA)$
UB:(XTS0-XA)^d *  (YTS0-YA)^(1-d)$
MUXB: -diff(UB,XA)$  MUYB: -diff(UB,YA)$ YACC :
  rhs((float(solve(MUXA/MUYA=MUXB/MUYB,YA)))[1]);
```

$$\frac{3.0574\,10^{14}\,XA}{1.1072\,10^{12}\,XA+6.0195\,10^{14}}$$

The figure below graphs the Consumption Contract Curve, $YACC$, along with the Edgeworth box and the initial indifference curves. That is, it adds $YACC$ to the previous figure.

```
wxdraw2d ( ... Some commands omitted ...
 points([XA0],[YA0]),key="",line_width=2,explicit(
 if XTsup < XTS0 then YTS0 else 0,XTsup,0,XTS0),
 key="ICA0, refer to SW origin",explicit(ICA0,XA,
 .01*XTS0,.99*XTS0),color=gray,key = "ICB0,
  refer to NE origin",explicit(ICB0,XA,.01*XTS0,
 .99*XTS0),key="Consumption Contract Curve",color=
 black,line_width=1, explicit(YACC,XA,0,.99*XTS0))$
```

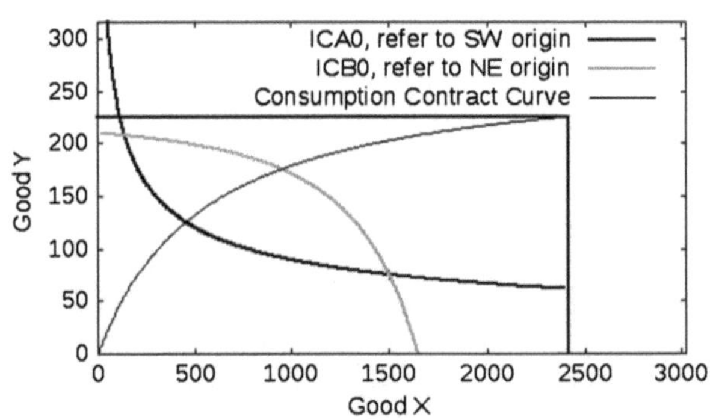

The two commands below show the values of XA at which the Consumption Contract Curve intersects Anderson's and Brooks's initial indifference curves. Selecting a value of XA between these two values and finding the resulting point on $YACC$ corresponds to increasing both Anderson's and Brooks's utility levels. Specifically, they increase to a point on the Consumption Contract Curve where the two indifference curves are tangent.

```
[find_root(ICA0 - YACC, XA, 200, 600),
find_root(ICB0 - YACC, XA, 500, 1500)];
```

$$[452.2, 947.09]$$

In the next cell, we select the value $XA = 600$, determine Anderson's and Brooks's new utility levels, and add the associated indifference curves to the previous graph.

```
XA1:600$  YACC1: subst(XA = XA1,YACC)$
[UA1,UB1]: subst([XA=XA1,YA=YACC1],[UA,UB])$
matrix(["UA0","UA1","UB0","UB1"],[UA0,UA1,UB0,UB1]);
matrix(["ICA1:",ICA1 :(UA1/XA^c)^(1/(1-c)),"ICB1:",
ICB1 : YTS0 - (UB1/(XTS0 - XA)^d)^(1/(1-d))]);
```

$$\begin{bmatrix} UA0 & UA1 & UB0 & UB1 \\ 184.23 & 221.88 & 532.84 & 713.24 \end{bmatrix}$$

$$\left[ICA1 : \frac{2247.0}{\left(XA^{0.3}\right)^{1.4286}} \quad ICB1 : 225.41 - \frac{3.24184\,10^{9}}{\left((2416.3-XA)^{0.7}\right)^{3.3333}}\right]$$

```
wxdraw2d( ... Some commands omitted ... key =
"ICA0, refer to SW origin",explicit(
  ICA0,XA,.01*XTS0,.99*XTS0),line_width=2,color=
  gray,key="ICB0, refer to NE origin",
 explicit(ICB0, XA, .01*XTS0, .99*XTS0) ,
 key = "Consumption Contract Curve",line_width=3,
 explicit(YACC, XA, 0, .99*XTS0),
 ... Some commands omitted ...   )$
```

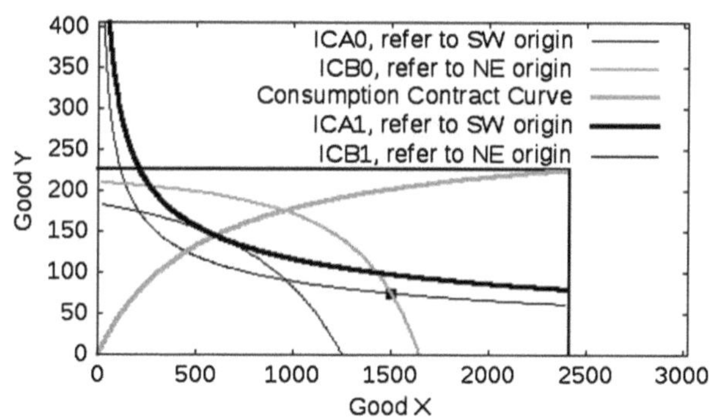

The graph above shows two important aspects of this model. First, any arbitrary distribution of goods among individuals is likely to be inefficient. Given the initial distribution, Anderson and Brooks can trade to mutual advantage, moving to some point inside the lens-shaped area. Improvements can continue until they reach a point on the contract curve, a point at which they have the same marginal willingness to exchange one good for the other.

The second point is that once the contract curve is reached, no further mutually advantageous moves exist. Now, Anderson's gain is Brooks's loss and vice versa. We could map the consumption contract curve into a graph like the PPC above, that would show the levels of utility for each of the two consumers consistent with their being at various points on the consumption contract curve.

This observation relates to one task that remains in developing this model. We require a way to "close" the model. That is, we need enough information to solve the system without arbitrarily setting the value of LX or PX.

One possibility is to posit the existence of an exogenous social welfare maximizer. This entity's objective function would take Anderson's and Brooks's utility levels (which the maximizer could evaluate—now units, known only to the welfare maximizer, would matter). The model is closed when one finds the values of XT and YT that the exogenous maximizer considers best. A classic treatment that employs this approach is [1].

The next section develops an alternative way of closing the model. The consumers of products X and Y are also owners of the labor and capital that are used to produce these goods. When w and r are determined, so are the producers' income levels.

11.6 Demand and Equilibrium

This section develops the demand curve for good X. It then sets demand equal to supply and, in doing so, provides the basis for determining all relevant values.[7]

11.6.1 Incomes and Demand

Before looking directly at the demand curve, we determine the income levels for Anderson and Brooks. Then we take advantage of an aspect of the Cobb–Douglas utility function to determine how the consumers allocate their income. With that information, we develop a set of graphs that summarizes much of what we have done to this point.

[7]We can limit attention to good X because of Walras's Law, which says that in a system of n markets equilibrium in $n - 1$ of the markets implies equilibrium in the nth market. Our system consists of four markets. We specify that $LX + LY = LT$ and $KX + KY = KT$. Thus, these two markets are in equilibrium. Therefore, equilibrium in the market for X is all that is required. The market for Y must also be in equilibrium.

We specify income levels by setting ownership levels for the two resources. In our illustrative example, Anderson owns 70 % of the labor and half of the capital. Thus, Anderson is the wealthier of the two producer/consumers. Anderson's money income level is $MA = w \cdot LA + r \cdot KA$, where LA and KA are determined as noted immediately above. Brooks's money income is determined in the same way. Because w and r are variable, depending on LX, so are the income levels.

For the Cobb–Douglas utility function, the share of income spent on each of the goods equals the exponent of that good. Therefore $c \cdot MA = PX \cdot XA$ and $d \cdot MB = PX \cdot XB$. We rearrange this relationship below, solving for XA and XB, which are labeled `XAD` and `XBD` to indicate demand. The price of X is `PXCC`, which corresponds to the slope of the PPC. This is the same price that appears in the supply curve above, because the price that sellers receive must equal the price that buyers pay.

Before showing the demand curve graphically, we relate a number of variables to LX. To save space, we suppress the printing of the functions below. All expressions are in terms of LX.

```
[alpha0,beta0]:[0.7,0.5]$ [LA,KA]:[alpha0*LT0,
 beta0*KT0]$     [w,r]:subst([KX=KXCC,alpha=alpha0,
 %beta=beta],[MPLY,MPKY])$
[MA,MB]:[w*LA+r*KA,w*(LT0-LA)+r*(KT0-KA)]$
[XAD, XBD] : subst([alpha=alpha0,%beta=beta0],
  [c*MA/PXCC, d*MB/PXCC])$     XD : XAD + XBD$
```

The graphs appear in four panels. The panels are created with the `grd2d` command, which has the same syntax as `wxdraw2d`. Instead of drawing a graph, this option creates a graphic image that can be bound to a name. The four graphs so generated are named, `inputPrices`, `xPrices`, `IncomeLevels`, and `Xvalues`.

Several things warrant attention:

- The first panel shows that the wage rate falls and the rental rate for capital increases as LX increases.
- The nominal price of good X increases as LX increases, reflecting the increasingly steep slope of the PPC.
- Because of the increasing price of good X, the nominal income levels increase as LX increases. One could create a "GDP deflater" by selecting any one value of LX as the base and computing an average price at other values.
- The quantity demanded of good X decreases as LX increases. This graph is *not* a demand curve for good X. The quantity demanded of good X at each value of LX depends on the money income levels of Anderson and Brooks and on the relative price of good X. That is, the curve consists of a set of quantities from various demand curves.

```
maxMT : subst(LX = 0.9999*LT0, MA + MB)$
 key="w",explicit(w,LX,.01*LT0,.99*LT0),
 key="r", ... explicit(r, LX,.01*LT0,.99*LT0))$
xPrices:gr2d(... title="Product Price, PXCC",
 ... explicit(PXCC,LX,.01*LT0,.99*LT0) )$
IncomeLevels:gr2d(...key="A",
 explicit(MA,LX,.01*LT0,.99*LT0), key="B", ...
 explicit(MB,LX,.01*LT0,.99*LT0),key="Total",
 ...explicit(MA+MB,LX,.01*LT0,.99*LT0) )$
Xvalues: gr2d( ...key="A",
 explicit(XAD,LX,.01*LT0,.99*LT0),key="B",
 explicit(XBD,LX,.01*LT0,.99*LT0),... key="Total",
 explicit(XD,LX,.01*LT0,.99*LT0))$
wxdraw(inputPrices, xPrices,IncomeLevels,
  Xvalues,columns=2,dimensions=[640,480])$
```

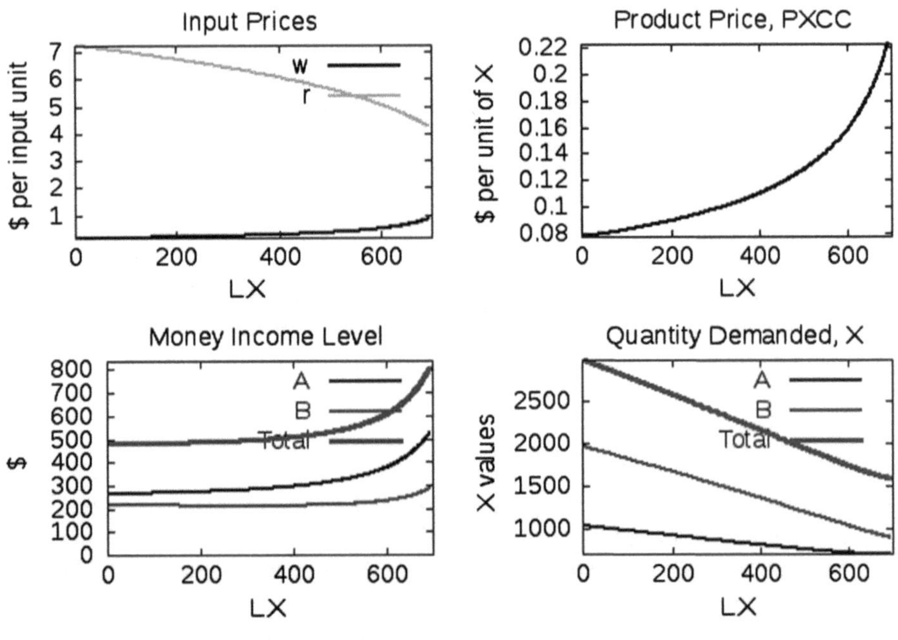

Anderson's and Brooks's demand curves and the market demand curve for good X appear below. The demand curves look much like the demand curves that are derived in partial-equilibrium analysis. They differ in one important respect, however: the income levels of the producers/consumers change as XT changes, so the usual "other things equal" condition must be amended. Now the other things that are equal include the total resource levels, their ownership, and the production functions. Thus, the same factors that affect supply also affect demand. This does not mean, however, that the two are not distinct.

```
maxPX:subst(LX=.99*LTO,PXCC)$  PX_QX: if X <
 XTSO then PXO else 0$   wxdraw2d(...
 Some commands omitted ...key="Supply Curve",
 parametric(XTS,PXCC,LX,.01*LTO,.99*LTO),key=
 "Demand Curve A",parametric(XAD,PXCC,LX,
 .01*LTO,.99*LTO),... key="Demand Curve B",
 parametric(XBD,PXCC,LX,.01*LTO,.99*LTO),...
 key="Market Demand",
  parametric(XD,PXCC,LX,.01*LTO,.99*LTO))$
```

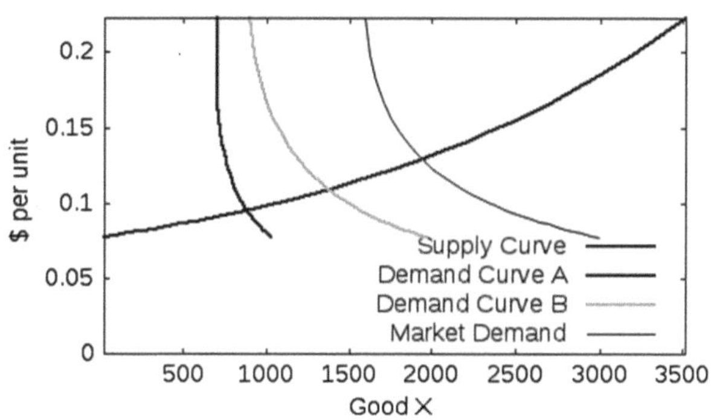

11.6.2 *Equilibrium Values*

We close the model. Once the quantity demanded of good X is determined, then all other values can be deduced. All that remains is to get a reasonably precise calculation of the value of XT at which the demand and supply curves intersect. Both demand and supply have been phrased as explicit functions of LX.

The `find_root` command below evaluates $XD - XTS$ over the range `L0`/100 to 0.99*`L0`, in order to avoid difficulties when either LX or LY equals zero. Given this value of `LX`, we can determine all other values.

```
LXE:find_root(  XD-XTS,LX,LT0/100,LT0*.99);
```
504.0

First, determine the equilibrium price and the quantities of the two goods by evaluating the relevant function given that $LX = LXE$. The equilibrium values are $PE = 0.129$, $XTE = 1942.1$ units, and $YTE = 291.34$ units. Product and input prices are now in dollars per unit.

```
[PXE,XTE,YTE]:subst(LX=LXE,
   [PXCC, XTS, YTS]);
```
[.12859, 1942.1, 291.34]

Next, determine the input price levels the same way. These values are $wE = 0.372$ and $rE = 5.6187$, again in dollar units.

```
[wE,rE]:subst(LX=LXE,[w, r]);
```

$[.37161, \ 5.6187]$

We now determine nominal GDP and nominal GDI. We find that $GDP = GDI = 541.06$ when the system is at equilibrium.

```
[GDPE:XTE*PXE+YTE,GDIE:wE*LT0+rE*KT0];
```

$[541.06, \ 541.06]$

Determine the income levels of Anderson and Brooks. As a check, we confirm that these sum to the GDP.

```
MAE: subst([alpha=alpha0,beta=beta0,LX=LXE],MA)$
MBE: subst([alpha=alpha0,beta=beta0,LX=LXE],MB)$
MtotalE : MAE + MBE$  matrix(["MAE","MBE","
  Total Income","GDPE"],[MAE,MBE,MtotalE,GDPE]);
```

$$\begin{bmatrix} \text{MAE} & \text{MBE} & \text{Total Income} & \text{GDPE} \\ 322.56 & 218.51 & 541.06 & 541.06 \end{bmatrix}$$

Next, determine the equilibrium quantity demanded of each good by each consumer and confirm that these sum to the equilibrium quantities of the two goods.

```
[XAE,YAE]:[c*MAE/PXE,(1-c)*MAE]$
[XBE, YBE]:[d*MBE/PXE, (1-d)*MBE]$
matrix(["","X","Y"],["Anderson",XAE,YAE],
  ["Brooks",XBE,YBE], ["Total",XAE+XBE,YAE+YBE]);
```

$$\begin{bmatrix} & X & Y \\ \text{Anderson} & 752.55 & 225.79 \\ \text{Brooks} & 1189.5 & 65.552 \\ \text{Total} & 1942.1 & 291.34 \end{bmatrix}$$

Finally, determine Anderson's and Brooks's utility levels. Recall that these two numbers are not comparable.

```
[UAE:XAE^c*YAE^(1-c),UBE:XBE^d*YAE^(1-d)];
```

$[324.01, 722.55]$

11.7 Summary

The material above applies *Maxima* to the examination of a general-equilibrium model. The model consists of three parts: production, consumption, and equilibrium. Production is represented by two Cobb–Douglas production functions, one for each of the two goods, X and Y. The resources used, labor and capital, are in fixed supply and are allocated to X and Y production, first arbitrarily and then efficiently. The latter involves development of a "contract curve" that demonstrates the requirement for productive efficiency.

Consumption is represented by two Cobb–Douglas utility functions, one for each of the two individuals, named Anderson and Brooks. A point on the PPC is selected and the output of the two goods is distributed arbitrarily to Anderson and Brooks. Next, the output is distributed efficiently. As with production, the latter involves developing the requirement for an efficient distribution between the consumers.

Finally, equilibrium is established. This is accomplished by allocating the ownership of resources to Anderson and Brooks and tracing the implications for their income. Once the income levels are established, the individual and market demand curves for good X can be determined. The supply curve is determined earlier, during the illustration of production. Together, these complete the model. Once the equilibrium quantity of good X is determined, all other values are determined.[8]

11.8 Summary of Specifications and Results

The next two cells summarize the model specifications and the results at equilibrium. These are provided in order to help with comparative statics analysis. The model is too complicated for such analysis to be incorporated directly. One can do the following: Highlight the output in the two cells that follow and paste it into a word processor. Then change any of the parameters and run the model again. A new set of values will appear below. Copy them and paste them into the same document for comparison.

The first cell reports the input values and parameters that have been used to this point.

```
matrix(  ["ResourceEndowment", LT0, KT0, "", ""],
["ProductionParameters", A, a, B, b],
["Utility Parameters", c, d, "",""],
["OwnershipParameters", alpha0, beta0,"",""]);
```

$$\begin{bmatrix} \text{ResourceEndowment} & 700 & 50 & & \\ \text{ProductionParameters} & 10 & 0.75 & 5 & 0.25 \\ \text{Utility Parameters} & 0.3 & 0.7 & & \\ \text{Ownership Parameters} & 0.7 & 0.5 & & \end{bmatrix}$$

The second cell shows the equilibrium values, given the input values above.

[8]These values are: the quantity of good Y; the price of good X with the price of good Y set at one; the per-unit prices of resources; and the allocation of resources between the two goods; and Anderson's and Brooks's income, consumption, and utility levels.

```
matrix(["wE","rE","MAE","MBE","GDIE"],
  [wE,rE,MAE,MBE,GDIE] ,["--","--","--",
  "--","--"],["PXE","XTE","YTE","GDPE"," "],
  [PXE, XTE,YTE,GDPE," "],["--","--","--",
  "--",""],["XAE","YAE","XBE","YBE" ," "],
  [XAE, YAE, XBE, YBE," "],["--","--",
  "","",""],["UAE", "UBE"," "," "," "],
  [UAE, UBE," "," "," "]);
```

$$\begin{bmatrix} wE & rE & MAE & MBE & GDIE \\ .37161 & 5.6187 & 322.56 & 218.51 & 541.06 \\ -- & -- & -- & -- & -- \\ PXE & XTE & YTE & GDPE & \\ .12859 & 1942.1 & 291.34 & 541.06 & \\ -- & -- & -- & -- & \\ XAE & YAE & XBE & YBE & \\ 752.55 & 225.79 & 1189.5 & 65.552 & \\ -- & -- & & & \\ UAE & UBE & & & \\ 324.01 & 722.55 & & & \end{bmatrix}$$

Reference

1. Bator FM (1957) The simple analytics of welfare maximization. Am Econ Rev 47:22–59

Chapter 12
Price-Searcher Markets

Abstract This chapter and the next two examine price-searching firms. Such firms inhabit one of the three types of industries: simple monopoly (one firm constitutes the industry), monopolistic competition (many firms inhabit an industry for which each firm's product differs slightly from those of other firms), and oligopoly (a few firms inhabit an industry, and the firms' products may be either identical or similar). This chapter addresses the first two cases.

12.1 Overview

Monopoly, monopolistic competition, and perfect competition share one important aspect: in both, no firm considers the effect that its actions have on the actions of other firms. With the simple monopoly, no such reactions are possible, for no other firms exist. With monopolistic competition, each firm is too small to appreciably affect other firms, so other firms do not react to any firm's actions. Therefore, neither firms that are simple monopolies nor firms that are in monopolistically competitive industries engage in *strategic* behavior; they do not "play games."

In contrast, every member of an oligopolistic industry must consider how other firms will react to its behavior, and these considerations affect the member's actions. Chapter 14 shows some ways to model the such industries.

The current chapter addresses a case that is simplified in one other respect, its treatment of the firms' pricing policies. The firm sells its product in a single market, and in that market each unit sells for the same price. Chapter 13 extends this analysis to the case in which the firm sells in different, separable markets with the product price varying across markets. Also, it considers ways that firms' pricing policy in a single market is more complex than simply charging the same price for each unit. Each extension represents a way in which the seller(s) can extract part of the consumer surplus that its product would provide in the simple case.

M.R. Hammock and J.W. Mixon, *Microeconomic Theory and Computation: Applying the Maxima Open-Source Computer Algebra System*,
DOI 10.1007/978-1-4614-9417-1_12, © Springer Science+Business Media New York 2013

Chapter 8 developed the conditions that must be met for a price-searching firm to maximize its profits: produce the quantity at which marginal revenue equals marginal cost and charge the price dictated by the inverse demand curve for that quantity. This condition is general in that it applies to cases in which the price-searcher is a monopolist (the only firm in an industry with no entry), an oligopolist (one of a few firms in an industry that has barriers to entry), or a monopolistic competitor (a price-searching firm in an industry with sufficient entry that the marginal firm or firms earn no economic profit).

12.2 Monopoly

Every price-searcher faces the trade-off between price and quantity that is implicit in the downward-sloping demand curve that faces its product. This trade-off and its implications are most easily developed in a model of pure monopoly.

Before addressing more formally the nature and implication of this trade-off, we consider what "monopoly" means. By definition, it means "one seller." If the firm leaves the industry, the industry no longer exists. Entry must be blocked, because entry implies a change in market structure. Thus, an essential aspect of monopoly is *barriers to entry*. These barriers may be natural, related either to geographic isolation or to prohibitively large-scale investments required to enter the industry. The barriers may be artificial, consisting of government-granted exclusivity or actions taken by the single firm to preclude entry. The latter can involve advertising or other reputation-enhancing actions to increase the cost that a would-be entrant must incur in order to attract buyers.

These observations suggest a structural definition of monopoly: a monopoly exists when a firm is the only seller in a market for which close substitutes do not exist. This definition has some merit, but the term "close substitutes" is not defined. Unless this term can be defined precisely, then the structural definition must remain imprecise.

An alternative approach to defining "monopoly" is behavioral. We may say that a monopoly exists when the demand curve for a firm's product slopes downward and is well-defined, and in which the incumbent firm can continue to earn profits even in the long run. In this case "well defined" means that if the monopolist changes its price by a certain amount, then a specific change in quantity will occur. This change is not conjectural, in the sense that it depends on how other firms might react to the price change. Again, Chap. 14 addresses situations in which the demand for a firm's product must be a matter of conjecture.

We assume that the monopolist, facing a given demand curve, selects the price/quantity pair that yields maximum profit. We can interpret this assumption in either of two ways. It may include the assumption that the monopolist knows the demand curve for its product, so that it predicts the monopolist's behavior. Alternatively, it may be an assertion that such a demand curve exists and that the monopolist searches for both the demand curve itself and for the profit-maximizing price, quantity pair. If so, then the model predicts tendencies in the monopolist's behavior. We leave this question open.

The second part of this chapter considers a variant of this model, in which entry is not restricted. With unrestricted entry, a large number of firms can sell similar products. Each firm's product is distinct in some way, so that each firm has some latitude in setting the price for its product. The market for these goods is in equilibrium when two conditions hold: Each firm produces a quantity that maximizes its profit, and the marginal firm(s) earn zero economic profit. An industry with these characteristics is referred to as *monopolistically competitive*.

12.2.1 Quantity, Price, and Marginal Revenue

The downward-sloping demand for a firm's product implies that it can sell more output only by reducing the price on all units sold. In turn, this implies that the marginal revenue, the change in total revenue per one-unit change in the quantity sold, must be below the price at each quantity.

This section illustrates this point using a linear demand curve and focusing on two points on that curve. If the firm increases output from $x = 8$ to $x = 15$, it must reduce the price from $p = 34$ to $p = 20$, so that total revenue increases from $tr = 272.0$ to $tr = 300.0$, so the revenue change is 28. Thus the change in revenue per one-unit change in quantity is $arc\_mr = \frac{300-272}{15-8} = 4$.[1]

The graph below shows the two effects of increasing output. The firm sells additional units at the reduced price, but the firm must sell each of the initial number of units for a lower price. The areas associated with these changes are labeled. An exercise: repeat this with x1 = 12 and x2 =16 and confirm that the resulting arc marginal revenue value is negative.

```
[x1,x2]:[8,15]$  [p1,p2]:[50.0-2*x1,50-2*x2]$
matrix(["x1","p1","x2","p2"],[x1,p1,x2,p2]);
wxdraw2d( ... Some commands omitted ...
 points_joined=true,points([[x1,0],[x1,p1],
 [0,p1]]),points([[x2,0],[x2,p2],[0,p2]]),
 ... key="Demand",explicit(50-2*x,x,0,30),
 label(["Gain, higher x",12,10]),
 label(["Loss,lower p",4,25]) )$
```

[1]We use arc_mr, arc marginal revenue, to indicate marginal revenue for a discrete quantity change. Most of the material below involves analysis of marginal revenue at a point: $mr = \frac{dtr}{dx}$.

$$\begin{bmatrix} x1 & p1 & x2 & p2 \\ 8 & 34.0 & 15 & 20 \end{bmatrix}$$

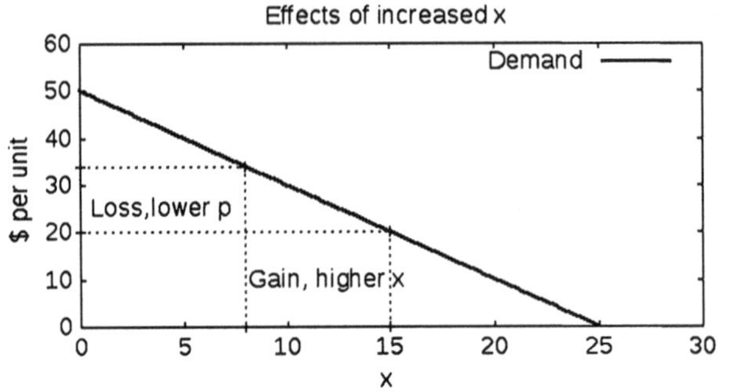

The cell below calculates the change in revenue per unit change in quantity. That is, it calculates the *marginal revenue* for a discrete change in quantity, Δx, based on the areas label in the graph.

```
[Gain,Loss]:[p2*(x2-x1),x1*(p1-p2)]$
NetGain:Gain-Loss$  PerUnitNetGain:NetGain/(x2-x1)$
matrix(["Gain","Loss","NetGain", "PerUnitNetGain"],
[Gain,Loss,NetGain,PerUnitNetGain]);
```

$$\begin{bmatrix} \text{Gain} & \text{Loss} & \text{NetGain} & \text{PerUnitNetGain} \\ 140 & 112.0 & 28.0 & 4.0 \end{bmatrix}$$

The graph below repeats the one above, but shows the value of the arc marginal revenue at the midpoint between the two quantities. It also contains the marginal revenue curve. For this linear illustration, the marginal revenue curve begins at the same point on the vertical axis as the demand curve and has twice the demand curve's slope. Accordingly, it lies half as far to the right as the demand curve. The graph confirms that for this example, the marginal revenue curve passes through the arc marginal revenue. This is true for all linear demand curves but not for demand curves in general.

An exercise: use *Maxima* to demonstrate that, in general, if the demand curve is $p = A + a \cdot x$, then $mr = A - 2 \cdot a \cdot x$.

```
wxdraw2d( ... Some commands omitted ...
 points([[x1,0],[x1,p1],[0,p1]]),points(
 [[x2,0],[x2,p2],[0,p2]]),points([[(x1+x2)/2,
 PerUnitNetGain]]),line_type=solid,line_width=2,
 key="Demand",explicit(50-2*x,x,0,30),line_width=1,
 key="Marginal Revenue",explicit(50-4*x,x,0,30))$
```

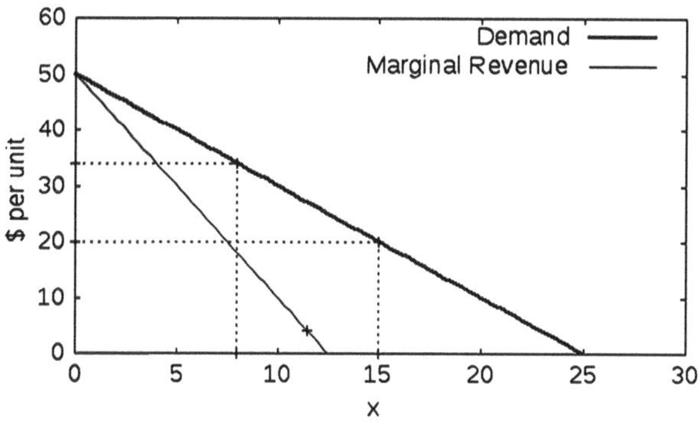

The relationship below is approximately correct for any functional form and exactly correct for the linear demand curve. The approximation becomes closer as $\Delta x \to 0$.

We rewrite the arc elasticity as follows: $arcEp = \frac{\Delta x}{\Delta p} \cdot \frac{\overline{p}}{\overline{q}}$, where $\overline{p}$ is the average price and $\overline{x}$ is the average quantity. The marginal revenue approximation using the arc formula is $arc\_mr = \overline{p} \cdot (1 + \frac{1}{arcEp})$.

```
arcEp:((x2-x1)/((x2+x1)/2))/((p2-p1)/((p2+p1)/2))$
arc_mr:0.5*(p1+p2)*(1 + 1/arcEp)$
matrix(["arc Elasticity","arc MR"],[arcEp,arc_mr]);
```

$$\begin{bmatrix} \text{arc Elasticity} & \text{arc MR} \\ -1.1739 & 4.0 \end{bmatrix}$$

The relationship between arc elasticity and arc marginal revenue is exact when $\overline{p}$ and $\overline{q}$ are replaced by the lower of the two values. In the example above, $arcEp = \frac{\Delta x}{\Delta p} \cdot \frac{p2}{x1}$. In this linear illustration, the two measures of arc elasticity yield the same value. In general, they do not, though the difference is small so long as the deviation from linearity is not large or the change in x is small.

```
(x2-x1)/((p2-p1))*(p2/x1);
p2*(1+1/%);
```

$-1.25 \quad 4.0$

12.2.2 *Profit Maximization Illustrations*

This section illustrates profit maximization using three shapes for demand curves: linear, constant-elasticity, and quadratic. The last of these, quadratic is defined so that $\frac{dp}{dx}$ is a decreasing function of x. The next cell defines the inverse demand curves and the implied marginal revenue curves. The uppercase L, CE, and Q indicate which demand curve is specified.

We define a set of parameters here, selected to ensure that the profit-maximizing quantity will lie between 0 and 5 units. Doing so simplifies the graphing below. We do not change these parameters below. Rather, we add a tax or subsidy that either shifts the marginal cost curve parallel to the original (excise) or changes its slope (ad valorem). Analyzing the effects of demand curve parameter changes is left as an exercise.

```
[a0,b0,m0]:[-4.0,-2.0,-2]$ [A0,B0,M0]:[-5*a0,
 4/10^b0,-25*m0]$   C0:5$   c0:(10-C0)/4.0$
 matrix(["A0","a0","B0","b0","M0","m0",
 "C0","c0"],  [A0,a0,B0,b0,M0,m0,C0,c0]);
```

$$\begin{bmatrix} \mathrm{A0} & \mathrm{a0} & \mathrm{B0} & \mathrm{b0} & \mathrm{M0} & \mathrm{m0} & \mathrm{C0} & \mathrm{c0} \\ 20.0 & -4.0 & 400.0 & -2.0 & 50 & -2 & 5 & 1.25 \end{bmatrix}$$

For each case, we find the quantity at which marginal revenue equals marginal cost. For the linear case *Maxima*'s `solve` command yields the solution. For the other two, we use `find_root`. Once the quantity is found, we determine the price, p, and profit level, P. Profit is determined by integrating the difference between marginal revenue and marginal cost from 0 to the maximizing quantity and subtracting the unspecified fixed cost, F. Values for the following are reported: quantity (x), price (p), average variable cost (avc), profit level (P), marginal revenue (mr) and marginal cost (mc). The last two must be equal.

```
/*Linear*/solve(mrL(xL,A0,a0)=mc(xL,C0,c0),xL)$
xL0: float(rhs(%[1]))$      pL0:pL(xL0,A0,a0)$
PL0:float(integrate(mrL(xL,A0,a0)-mc(xL,C0,c0),xL,
   0,xL0)-F)$ matrix(   ["xL0","pL0","avcL0","PL0",
"mrL0","mcL0"],[xL0,pL0,C0+.5*c0*xL0,PL0,
   mrL(xL0,A0,a0),mc(xL0,C0,c0)]);
```

$$\begin{bmatrix} \mathrm{xL0} & \mathrm{pL0} & \mathrm{avcL0} & \mathrm{PL0} & \mathrm{mrL0} & \mathrm{mcL0} \\ 1.6216 & 13.514 & 6.0135 & 12.162 - 1.0\,F & 7.027 & 7.027 \end{bmatrix}$$

```
/*CE*/ xCE0:find_root(mrCE(xCE,B0,b0)=mc(xCE,C0,c0),
  xCE,.1,5)$  pCE0:pCE(xCE0,B0,b0)$    PCE0:
  (pCE0-(C0+c0*xCE0/2))*xCE0-F$ float(integrate(
  mrCE(xCE,B0,b0)-mc(xCE,C0,c0),xCE,0,xCE0)-F)$
  matrix(["xCE0","pCE0","avcCE0",   "PCE0",
  "mrCE0","mcCE0"],[xCE0,pCE0,C0+.5*c0*xCE0,
  PCE0, mrCE(xCE0,B0,b0),mc(xCE0, C0,c0)]);
```

$$\begin{bmatrix} \mathrm{xCE0} & \mathrm{pCE0} & \mathrm{avcCE0} & \mathrm{PCE0} & \mathrm{mrCE0} & \mathrm{mcCE0} \\ 1.8623 & 14.656 & 6.1639 & 15.814 - F & 7.3279 & 7.3279 \end{bmatrix}$$

```
/*Quadratic*/ xQ0:find_root(mrQ(xQ,M0,m0)=
  mc(xQ,C0,c0),xQ,0,5)$ pQ0:pQ(xQ0,M0,m0)$
  PQ0:float(integrate(mrQ(xQ,M0,m0)-mc(xQ,C0,c0),
  xQ,0,xQ0)-F)$
matrix(["xQ0","pQ0","avcQ0","PQ0","mrQ0","mcQ0"],
  [xQ0,pQ0,C0+.5*c0*xQ0, PQ0,mrQ(xQ0,M0,m0),
  mc(xQ0,C0,c0)]);
```

$$\begin{bmatrix} \text{xQ0} & \text{pQ0} & \text{avcQ0} & \text{PQ0} & \text{mrQ0} & \text{mcQ0} \\ 2.6364 & 36.099 & 6.6478 & 77.645 - 1.0\,F & 8.2955 & 8.2955 \end{bmatrix}$$

The graph below shows the three cases.[2]

```
Linear:gr2d( ... Some commands omitted ...
 points([[xL0,0],[xL0,pL0],[0,pL0]]),
 key="D",explicit(pL(xL,A0,a0),xL,0,5),line_width
 =2,key="MR",explicit(mrL(xL,A0,a0),xL,0,5),
 color=gray,key="MC",explicit(mc(X,C0,c0),X,0,5))$
ConstElas: gr2d( ... Some commands omitted ...
 key="MR",explicit(mrCE(xCE,B0,b0),xCE,0.1,5), ...
 key="MC",explicit(mc(X,C0,c0),X,0,5) )$
Quadratic: gr2d(   ... Some commands omitted ...
 key="MC",explicit(mc(X,C0,c0),X,0,5))$
 wxdraw(Linear,ConstElas,Quadratic,
   columns=3,dimensions=[700,360])$
```

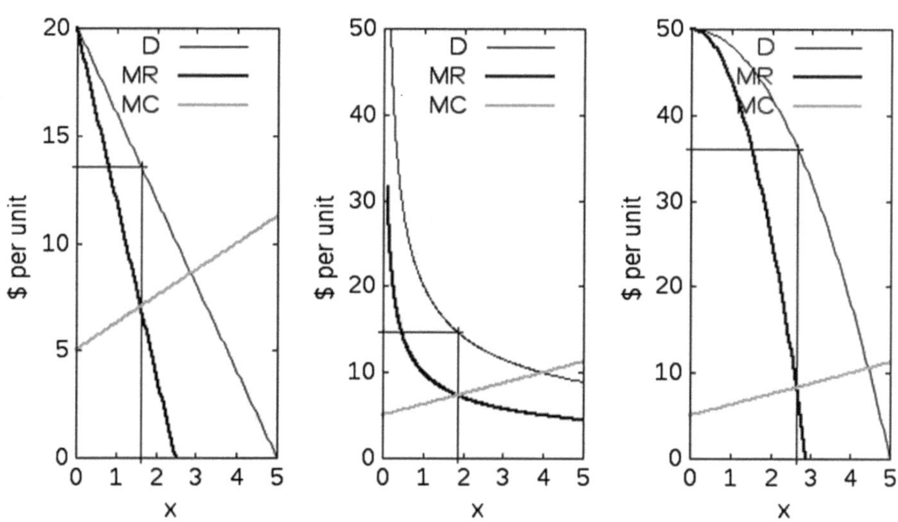

[2]Exercise: Create three separate graphs, replacing `gr2d` with `wxdraw2d`. Add the average cost and average variable cost curves to these graphs.

12.2.2.1 Elasticities and Marginal Revenue

Because $mr = p(1 + 1/E)$, where E is the price elasticity of demand and $mr = mc$ if profits are to be maximized, this important mark-up feature must pertain: $p = \frac{mc}{1+1/E} = mc\left(\frac{E}{1+E}\right)$. The next cell confirms this result in these three examples. Elasticity is $E = \frac{dx}{dp} \cdot \frac{p}{x}$. In the linear case, this equals $\frac{1}{a0} \cdot \frac{pL0}{xL0}$. For the constant-elasticity case, it equals $b0$. Calculating the elasticity for the quadratic demand curve requires evaluating the slope $xQ0$ and then computing the elasticity at that value of x.

In all cases, the prices implied below equal those in the table above. The least elastic demand curve, the quadratic curve, results in the largest markup of price over marginal cost, as it must.

```
EpL : (1/a0)*pL0/xL0$        EpCE : b0$
diff(pQ(xQ,M0,m0),xQ)$  EpQ :(1/%)*(pQ0/xQ0),xQ=xQ0$
matrix(["EpL", "pL", "EpCE", "pCE", "EpQ", "pQ"] ,
[EpL,(C0+c0*xL0)/(1+1/EpL),EpCE,(C0+c0*xCE0)/
 (1+1/EpCE), EpQ, (C0 + c0*xQ0)/(1+1/EpQ) ]);
```

$$\begin{bmatrix} \text{EpL} & \text{pL} & \text{EpCE} & \text{pCE} & \text{EpQ} & \text{pQ} \\ -2.0833 & 13.514 & -2.0 & 14.656 & -1.2984 & 36.099 \end{bmatrix}$$

12.2.3 *Consumer Surplus and Deadweight Loss*

The table below shows the surplus that consumers realize. It also shows the deadweight loss that results from the fact that the monopolist produces a quantity that is less than the quantity at which price equals marginal cost. We refer to that quantity as the efficient quantity, using the notation `xLe`, `xCEe`, and `xQe` to indicate the efficient quantities.

In the graphs above, consumer surplus corresponds to the areas between the demand curves and the prices up to the quantities the monopolist would sell under each set of demand conditions. The deadweight loss corresponds to the areas between the demand curves and the marginal cost curves over the quantities ranging from the quantities that would be sold to the quantities at which marginal costs equal the demand prices, where the demand prices are the demand curves' heights.

The notation in the preceding paragraph pertains to the linear illustration. For constant-elasticity demand substitute `CE` for `L`, and for quadratic demand substitute `Q` for `L`. The initial quantity of x is `xL0`, and `xLe0` is the efficient quantity.

The initial consumer surplus is CSL01, and DL0 is the deadweight loss that results from xL0 < xLe0.

```
xLe0:find_root(pL(xL, A0, a0) -mc(xL,C0,c0),
 xL,0,5)$CSL0:float(integrate(pL(xL, A0, a0)-
 pL0,xL,0,xL0))$  DL0:float(integrate(pL(xL,A0,a0)-
 mc(xL,C0,c0),xL,xL0,xLe0))$
xCEe0 : find_root(pCE(xCE,B0,b0)-mc(xCE,C0,c0),
 xCE,0.1,5)$  CSCE0:float(integrate(pCE(xCE,B0,b0)-
 pCE0,xCE,0,xCE0))$ DCE0:float(integrate(pCE(xCE,
 B0,b0)- mc(xCE,C0,c0),xCE,xCE0,xCEe0))$
xQe0 : find_root(pQ(xQ,M0,m0)-mc(xQ,C0,c0),
 xQ,0,5)$CSQ0:float(integrate(pQ(xQ,M0,m0)-
 pQ0,xQ,0,xQ0))$ DQ0:float(integrate(pQ(xQ,M0,m0)-
 mc(xQ,C0,c0),xQ,xQ0,xQe0))$
```

```
matrix(["xL0","xLe0","CSL0","DL0"],[xL0,xLe0,CSL0,
 DL0],["--","--","--","--"],["xCE0","xCEe0",
"CSCE0","DCE0"],[xCE0,xCEe0,CSCE0,DCE0], ["--",
"--","--","--"],["xQ0","xQe0","CSQ0","DQ0"],
   [xQ0, xQe0, CSQ0, DQ0]   );
```

xL0	xLe0	CSL0	DL0
1.6216	2.8571	5.2593	4.0071
–	–	–	–
xCE0	xCEe0	CSCE0	DCE0
1.8623	4.0	27.293	6.8928
–	–	–	–
xQ0	xQe0	CSQ0	DQ0
2.6364	4.4412	24.434	27.049

12.2.4 Shifting Marginal Cost: An Excise Tax

This section considers shifts in the marginal cost function. In our example, an excise tax is nominally placed on the selling of the good. Effects of demand curve shifts are left as exercises. When the shift in the firm's cost curve is due to a tax, the efficient quantity remains unchanged. A tax (or subsidy) does not affect the (social) opportunity cost of producing additional units, just the firm's perception of that cost.

The new equilibrium values, after a tax of $1 appear below. They are named `xL1`, and so on.

```
xL1:find_root(mrL(xL,A0,a0)=mc(xL,C0+1,c0),xL,0,5)$
pL1:pL(xL1,A0,a0)$
matrix(["xL0","pL0","xL1","pL1"],[xL0,pL0,xL1,pL1]);
xCE1:find_root(mrCE(xCE,B0,b0)=mc(xCE,C0+1,c0),
 xCE,0.1,5)$  pCE1:pCE(xCE1,B0,b0)$  matrix(["xCE0",
 "pCE0","xCE1","pCE1"],[xCE0,pCE0,xCE1,pCE1]);
xQ1:find_root(mrQ(xQ, M0,m0)=mc(xQ,C0+1,c0),xQ,0,5)$
pQ1 : pQ(xQ1, M0, m0)$
matrix(["xQ0","pQ0","xQ1","pQ1"],[xQ0,pQ0,xQ1,pQ1]);
```

$$\begin{bmatrix} \text{xL0} & \text{pL0} & \text{xL1} & \text{pL1} \\ 1.6216 & 13.514 & 1.5135 & 13.946 \end{bmatrix}$$

$$\begin{bmatrix} \text{xCE0} & \text{pCE0} & \text{xCE1} & \text{pCE1} \\ 1.8623 & 14.656 & 1.5749 & 15.937 \end{bmatrix}$$

$$\begin{bmatrix} \text{xQ0} & \text{pQ0} & \text{xQ1} & \text{pQ1} \\ 2.6364 & 36.099 & 2.6058 & 36.419 \end{bmatrix}$$

The table above shows that for the constant-elasticity case, the parameters result in a price increase of more than 1.0. That is, the price rises by more than the tax.[3] The possibility that the price can rise by more than the tax calls into question the meaning of tax incidence when the demand curve for a firm's product slopes downward. Both the buyers and the seller lose because of the tax, but no neat allocation of the tax's (or subsidy's) effect can be defined.

12.2.4.1 Tax-Induced Deadweight Loss

The table below shows the deadweight loss due to the combination of monopoly and taxation. The last column shows the deadweight loss change due directly to the tax. The tax is 1.0, so the tax revenue equals the post-tax quantity, and we can directly compare the tax revenue with the resulting deadweight loss increase. In all cases, the added deadweight loss per dollar of tax revenue is large: 0.484 ($= \frac{0.732}{1.5135}$) for the linear demand curve, 1.4823 ($= \frac{2.3344}{1.5749}$) for the constant-elasticity demand curve, and 0.328 ($= \frac{0.856}{2.6058}$) for the quadratic demand curve.

[3] Exercise: Demonstrate that this result must hold if $c = 0$, that is if marginal cost is constant and the demand curve exhibits constant elasticity.

```
CSL1:float(integrate(pL(xL,A0,a0)-pL1,xL,0,xL1))$
DL1:float(integrate(pL(xL,A0,a0)-mc(xL,C0,c0),xL,
 xL1,xLe0))$ CSCE1:float(integrate(pCE(xCE,B0,b0)-
 pCE1,xCE,0,xCE1))$ DCE1:float(integrate(pCE(xCE,B0,
 b0)-mc(xCE,C0,c0),xCE,xCE1,xCEe0))$
CSQ1:float(integrate(pQ(xQ,M0,m0)-pQ1,xQ,0,xQ1))$
DQ1:float(integrate(pQ(xQ,M0,m0)-mc(xQ,C0,c0),xQ,
 xQ1,xQe0))$
matrix( ["xL1","CSL1","DL1","ChgDL"],
 [xL1,CSL1,DL1,DL1-DL0],["--","--","--","--"],
 ["xCE1","CSCE1","DCE1","ChgDCE"], [xCE1,CSCE1,
 DCE1,DCE1-DCE0],["--","--","--","--"],["xQ1",
"CSQ1","DQ1","ChgDQ"],[xQ1,CSQ1,DQ1,DQ1-DQ0]);
```

xL1	CSL1	DL1	ChgDL
1.5135	4.5814	4.739	.73192
–	–	–	–
xCE1	CSCE1	DCE1	ChgDCE
1.5749	25.099	9.2271	2.3343
–	–	–	–
xQ1	CSQ1	DQ1	ChgDQ
2.6058	23.593	27.904	.85564

As these illustrative numbers suggest, taxes on monopolists' products tend to cause relatively large increases in deadweight loss. The graph below, showing the striking case of the constant-elasticity demand curve, demonstrates why this loss is relatively large. The monopolist has already created deadweight loss as shown by the area between the demand and marginal cost curves to the right of the initial quantity. Now, imposing the tax adds a loss indicated by the trapezoid defined by the area between the demand and original marginal cost curves from the post-tax quantity to the initial quantity.

```
xmax: 1.25*max(xCE0, xCE1)$
wxdraw2d( ... Some commands omitted...
 key="Initial x and p",points([[xCE0,0],[xCE0,
 pCE0],[0,pCE0]]),... points([[xCE1,0],[xCE1,
 pCE1],[0,pCE1]]), ... explicit(pCE(xCE,B0,b0),xCE,
 0.1,xmax), ... explicit(mrCE(xCE,B0,b0),xCE,0.1,
 xmax), ... explicit(mc(X,C0,c0),X,0,xmax),...
 explicit(1+mc(X,C0,c0),X,0,xmax), ...fill_color=
 gray40,polygon([[0,0],[xCE1,0],[xCE1,1],[0,1]]))$
```

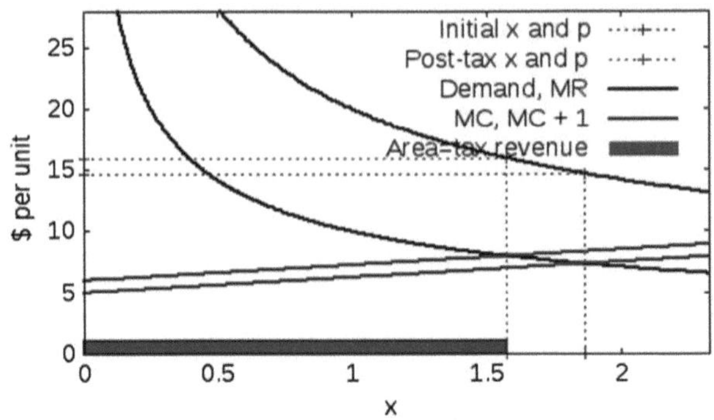

Consider the distributional effects. Consumers pay an additional 2.018 (1.2814 · xCE1) for the units purchased at the higher price. The seller's profit falls by 1.7147. Thus private losses equal 4.5442, in this case much more than the tax revenue raised. Be aware that the difference between private-sector losses and tax revenue generated does not correspond directly to the additional deadweight loss as it can in perfectly competitive markets.

```
PCE1 : (pCE1 - (C0 +1 + c0*xCE1/2))*xCE1 - F$
transpose(matrix( ["Price Change",
"Increased spending on post-tax units",
"Tax Revenue","Consumer Surplus Change",
"Profit Change"],[(pCE1-pCE0),(pCE1-pCE0)*xCE1,
 xCE1,CSCE1-CSCE0,PCE1- PCE0]));
```

Price Change	1.2814
Increased spending on post-tax units	2.018
Tax Revenue	1.5749
Consumer Surplus Change	−2.1945
Profit Change	−1.7147

The cell below shows the corresponding values for the quadratic illustration.

```
PQ1:(pQ1-(C0+1+c0*xQ1/2))*xQ1-F$
transpose(matrix(["Price Change",
 "Increased spending on post-tax units",
 "Tax Revenue","Consumer Surplus Change",
 "Profit Change"],[(pQ1-pQ0),(pQ1-pQ0)*xQ1,xQ1,
 CSQ1-CSQ0,PQ1- PQ0]));
```

Price Change	.32059
Increased spending on post-tax units	.83542
Tax Revenue	2.6058
Consumer Surplus Change	−.84033
Profit Change	−2.6212

Finally, the cell below shows the corresponding values for the linear illustration. In both the quadratic and linear illustrations, the sum of buyers' and seller's losses exceeds the tax revenue raised, though the losses are proportionately less than in the constant-elasticity illustration.[4]

```
PL1 : (pL1 - (C0 +1 + c0*xL1/2))*xL1 - F$
transpose(matrix( ["Price Change",
 "Increased spending on post-tax units",
 "Tax Revenue","Consumer Surplus Change",
 "Profit Change"],[(pL1-pL0),
 (pL1-pL0)*xL1, xL1, CSL1-CSL0, PL1- PL0]));
```

$$\begin{bmatrix} \text{Price Change} & .43243 \\ \text{Increased spending on post-tax units} & .65449 \\ \text{Tax Revenue} & 1.5135 \\ \text{Consumer Surplus Change} & -.67787 \\ \text{Profit Change} & -1.5676 \end{bmatrix}$$

12.3 Upward-Sloping Marginal Revenue

For the examples above, the marginal revenue curve for the monopolist's product uniformly slopes downward. In such cases, a horizontal or upward-sloping marginal cost curve ensures that the second-order condition for the quantity at which $mr = mc$ is the profit-maximizing quantity.

Suppose, however, that the marginal revenue curve does not slope downward throughout its range. Then, finding a quantity at which $mr = mc$ can correspond to finding a quantity that results in local minimum profit. Consider a relatively simple inverse demand curve, $p = a \cdot x^3 + b \cdot x^2 + c \cdot x + d$, with the following conditions applying: $a < 0$, $b > 0$, $c < 0$, and $d > 0$. For some parameter values, the demand curve can slope downward throughout the range of positive x values, but can generate an associated marginal revenue curve that slopes upward over some range.

```
kill(all)$  p(x,a,b,c,d):=a*x^3 + b*x^2 +c*x +d;
```

$$\mathrm{p}(x,a,b,c,d) := a\,x^3 + b\,x^2 + c\,x + d$$

Consider the following parameter values (adapted from [3]). For $0 \leq x \leq 79.917$, the price is positive and the demand curve has a negative slope.

[4]Bishop [1] provides a more general discussion of the relevance of the demand curve's curvature to deadweight loss.

```
[a0,b0,c0,d0] : [-0.018, +1.1, -23.0, 4000.0]$
xmax0: find_root(p(x,a0,b0,c0,d0), x,0,500);
wxdraw2d(xlabel="x", ylabel="$ per unit",
 explicit(p(x,a0,b0,c0,d0),x, 0, xmax0))$
```

79.917

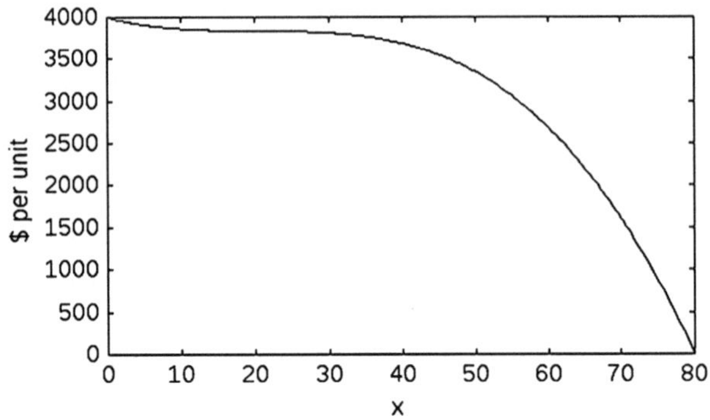

The next cell creates the total revenue and marginal revenue functions.

```
tr(x,a,b,c,d):=''(radcan(p(x,a,b,c,d)*x));
 mr(x,a,b,c,d):=''(diff(tr(x,a,b,c,d),x));
```

$$\operatorname{tr}(x,a,b,c,d) := a\,x^4 + b\,x^3 + c\,x^2 + d\,x$$

$$\operatorname{mr}(x,a,b,c,d) := 4\,a\,x^3 + 3\,b\,x^2 + 2\,c\,x + d$$

The graph below shows a segment of the demand curve and the marginal revenue curve over a subset of the range defined above. The price axis is truncated in order to emphasize the curvature of the demand and marginal revenue curves.

```
wxdraw2d(yrange=[mr(30,a0,b0,c0,d0),d0],xlabel="x",
 ylabel="$ per unit", key = "Demand", explicit(
 p(x,a0,b0,c0,d0),x,0,30),key="Marginal Revenue",
 line_width=2,explicit(mr(x,a0,b0,c0,d0),x,0,30))$
```

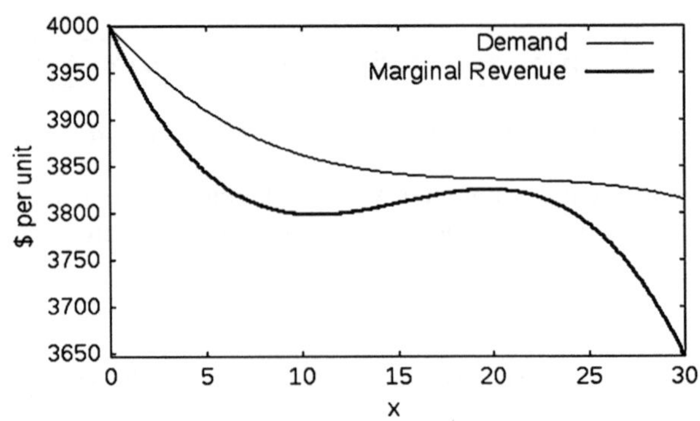

To find the values for which the marginal revenue curve achieves a local minimum and a local maximum, solve $\frac{dmr}{dx}$. At $x \approx 10.8$, $mr \approx 3797.41$, a local minimum value. At $x \approx 19.8$, $mr \approx 3824.04$, a local maximum. Between the two values of x, mr is increasing even though the demand curve continues to slope downward.

```
[soln2: float(solve(diff(mr(x,a0,b0,c0,d0),x), x)),
[mr1: float(mr(rhs(soln2[1]), a0, b0, c0, d0)),
 mr2: float(mr(rhs(soln2[2]), a0, b0, c0, d0))]];
```

$$[[x = 10.756, \quad x = 19.8], \quad [3797.4, \quad 3824.0]]$$

We select a value for marginal cost $mc = 3{,}810$, between the two values above. Given constant marginal cost, $mr = mc$ occurs at three values of x. The largest and smallest x values result in local profit maxima; the middle value results in a local profit minimum.

```
mc0:3810$    wxdraw2d( ... Some commands omitted ...
 key="Demand",explicit(p(x,a0,b0,c0,d0),x,0,30),
 key="MR",line_width=2,explicit(mr(x,a0,b0,c0,d0),
 x,0,30), ... explicit(mc0,x,0,30))$
```

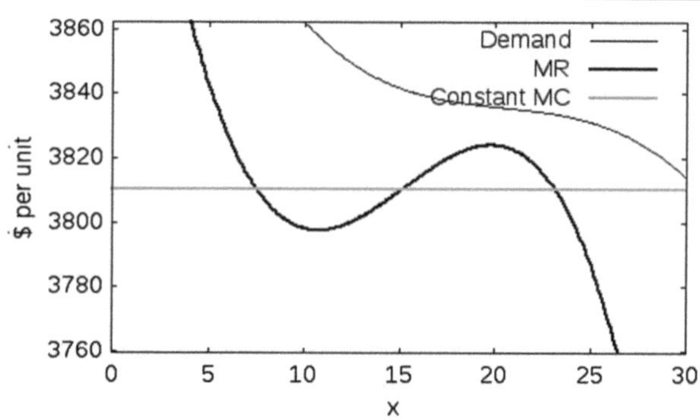

The three values of x are determined below. If the firm knows its demand curve, the upward-sloping marginal revenue segment presents no difficulty. The firm would be able to determine the profit level at each value of x and choose between $x \approx 7.53$ and $x \approx 23.19$, depending on which local maximum corresponds to the global maximum.

Consider a firm that is producing $x \approx 15.11$ units. Either increasing or decreasing x would result in increased profits, so the firm might not have enough information to determine the profit-maximizing quantity.

```
[x1:find_root(mr(x,a0,b0,c0,d0)-mc0,x,0,10),
x2:find_root(mr(x,a0,b0,c0,d0)-mc0,x,x1+1,20),
x3:find_root(mr(x,a0,b0,c0,d0)-mc0,x,x2+1,50)]$
matrix(["x1","x2","x3"],[x1,x2,x3]);
```

$$\begin{bmatrix} \text{x1} & \text{x2} & \text{x3} \\ 7.5291 & 15.113 & 23.191 \end{bmatrix}$$

We determine the profit level at each output rate.

```
[P1: tr(x1,a0,b0,c0,d0)-(mc0*x1+F),
P2: tr(x2,a0,b0,c0,d0)-(mc0*x2+F),
P3:tr(x3,a0,b0,c0,d0)-(mc0*x3+F) ]    matrix(
["Profit1","Profit2","Profit3"],[P1,P2,P3]);
```

$$\begin{bmatrix} \text{Profit1} & \text{Profit2} & \text{Profit3} \\ 538.36 - F & 476.22 - F & 549.72 - F \end{bmatrix}$$

From the firm's viewpoint, the result of selling x1 rather than x3 is not very large, given this model's parameters. The social implication, however, is another matter. The efficient quantity is $x = xe \approx 30.67$ units. If the firm selects output $x = x1$, the resulting deadweight loss is $dwl \approx 685.71$, versus $dwl \approx 113.38$ for $x = x3$.

```
[xeff:find_root(p(x,a0,b0,c0,d0)-mc0,x,0,100), DWL1:
 float(integrate(p(x,a0,b0,c0,d0)-mc0,x,x1,xeff)),
 DWL3:float(integrate(p(x,a0,b0,c0,d0)-mc0,x,x3,
 xeff))]$  matrix(["Efficient x","DWL, x1",
 "DWL, x3"],[xeff,DWL1,DWL3]);
```

$$\begin{bmatrix} \text{Efficient x} & \text{DWL, x1} & \text{DWL, x3} \\ 30.671 & 685.71 & 113.38 \end{bmatrix}$$

The implications of an upward-sloping segment of the marginal revenue function might not be as consequential as the analysis above suggests, for at least two reasons. First, marginal cost curves often have positive slopes. Even if the marginal revenue curve slopes upward, a single profit-maximizing value of x results if the marginal cost's slope exceeds that of the marginal revenue. Furthermore, even if marginal cost is constant, the difficulty occurs only if it falls within the range indicated above. For our illustration, $mc > 3824.04$ results in $x < 10.8$, and $mc < 3797.41$ results in $x > 19.8$.

12.4 Monopolistic Competition

So far, we have analyzed two polar cases. Previous chapters considered industries that consist of many price-taking firms. The long-run behavior of such an industry is determined by the entry or exit of firms depending on profits. In the long run, no economic profits exist for the (marginal) firm(s). At the other extreme, a price searcher in an industry in which entry cannot occur must determine both price and quantity. If this market structure pertains, then the firm can earn economic profits in the long run.[5]

This section considers cases in which the firms in an industry sell products that are differentiated, so that each firm's product faces a downward-sloping demand curve. Like a pure monopolist, each of the firms must select a price/quantity pair. This choice is constrained, however, by entry or exit. In this way, the industry is like a purely competitive industry. The hybrid nature of this industry type is reflected in its name, monopolistic competition.

As in Chap. 9, we consider three cases: industries consisting of identical firms whose cost curves are independent of the number of firms in the industry (constant-cost industries), industries consisting of identical firms whose cost curves are directly affected by the number of firms in the industry (increasing-cost industries), and industries consisting of firms that have different cost curves.

In an industry that consists of price-taking firms, a constant-cost industry's long-run supply curve is horizontal. That is, the supply price is the same for all quantities. In monopolistically competitive industries, a constant cost industry's price actually *decreases* as a result of increased industry output. In increasing-cost monopolistically competitive industries with identical firms whose cost is directly related to the number of firms in the industry, the relationship between product price and industry quantity can be either direct or inverse, depending on the system's parameters.

12.4.1 A Constant-Cost Industry

The cost function in the cell below applies to each firm in a constant-cost industry. The average cost function is U-shaped, with its minimum value occurring at $x = 1$. That minimum value is $ac = 1$.[6] The marginal cost curve is $mc = x^c$, where x is the firm's output per period, and c is the elasticity of marginal cost with respect to output. This function implies the variable cost function shown in the table. We define a per-period fixed cost such that total cost equals 1 when $x = 1$. Thus, the total cost function is defined in terms of the single parameter c, as are the average cost function and the marginal cost function.[7]

[5]Exit is possible, of course. If the monopolist exits the industry, however, nothing remains to be analyzed in this partial-equilibrium framework.

[6]This specification is more general than it might appear. We can select product units such that $x = 1$ at the average cost curve's minimum point, and we can select monetary units such that the height of the curve at that point is 1.

[7]The table shows that the elasticity of the total variable cost function equals $c + 1$. The elasticity of total cost with respect to output approaches $c + 1$ from below and that the elasticity of average cost with respect to output approaches c from below.

```
assume(c+1>0)$  mc(x,c):=x^c$ VC:integrate(mc(x,c),
 x)$   tc(x,c):= ''(ratsimp(subst(x=1,1-VC)))+VC$
 ac(x,c):=''(ratsimp(tc(x,c)/x))$    matrix(["MC",
  "Variable Cost","Total Cost","Average Cost"],
  [mc(x, c), VC, ratsimp(tc(x, c)), ac(x, c)]);
```

$$\begin{bmatrix} \text{MC} & \text{Variable Cost} & \text{Total Cost} & \text{Average Cost} \\ x^c & \frac{x^{c+1}}{c+1} & \frac{x^{c+1}+c}{c+1} & \frac{x^{c+1}+c}{(c+1)\,x} \end{bmatrix}$$

Firms in this industry must neither earn economic profits nor suffer losses. This condition, due to easy entry and exit, is expressed in `zeroP`, where `P` denotes profit. The term *pavg* represents the average price in the industry, and *n* represents the number of firms. The values of last two terms have not yet been determined.

```
zeroP(n,pavg,A,a,c):=''(pavg-ac(x,c));
```

$$\text{zeroP}\,(n, pavg, A, a, c) := pavg - \frac{x^{c+1}+c}{(c+1)\,x}$$

The next cell derives the inverse demand (price) expression and the marginal revenue expression from the demand curve that faces the individual firm. We use a constant-elasticity specification. The parameters of the demand curve are these: (1) the constant A; (2) a, the response if all firms' price rise together; and (3) b, the effect on the firm's quantity demanded of a one-unit deviation of the firm's price from the industry average.

```
x(p,n,A,a,b,pavg):=A*pavg^a*(1/n-b*(p-pavg))$
solve(x(p, n,A,a,b,pavg)-x,p)$  p(x,n,A,a,b,pavg):=
  ''(rhs(%[1]))$   mr(x,n,A,a,b,pavg):=
  ''(ratsimp(diff(p(x,n,A,a,b,pavg)*x,x)))$
transpose(matrix( ["Demand","Price",
  "Marginal Revenue"],[x(p,n,A,a,b,pavg),
  p(x,n,A,a,b,pavg),mr(x,n,A,a,b,pavg)]));
```

$$\begin{bmatrix} \text{Demand} & \left(\frac{1}{n} - b\,(p - pavg)\right)\,pavg^a\,A \\ \text{Price} & \frac{\left(b\,n\,pavg^{a+1}+pavg^a\right)A-n\,x}{b\,n\,pavg^a\,A} \\ \text{[4pt]Marginal Revenue} & \frac{pavg^a\,(b\,n\,pavg+1)\,A-2\,n\,x}{b\,n\,pavg^a\,A} \end{bmatrix}$$

The next figure shows a demand curve and the associated marginal revenue curve for a set of parameter values, along with the average and marginal cost curves. This illustration does not represent equilibrium, because the ten firms in the industry earn profits.

```
[n0,A0,a0,b0,pavg0,c0]:[10,15,-0.25,0.5,1.2,0.5]$
wxdraw2d( ... Some commands omitted ...
 points([[0,1],[1,1],[1,0]]),line_type = solid,
 key="Average Cost and Marginal Cost",explicit(
 mc(x,c0),x,0,2),key="",explicit(ac(x,c0),x,0,2),
 line_width=2,key="Demand and Marginal Revenue",
 explicit(p(x,n0,A0,a0,b0,pavg0),x,0,2),key="",
explicit(mr(x,n0,A0,a0,b0,pavg0),x,0,2) )$
```

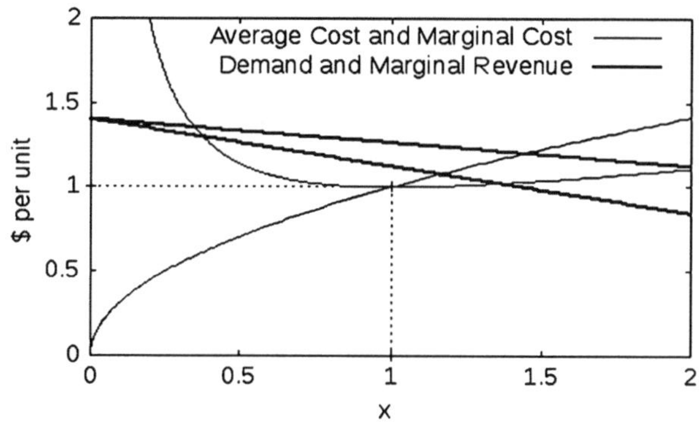

The graph suggests the second of the three conditions for equilibrium: Each firm must produce the quantity at which its marginal revenue equals its marginal cost. The expression *xeq* states that condition as an implicit equation of x and $pavg$.

```
solve(mr(x,n,A,a,b,pavg)= mc(x, c), x)$
xeq(x, n, pavg, A, a, b, c) := ''(x - rhs(%[1]));
```

$$\mathrm{xeq}\,(x,n,pavg,A,a,b,c) := \frac{b\,n\,pavg^a\,x^c\,A - b\,n\,pavg^{a+1}\,A - pavg^a\,A}{2\,n} + x$$

The third condition for equilibrium is that the individual output decisions must add up: $n \cdot x = A \cdot pavg^a$, because the market quantity demanded as a constant-elasticity function of the market quantity X. A and a are the demand curve's parameters. Alternatively stated, each identical firm produces an equal share of total output. This expression is to be solved for n, the equilibrium number of firms.

```
Xeq(x,n,pavg,A,a):=n*x-A*pavg^a;
```

$$\mathrm{Xeq}\,(x,n,pavg,A,a) := n\,x - A\,pavg^a$$

12.4.2 Equilibrium

We use `mnewton` to determine equilibrium values. The table below shows the equilibrium values of x, n, and $pavg$. It also shows the price that each firm charges and each firm's average cost, each of which must equal *pavg* if equilibrium is attained. In addition, marginal revenue and marginal cost are shown.

```
load(mnewton)$    mnewton([Xeq(x,n,pavg,A0,a0),
 xeq(x,n,pavg,A0,a0,b0,c0),
zeroP(n,pavg,A0,a0,c0)],[x,n,pavg], [1,15,1])$
[x1,n1,pavg1]:[rhs(%[1][1]),rhs(%[1][2]),
 rhs(%[1][3])]$ p1:p(x1,n1,A0,a0,b0,pavg1)$
 matrix(["x1","n1","pavg","p1","mr1","mc1","ac1"],
  [x1,n1,pavg1,p1,mr(x1,n1,A0,a0,b0,pavg1),
  mc(x1,c0),ac(x1,c0)]);
```

$$\begin{bmatrix} \text{x1} & \text{n1} & \text{pavg} & \text{p1} & \text{mr1} & \text{mc1} & \text{ac1} \\ .81392 & 18.379 & 1.011 & 1.011 & .90217 & .90217 & 1.011 \end{bmatrix}$$

12.4.2.1 Efficiency

The table above contains evidence of two sources of inefficiency. First, $p > mc$, signaling deadweight loss due to underproduction. Also, $mc < ac$ implies that economies of scale are not fully exploited.

The cell below shows the deadweight loss due to underproduction ($p > mc$) in each firm. It also shows the consumer surplus generated by each firm's production. To derive industry values, multiply the values below by $n1$.

```
xeff1:find_root(p(x,n1,A0,a0,b0,pavg1)-mc(x,c0),
 x,0.0001,2)$     dwl1:float(integrate(p(x,n1,A0,
 a0,b0,pavg1)-mc(x,c0),x,x1,xeff1))$   cs1:float(
 integrate(p(x,n1,A0,a0,b0,pavg1)-p1,x,0,x1))$
matrix(["Efficient x","cs","dwl"],[xeff1,cs1,dwl1]);
```

$$\begin{bmatrix} \text{Efficient x} & cs & dwl \\ .97818 & .044285 & .0088283 \end{bmatrix}$$

A second, new source of inefficiency is that each firm is too small to exploit economies of scale. Chapter 9 shows that in a constant-cost competitive industry, each firm is driven to produce at a quantity such that it reaches the minimum point on its long-run average cost curve. In a constant-cost monopolistically competitive industry, as the next graph demonstrates, firms cannot produce at the minimum point on their cost curve.

To determine the excess cost due to overcapacity (too many small firms), compute each firm's total cost when $x = x1$ and multiply times $n1$. By construction, the minimum-cost output is $x = 1$, at which level $ac = 1$ so the minimum total cost equals the total number of units produced: $X = n1 \cdot x1$. For the parameters used here, the cost due to undersized firms is around 1.1 % of total cost (0.164/15.123), a good bit larger than the dwl but not very large.

```
TC1:n1*ac(x1,c0)*x1$ minTC1:n1*x1$ excesscost1:
 TC1-minTC1$    matrix(["Cost", "Minimum Cost",
   "Excess Cost"],[TC1, minTC1, excesscost1]);
```

$$\begin{bmatrix} \text{Cost} & \text{Minimum Cost} & \text{Excess Cost} \\ 15.123 & 14.959 & .16442 \end{bmatrix}$$

The graph below reproduces the earlier graph but with equilibrium values of n and $pavg$ replacing the values shown in the previous graph. The vertical scale is exaggerated. The graph shows why no firm can produce at $x = 1$: The price for which each firm can sell this output rate is below the firm's average cost.

```
wxdraw2d( ... Some commands omitted ...
 points([[0,p1],[x1,p1],[x1,0]]),line_type = solid,
 key="Average Cost and Marginal Cost", explicit(
 mc(x,c0),x,0,2),key="",explicit(ac(x, c0),x,0,2),
 ... explicit(p(x,n1, A0,a0,b0, pavg1),x,0,2), ...
 explicit(mr(x,n1,A0,a0,b0,pavg1), x, 0, 2)  )$
```

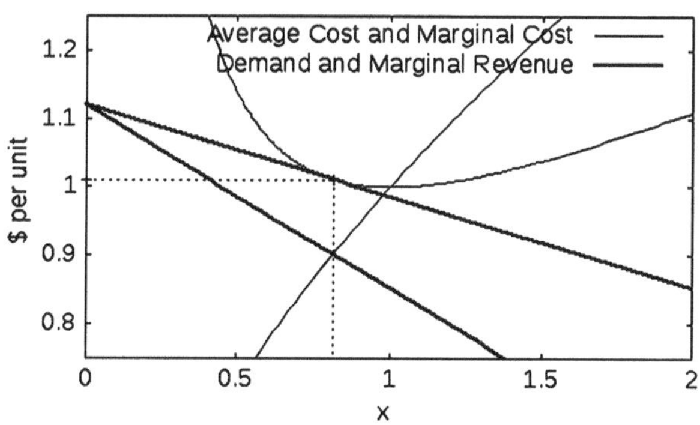

12.4.2.2 Price Elasticity of Demand for a Firm's Product

One parameter of the model is the market demand elasticity. As we should expect, the demand for the product of any individual firm is more elastic than the demand at the market level (unless $b = 0$). In the present illustration, the market demand curve has a price elasticity of $a = -0.25$; the demand elasticity for each firm's product, at equilibrium values of x and p, is -9.2906.

```
matrix(["Demand elasticity for each firm's product ="
 ,diff(x(p, n1,A0,a0,b0,pavg1),p)*p1/x1]);
```

$$[\text{Demand elasticity for each firm's product} = -9.2906]$$

The elasticity of demand for each firm's product is one reason that the increased cost due to overcapacity is so small. Each firm produces to the left of the point that yields minimum average cost, but not far to the left.

12.4.2.3 Effects of Market Size

In a perfectly competitive constant-cost industry, increased demand elicits increased output but no price change at long-run equilibrium. In its monopolistically competitive counterpart, increased demand causes price to fall in the long run. The next cell computes the equilibrium values when demand doubles ($2 \cdot A$ replaces A in all expressions). The number of firms increases from 18.379 to 33.64, less than doubling. Output per firm increases, and price decreases. As a result of the price decrease, the new industry quantity is more than twice the initial value.

```
mnewton([Xeq(x,n,pavg,2*A0,a0),xeq(x,n,pavg,2*A0,
  a0,b0,c0),zeroP(n,pavg,2*A0,a0,c0)],[x,n,pavg],
  [1,30,1])$ [x2,n2,pavg2]:[rhs(%[1][1]),
  rhs(%[1][2]),rhs(%[1][3])]$
  p2:p(x2,n2,2*A0,a0,b0,pavg2)$   matrix(["x1",
  "x2","p1","p2","n1","n2"],[x1,x2,p1,p2,n1,n2]);
```

$$\begin{bmatrix} \text{x1} & \text{x2} & \text{p1} & \text{p2} & \text{n1} & \text{n2} \\ .81392 & .89103 & 1.011 & 1.0034 & 18.379 & 33.64 \end{bmatrix}$$

This analysis implies that removing trade barriers can enhance efficiency even if the trading countries have similar technologies and resource endowments. As Krugman and Obstfeld [2, p. 125] report, "Each country can [produce] a narrower range of products..., yet by buying goods that it does not make from other countries, each ... [increases] the variety of goods available to its consumers." The example above would coincide with two identical countries that open to trade. As a result, the number of producers in each country falls to $\frac{33.64}{2}$ but each country's consumers have access to the products of 33.64 firms rather than those of the 18.379 firms that would exist in each country in the absence of trade. Further, as noted, the total number of units consumed increases and the average price level falls.

Exercises: (1) Solve the system for $a = 0$; report and discuss the changes in pertinent values. (2) Solve the system for $a = -5.0$ or some other large value for the elasticity; report and discuss the changes in pertinent values. (3) Examine and discuss implications of changes in c. (4) Solve the system for $b = 0$. What does this value imply about the degree to which the firms' products are substitutes for each other? (5) Suppose that the industry is operated by a single seller. This seller can treat each of the erstwhile firms as a plant. Explain the following: The single seller's long-run average cost curve is scalloped with minimum points $ac = 1$ at integer values of its output, and its marginal cost curve is a discontinuous series of segments. Given a value of c, determine the minimum price and maximum price that such a firm would charge, assuming that it can sell fractional units of its product.

12.4.3 An Increasing-Cost Industry

As with competitive industries, the number of firms can affect the cost of member firms. This section continues to consider an industry composed of identical firms, but each firm's cost curves reflect the number of firms. Specifically, we multiply the marginal cost curve by the term n^e where n is the number of firms and e is the elasticity of response of the marginal cost with respect to n. The total cost function is also modified, so that the total cost at $x = 1$ is n^e, rather than 1. Otherwise, the analysis proceeds as before.

```
kill(all)$assume(c+1>0)$    mc(x,c,e,n):=(x^c)*n^e$
VC: integrate(mc(x,c,e,n), x)$   tc(x,c,e,n) :=
 ''(ratsimp(subst(x=1,n^e - VC)) + VC)$
ac(x,c,e,n):=''(ratsimp(tc(x,c,e,n)/x))$
matrix( ["MC","Variable Cost","Total Cost",
 "Average Cost"],[mc(x,c,e,n), VC,
 ratsimp(tc(x,c,e,n)), ac(x,c,e,n)]);
```

$$\begin{bmatrix} \text{MC} & \text{Variable Cost} & \text{Total Cost} & \text{Average Cost} \\ n^e\,x^c & \frac{n^e\,x^{c+1}}{c+1} & \frac{n^e\,x^{c+1}+c\,n^e}{c+1} & \frac{n^e\,x^{c+1}+c\,n^e}{(c+1)\,x} \end{bmatrix}$$

Firms in this industry must neither earn economic profits nor suffer losses. This condition, due to easy entry and exit, is expressed in `zeroP`, where `P` denotes profit.

```
zeroP(n,pavg,A,a,c,e):=''(pavg-ac(x,c,e,n));
```

$$\text{zeroP}\,(n, pavg, A, a, c, e) := pavg - \frac{n^e\,x^{c+1}+c\,n^e}{(c+1)\,x}$$

The next cell derives the inverse demand (price) expression and the marginal revenue expression from the demand curve that faces the individual firm. We use the same specification as in the case of constant-cost industries.

```
x(p,n,A,a,b,pavg):=A*pavg^a*(1/n-b*(p-pavg))$ solve
 (x(p,n,A,a,b,pavg)-x, p)$  p(x,n,A,a,b,pavg):=
 ''(rhs(%[1]))$ mr(x,n,A,a,b,pavg):=
 ''(ratsimp(diff(p(x,n,A,a,b,pavg)*x,x)))$
transpose(matrix(["Demand","Price",
 "Marginal Revenue"],[x(p,n,A,a,b,pavg),
 p(x,n,A,a,b,pavg), mr(x,n,A,a,b,pavg)]));
```

$$\begin{bmatrix} \text{Demand} & \left(\frac{1}{n} - b\,(p - pavg)\right)\,pavg^a\,A \\ \text{Price} & \frac{\left(b\,n\,pavg^{a+1} + pavg^a\right)A - n\,x}{b\,n\,pavg^a\,A} \\ \text{Marginal Revenue} & \frac{pavg^a\,(b\,n\,pavg+1)\,A - 2\,n\,x}{b\,n\,pavg^a\,A} \end{bmatrix}$$

The next figure shows a demand curve and the associated marginal revenue curve for a set of parameter values, along with the average and marginal cost curves. This illustration does not represent equilibrium, because the ten firms in the industry earn profits. This is like its counterpart in the preceding section, except that the per-unit cost curves are higher.

```
[n0,A0,a0,b0,pavg0,c0,e0]:
[10,15,-0.25,0.5,1.2,0.5,0.05]$  wxdraw2d(
  ... Some commands omitted ...
  points([[0,1],[1,1],[1,0]]),line_type=solid,key=
 "Average Cost and Marginal Cost",explicit(mc(x,c0,
 e0,n0),x,0,2),key = "",explicit(ac(x,c0,e0,n0),
 x,0,2),line_width=2,key =
 "Demand and Marginal Revenue", ...) )$
```

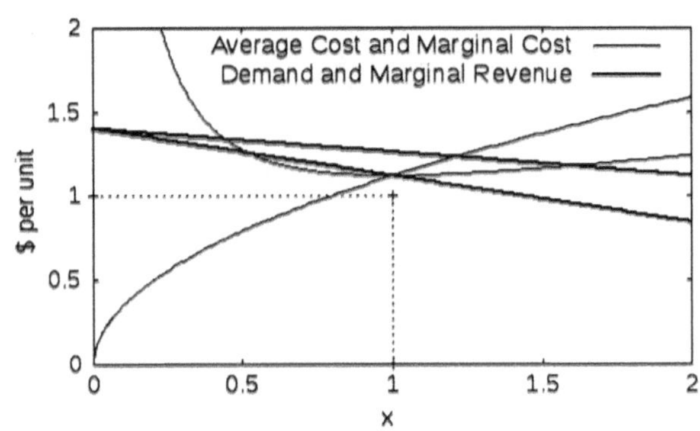

As before, each firm must produce the quantity at which its marginal revenue equals its marginal cost. The expression *xeq* states that condition as an implicit equation of x and $pavg$.

```
solve(mr(x,n,A,a,b,pavg)= mc(x,c,e,n), x)$
xeq(x,n,pavg,A,a,b,c,e)  := ''(x - rhs(%[1]));
```

$$\mathrm{xeq}\,(x, n, pavg, A, a, b, c, e) := \frac{b\,n^{e+1}\,pavg^a\,x^c\,A - b\,n\,pavg^{a+1}\,A - pavg^a\,A}{2\,n} + x$$

The third condition for equilibrium is identical to its counterpart in the preceding section.

```
Xeq(x, n, pavg, A, a,b,c,e)  := ''(n*x - A*pavg^a);
```

$$(\%o17)\ \mathrm{Xeq}\,(x, n, pavg, A, a, b, c, e) := n\,x - pavg^a\,A$$

12.4.4 Equilibrium

As before, we cannot derive expressions for the equilibrium values, but `mnewton` can determine their values. The table below shows the equilibrium values of x, n, and $pavg$. It also shows the price that each firm charges and each firm's average cost, each of which must equal $pavg$ if equilibrium is attained. In addition, marginal revenue and marginal cost are shown.

```
mnewton([Xeq(x,n,pavg,A0, a0,b0,c0,e0),
  xeq(x,n,pavg,A0,a0,b0,c0,e0),zeroP(n,pavg,
  A0,a0,c0,e0)],[x,n,pavg],[1,15,1] )$
[x1,n1,pavg1]:[rhs(%[1][1]),rhs(%[1][2]),
 rhs(%[1][3])]$  p1:p(x1,n1,A0,a0,b0,pavg1)$
matrix(["x1","n1","pavg","p1","mr1","mc1",
"ac1"],[x1,n1,pavg1,p1,mr(x1,n1,A0,a0,b0,
  pavg1), mc(x1,c0,e0,n1),ac(x1,c0,e0,n1)]);
```

$$\begin{bmatrix} \text{x1} & \text{n1} & pavg & \text{p1} & \text{mr1} & mc1 & \text{ac1} \\ .82782 & 17.444 & 1.1643 & 1.1643 & 1.0497 & 1.0497 & 1.1643 \end{bmatrix}$$

12.4.4.1 Efficiency

The cell below shows the deadweight loss due to underproduction ($p > mc$) in each firm. It also shows the consumer surplus generated by each firm's production. To derive industry values, multiply the values below by $n1$.

```
xeff1:find_root(p(x,n1,A0,a0,b0,pavg1)-mc(x,c0,
 e0,n1),x,0.0001,2)$    dwl1:float(integrate(p(
  x,n1,A0,a0,b0,pavg1)-mc(x,c0,e0,n1),x,x1,xeff1))$
cs1:float(integrate(p(x,n1,A0,a0,b0,pavg1)-p1,
 x,0,x1))$   matrix(["Efficient x","cs","dwl"],
 [xeff1,cs1,dwl1]);
```

$$\begin{bmatrix} \text{Efficient x} & cs & dwl \\ .98161 & .047457 & .0087148 \end{bmatrix}$$

The inefficiency due to overcapacity is similar to that in the constant-cost illustration above. Assign each of `n1` firms an output of 1 unit. Then total cost equal `n1` times 1.1537 (the cost per unit when each firm produces $x = 1$). As before, given the current parameters, it is quite large when compared to the deadweight loss above but a small fraction of total cost.

```
TC1:n1*ac(x1,c0,e0,n1)$    minTC1:n1*tc(1,c0,e0,n1)$
excesscost1:TC1-minTC1$  matrix(["Cost","Minimum
 Cost","Excess Cost"],[TC1, minTC1, excesscost1]);
```

$$\begin{bmatrix} \text{Cost} & \text{Minimum Cost} & \text{Excess Cost} \\ 20.31 & 20.124 & .18571 \end{bmatrix}$$

12.4.4.2 Effects of Market Size

Unlike constant-cost industries, increased market size can drive up the price in increasing-cost industries. Shifting the demand curve rightward by replacing `A0` with `2*A0` results in both an increased number of firms and more output, but the price rises slightly.

```
mnewton([Xeq(x,n,pavg,2*A0,a0,b0,c0,e0),
 xeq(x,n,pavg,2*A0,a0,b0,c0,e0),
 zeroP(n,pavg,2*A0,a0,c0,e0)],[x,n,pavg],[1,30,1])$
[x2,n2,pavg2]:[rhs(%[1][1]),rhs(%[1][2]),
 rhs(%[1][3])]$p2:p(x2,n2,2*A0,a0,b0,pavg2)$
matrix(["x1","x2","p1","p2","n1","n2"],
[x1,x2,p1,p2,n1,n2]);
```

$$\begin{bmatrix} \text{x1} & \text{x2} & \text{p1} & \text{p2} & \text{n1} & \text{n2} \\ .82782 & .90218 & 1.1643 & 1.1921 & 17.444 & 31.824 \end{bmatrix}$$

12.4.5 Firm-Specific Cost

So far, all firms have identical cost curves. We now let each firm have a unique cost function, although all functions are in the same family. The cost functions are constructed much as those in the previous section with the notation i replacing n. In the earlier case, each of n firms has the same cost functions, which depend on n's value. Here, each firm, from $i = 1$ to $i = n$, has its own cost functions.

The difference among firms' cost functions introduces a third source of inefficiency: firms' marginal costs may no longer be equal to each other. This would imply that even if the number of firms were optimal, total cost is higher than it would be if a given output could be distributed among firms in amounts such that the value of marginal cost is the same in all firms. As before, the output of the marginal firm(s) is too small to exploit economies of scale, so the problem of excess capacity is found in this case as well.

We find, however, that both the cost due to unequal marginal costs and the cost due to excess capacity are quite small, given our model's parameters. The increased cost of the firms that are near the margin cause them to reduce output to an extent that their marginal costs do not differ greatly from those of more efficient firms. Furthermore, the demand curve becomes sufficiently elastic in their ranges of production that the equilibrium average cost is quite close to the minimum average cost.

```
assume(c+1>0)$     mc(x,c,e,i):=x^c*i^e$   VC:
 integrate(mc(x,c,e,i),x)$    FC: subst(x=1,i^e-VC)$
tc(x,c,e,i):=''(radcan(FC+VC))$    ac(x,c,e,i):=  ''
 (radcan(tc(x,c,e,i)/x))$   matrix(["Marginal Cost",
  "Variable Cost","Total Cost","Average Cost"],
    [mc(x,c,e,i), VC, tc(x,c,e,i), ac(x,c,e,i)]);
```

$$\begin{bmatrix} \text{Marginal Cost} & \text{Variable Cost} & \text{Total Cost} & \text{Average Cost} \\ i^e\, x^c & \frac{i^e\, x^{c+1}}{c+1} & \frac{i^e\, x^{c+1}+c\, i^e}{c+1} & \frac{i^e\, x^{c+1}+c\, i^e}{(c+1)\, x} \end{bmatrix}$$

The demand and revenue functions are the same as in previous sections. This specification is not quite appropriate here: We determine below that as i increases, firms' output decreases. That is, higher-cost firms produce less and sell their products for a higher price. Thus, we might expect the term $1/n$ to overstate the effect of the number of firms on each firm's demand. We keep the approximation used here because it is tractable.

```
x(p, i,A,a,b,pavg,n):=A*pavg^a*(1/n - b*(p-pavg));
solve(x(p,i,A,a,b,pavg,n)-x,p)$
p(x,i,A,a,b,pavg,n):=''(rhs(%[1]));
mr(x,i,A,a,b,pavg,n):=''(ratsimp(diff(
  p(x,i,A,a,b,pavg,n)*x,x)));
```

$$\mathrm{x}\,(p,i,A,a,b,pavg,n) := A\, pavg^a\, \left(\tfrac{1}{n} - b\,(p-pavg)\right)$$

$$\mathrm{p}\,(x,i,A,a,b,pavg,n) := \frac{\left(b\,n\,pavg^{a+1}+pavg^a\right)A - n\,x}{b\,n\,pavg^a\,A}$$

$$\mathrm{mr}\,(x,i,A,a,b,pavg,n) := \frac{pavg^a\,(b\,n\,pavg+1)\,A - 2\,n\,x}{b\,n\,pavg^a\,A}$$

The next cell derives expressions for the profit-maximizing quantity of each firm. The expression `exprxi` is a general statement of the ith firm's profit-maximizing output as a function of the system's parameters and variables. The name `xeq` is bound to the condition that x equals the specified quantity. This is expressed as an implicit function because an explicit function cannot be derived. The equilibrium condition is for the marginal firm. We derive the other firm's output rates after equilibrium values for the system are defined.

The output line shows the equilibrium condition given the system's parameters. It is used in the application of `mnewton` below. Simulations should place some limits on two of the parameters. Values of $c0$ should be less than 0.5. Also $e0$ should be kept within the range $0.1 \le e0 \le 0.25$. Values outside these ranges can cause the `mnewton` procedure to fail to converge.

```
xeq: solve(mr(x,i,A,a,b,pavg,n)=mc(x,c,e,i),x)[1];
xeq0:subst([A=A0,a=a0,b=b0,c=c0,e=e0, m=m0,i=n],xeq);
exprxi:rhs(xeq);
```

$$x = -\frac{b\, i^e\, n\, pavg^a\, x^c\, A - b\, n\, pavg^{a+1}\, A - pavg^a\, A}{2\, n}$$

$$x = -\frac{\frac{15.0\, n^{1.1}\, x^{0.5}}{pavg^{0.25}} - 15.0\, n\, pavg^{0.75} - \frac{150}{pavg^{0.25}}}{2\, n}$$

$$-\frac{b\, i^e\, n\, pavg^a\, x^c\, A - b\, n\, pavg^{a+1}\, A - pavg^a\, A}{2\, n}$$

As in the previous sections, we derive the condition for the number of firms by setting average cost equal to price, but now the average cost of the marginal firm must equal price. The average costs of other firms will be less than the price. As with x, we derive a general expression and then apply the system's parameters to determine an expression to which numerical analysis can be applied. The output below shows only the expression stated in terms of the system's parameters.

```
solnn: solve(ac(x,c,e,n)-p(x,n,A,a,b,pavg,n),n)$
nexpr : rhs(solnn[1])$
neq0:n-(subst([A=A0,a=a0,b=b0,c=c0,e=e0],nexpr));
```

$$n - \frac{0.1\, n^{1.1}\left(\frac{150\, x^{1.5}}{pavg^{0.25}} + \frac{75.0}{pavg^{0.25}}\right) - \frac{225.0\, x}{pavg^{0.25}}}{22.5\, pavg^{0.75}\, x - 1.5\, x^2}$$

Finally, we derive an expression for equilibrium industry output, X. This expression yields a third equation, all expressed in terms of these three variables: the average price ($pavg$), the number of firms (n), and the marginal firm's output (x). Given solutions for $pavg$ and n, we can determine the output rates, prices, and related values for the inframarginal firms. Among other things, we find that some firms in this industry operate beyond the range of decreasing average cost.

To derive an expression for X integrate the expression for firms' output over i from $i = 1$ to $i = n$. Set the resulting expression equal to $A \cdot p^a$. The output is suppressed because the expressions are too long to fit on a page. They involve the same three variables as the preceding two equations along with the system's parameters.

```
assume(n>1, e >0)$  exprX:integrate(exprxi,i,1,n)$
Xeq:A*pavg^a - %$
Xeq0: subst([A=A0,a=a0,b=b0,c=c0,e=e0],Xeq)$
```

We now establish the required equilibrium values. The average price is 1.4844, and the number of firms is 76.949. The marginal firm ($i = 76.946$) produces 0.918 units of output.

```
solnpnx:mnewton( [neq0,Xeq0,xeq0], [pavg,n,x],
 [1.0,1/e0,.8]);pavg1: rhs(solnpnx[1][1])$
n1: rhs(solnpnx[1][2])$ xn1: rhs(solnpnx[1][3])$
```

$$[[pavg = 1.4844, n = 76.949, x = .91797]]$$

The next cell determines the equilibrium output level of the ith firm given the system's parameters. The first command returns an expression for the equilibrium value that is equivalent to the implicit function $x = -0.0065 * (1045.7 * i^0.1 * x^0.5 - 1688.1)$. The function $f(i)$ uses the `find_root` command to determine the requisite x value, given i. The variable $n1f$ is created by using the `floor` command to round $n1$ to the next lowest integer. A list of x values for all integer values of i is created in two steps: A list of i values is created, and $f(i)$ is mapped onto that list. The table reports four of these values.

```
xeqi:subst([A=A0,a=a0,b=b0,c=c0,e=e0,n=n1,pavg=
  pavg1], xeq)$    f(i):=find_root(xeqi,x,0.01,100)$
n1f: floor(n1)$    ilist:makelist(i,i,1,n1f)$
xlist: map(f, ilist)$  matrix([ilist[1],xlist[1]],
 [ilist[2],xlist[2]],["...","..."],[ilist[n1f-1],
 xlist[n1f-1]],[ilist[n1f],xlist[n1f]]);
```

$$\begin{bmatrix} 1 & 1.815 \\ 2 & 1.6407 \\ ... & ... \\ 75 & .92196 \\ 76 & .91989 \end{bmatrix}$$

We determine total output `X1` and the associated total cost `TC1` by using the `sum` command for all integer values of i and then adding this sum to the values for the marginal firm.

```
X1:sum(xlist[i],i,1,n1f)+xn1$ TClist:tc(xlist,c0,e0
 ,ilist)$TC1:tc(xn1,c0,e0,n1)+sum(TClist[i],i,1,n1f)$
matrix(["X1", "TC1"],[X1, TC1]);
```

$$\begin{bmatrix} X1 & TC1 \\ 83.695 & 117.06 \end{bmatrix}$$

The graph below the table shows the marginal firm's output and associated cost and revenue curves. The graph points to two important facts: At their selected output levels (where mc intersects mr) marginal cost values are close to each other, and the equilibrium quantity of the marginal firm is near the quantity at which average cost reaches its minimum value.

```
wxdraw2d( ... Some commands omitted ... explicit(
 p(x,1,A0,a0,b0,pavg1,n1),x,.1,1.25*xlist[1]),
 key="",explicit(mr(x,1,A0,a0,b0,pavg1,n1),x,.1,
  1.25*xlist[1]),   ...   explicit(
 ac(x,c0,e0,1),x,.1,1.25*xlist[1]),key="",explicit(
 mc(x,c0,e0,1),x,.1,1.25*xlist[1]),  ...
 explicit(ac(x,c0,e0,n1),x,.1,2),
 key ="", explicit(mc(x,c0,e0,n1),x,.1,2))$
```

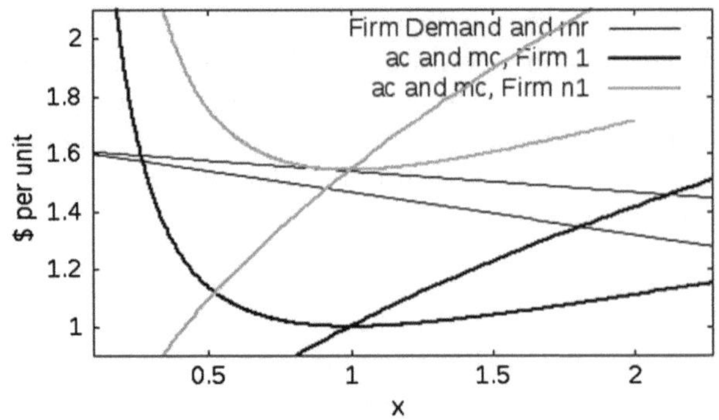

Information regarding prices can be derived directly. The table below shows the expression for the price of a firm's product in terms of the firm's output. It also shows the price charged by the lowest-cost firm and by the marginal firm. The former is lower because the firm with the lowest marginal cost produces more and sells at a lower price. A graph like the one above could be drawn for prices, but the fact that all firms face the same demand curve implies that its shape would mirror the shape in the preceding graph, providing no additional insight.

```
price :p(x,i,A0,a0,b0,pavg1,n1)$ pi1:p(xlist[1],i,
  A0,a0,b0, pavg1,n1)$ pi1n1:p(xn1,i,A0,a0,b0,pavg1,
  n1)$  float(matrix(["Firm Price","Firm 1 Price",
 "Firm n1 Price"],[price,pi1,pi1n1]));
```

$$\begin{bmatrix} \text{Firm Price} & \text{Firm 1 Price} & \text{Firm n1 Price} \\ 9.56298\,10^{-4}\,(1688.1 - 76.949\,x) & 1.4808 & 1.5468 \end{bmatrix}$$

The demand for the marginal firm's product is quite elastic (Epn1 $= -22.898$) even though the market demand is quite inelastic ($a = -0.25$). The large elasticity value accounts for the rather small difference between xn1 (0.918 units) and the cost minimizing output for the marginal firm (1 unit).

```
Epn1: diff(x(p,n1,A0,a0,b0,pavg1,n1),p)*
      pi1n1/xn1;
```

$$-22.898$$

12.4.6 Economic Efficiency

As before, this industry is subject to allocative inefficiency because each firm sells a quantity less than the quantity at which $p = mc$. The industry is also subject to productive inefficiency. As before, the industry has too many firms, in the sense that output could be allocated to fewer, smaller firms with resulting saving. In addition to this source of inefficiency, which we encountered in the previous two sections, this industry in which firms' cost curves differ suffers potential productive inefficiency from a second source, the individual firms do not select quantities such that marginal costs are equalized across firms.

In this section, we do not deal with allocative inefficiency, for two reasons. First, the preceding sections treat these issues. Second, as noted previously, the specification for individual firms' demand curves is only approximately appropriate, so the measures of allocative inefficiency would also be only approximately indicative of the values that the system's parameters should imply.

12.4.6.1 Inter-firm Output Distribution

We address the question of productive efficiency (least-cost production) in two steps. We first consider the effect on total cost of redistributing the equilibrium output among the $n1$ firms in quantities such that each firm has the same marginal cost. Then we allow the number of firms to vary in order to determine the saving from reducing the number of firms.

The cell below computes total cost at equilibrium. Its value is compared to those below in order to determine the savings from equalizing marginal cost and reducing the number of firms. The `romberg` numerical method replaces `integrate` because the latter failed to compute a value for the integral of the total cost function.

12.4.6.2 Equalizing Marginal Costs

The cell below defines a variable MC that is the common marginal cost of all firms. Each firm's marginal cost is set equal to MC, the value of which is to be determined. The resulting output function $xiMC$ is defined. The result is an expression stated in terms of i. The output functions are summed from $i = 1$ to $i = n1f$ and the sum is added to the marginal firms output, generating an expression that states X in terms of MC. Solving $X =$ `X1` yields the value $MC = 1.185$.

```
assume(MC>0, i>1)$   solve(mc(x,c0,e0,i)-MC, x)$
xMC(i,MC):=''(rhs(%[1]));  xMClist:xMC(ilist,MC)$
XMC:float(xMC(n1,MC)+sum(xMC(i,MC),i,1,n1f))$ matrix(
  ["Industry output as function of common mc =",XMC]);
```

$$\text{xMC}(i, MC) := \frac{MC^2}{i^{\frac{1}{5}}}$$

$$\left[\text{Industry output as function of common mc} = \;\; 39.85\, MC^2\right]$$

Solving $XMC =$ `X1` yields the value $MC = 1.4492$, as is shown below. As a check, the cell below also shows that this MC value does result in $X =$ `X1`.

The remainder of the cell generates values of x for selected i values. The second column shows the equilibrium values of x, and the third column shows the costs if equal marginal costs are imposed. As expected, equalizing marginal cost results in redistributing output allotments from high-cost firms to lower-cost firms.

```
MC1:find_root(XMC-X1,MC,0.1,10)$ matrix(
  ["MC1=",MC1]);  xMClist2:float(
  subst(MC=MC1,xMClist))$
Xcheck:float(xMC(n1,MC1)+sum(xMC(i,MC1),i,1,n1f))$
matrix(["Xcheck =", Xcheck, "=", X1, "= X1"]);
numbers1:[ilist[1], xlist[1], xMClist2[1] ]$
numbers2:[ilist[2], xlist[2], xMClist2[2] ]$
numbers_2:[ilist[n1f-2],xlist[n1f-2],
 xMClist2[n1f-2]]$ numbers_1:[ilist[n1f-1],
 xlist[n1f-1],xMClist2[n1f-1]]$numbers_0:
  [ilist[n1f],xlist[n1f],xMClist2[n1f] ]$
numbers: [n1, xn1, xMC(n1,MC1) ]$
```

```
matrix(["i","x","xMC"],numbers1,numbers2,["...",
 "...","..."],numbers_2,numbers_1,numbers_0,numbers);
```

$$[\mathrm{MC1} = 1.4492] \qquad [\mathrm{Xcheck} = 83.695 = 83.695 = \mathrm{X1}]$$

$$\begin{bmatrix} i & x & xMC \\ 1 & 1.815 & 2.1003 \\ 2 & 1.6407 & 1.8284 \\ \ldots & \ldots & \ldots \\ 74 & .92405 & .88803 \\ 75 & .92196 & .88565 \\ 76 & .91989 & 0.8833 \\ 76.949 & .91797 & .88111 \end{bmatrix}$$

Finally, we determine the saving from output redistribution by computing a second total cost function. The result is that total cost is reduced but only by about 0.01%.

```
tclist: tc(xlist,c0,e0,ilist)$    TC2 :
  tc(xMC(n1,MC1),c0,e0,n1)+sum(tclist[i],i,1,n1f)$
matrix(["TC2", "TC1", "Percent Change"],
  [TC2, TC1, 50*(TC2-TC1)/(TC1+TC2)]);
```

$$\begin{bmatrix} \mathrm{TC2} & \mathrm{TC1} & \text{Percent Change} \\ 117.0 & 117.06 & -.011527 \end{bmatrix}$$

12.4.6.3 Reducing the Number of Firms

In previous sections, we identified overcapacity as a source of inefficiency in monopolistically competitive industries. Here we consider the implications of varying cost functions for the added cost due to overcapacity. We reduce the number of firms, equalize the marginal cost among these firms and produce the equilibrium quantity in the new configuration.

As before, we define a new variable for the common marginal cost. MCI is the marginal cost given the number of firms, I, where $I \leq n1f$. We use a `block` that allows the construction for an MCI function and value for each I. The MCI values imply a set of x values and a related set of total cost values for firms. The firms' total cost values are summed to generate the industry total cost for the specific I.[8]

The output consists of three parts. The first shows the function for the ith firm's output, given the number of the firm and the implied marginal cost. The second shows the previously computed total cost values, for comparison. The last output line shows total cost for $I = nif$ and for $I = n1f - 1$. Given this system's parameters, removing the marginal firm and reducing I to the next highest integer value reduces total cost, albeit very slightly. Any further reduction in the number of firms, however, causes total cost to increase. Furthermore, as with equalizing marginal cost, optimizing the number for firms affects industry cost only very slightly.

```
assume(I>1,MCI>0,i>0)$    solve(mc(x,c0,e0,i)=MCI,x)$
xMCI(i, MCI):=''(rhs(%[1]));
/*start block*/  F(I)  :=  block(
 Ilist:makelist(i,i,1,I),XMCIlist:xMCI(Ilist,MCI),
 XMCI:sum(xMCI(i,MCI),i,1,I),
 MCI1: find_root(XMCI-X1,MCI,0,10), xMCI1list:
 xMCI(Ilist, MCI1),TC3:sum(tc(xMCI(i,MCI1),
 c0,e0,i),i,1,n1))$ /*end block*/  matrix(["TC1 =",
 TC1, "and TC2 =", TC2]);  matrix(["TC_n1f = ",
 F(floor(n1))," and TC_n1f-1 =",F(floor(n1)-1)]);
```

$$\text{xMCI}(i, MCI) := \frac{MCI^2}{i^{\frac{1}{5}}}$$

$$\left[\text{TC1} = \ 117.06 \text{ and TC2} = \ 117.0\right]$$

$$\left[\text{TC\_n1f} = \ 116.92 \text{ and TC\_n1f-1} = \ 118.24\right]$$

```
TCMCIlist:map(F,ilist)$    wxdraw2d(
 ... Some commands omitted ...
point_type=circle, points(ilist,TCMCIlist))$
```

[8]One line of input creates a list of x values. We do not use that list, but *Maxima* retains the values.

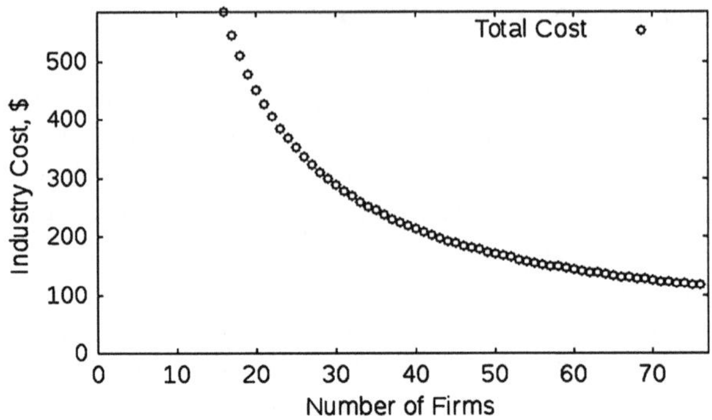

References

1. Bishop RL (1968) The effects of specific and ad valorem taxes. Q J Econ 82:198–218
2. Krugman PR, Obstfeld M (2004) International economics: theory and policy, 8th edn. Prentice-Hall, Saddle River
3. Wainwright K, Chiang AC (2004) Fundamental methods of mathematical economics, 4th edn. McGraw-Hill, New York

Chapter 13
Nonlinear Pricing: Capturing Consumer Surplus

Abstract Chapter 12 shows how a price-searching firm maximizes profits if its product faces a well-defined downward-sloping demand curve, given that the firm must sell all units of its product for the same price and that the per-unit price is the only charge that the seller can impose on the product's buyers. This pricing regime is often called *linear pricing*. This chapter analyzes deviations from linear pricing, which potentially increase the seller's profit.

13.1 Review and Introduction

We begin with a review of linear pricing, using the following simple representation of the relationships that dictate a price-taking firm's behavior. The firm's product faces a constant-elasticity demand curve, and its marginal cost curve is linear.

```
p(x,A,a):=(x/A)^(1/a)$
mr(x,A,a):=(1+1/a)*(x/A)^(1/a)$
mc(x,C,c):=C+c*x$  [A0,a0,C0,c0]:[5000,-1.5,5,2]$
```

The `find_root` command yields the profit-maximizing quantity and the quantity at which the inverse demand curve intersects the marginal cost curve. The prices at these two quantities are determined.

```
x0:find_root(mr(x,A0,a0)-mc(x,C0,c0),x,.01,200)$
p0:p(x0,A0,a0)$
xe0: find_root(p(x,A0,a0)-mc(x,C0,c0),x,.01,200)$
pe0:p(xe0,A0,a0)$
matrix(["x0","p0","xe0","pe0"],[x0,p0,xe0,pe0]);
```

$$\begin{bmatrix} \text{x0} & \text{p0} & \text{xe0} & \text{pe0} \\ 8.8717 & 68.23 & 18.444 & 41.887 \end{bmatrix}$$

M.R. Hammock and J.W. Mixon, *Microeconomic Theory and Computation: Applying the Maxima Open-Source Computer Algebra System*,
DOI 10.1007/978-1-4614-9417-1_13,

The graph below shows the profit-maximizing price and quantity for the price-searching firm that practices linear pricing. It also shows the areas designated by the labels "Consumer Surplus" and "Deadweight Loss." From the seller's viewpoint, these represent unexploited surpluses: Consumer Surplus is an amount that consumers would be willing and able to pay without being worse off than if they were not to buy x0 units of the seller's product, and Deadweight Loss is a surplus of value over cost that the seller does not gain.

```
wxdraw2d ( ... Some commands omitted ...
 label_alignment=left,label(["Consumer Surplus",
 0, 1.1*p0]), label(  ["Deadweight Loss", x0, pe0]),
 points_joined=true, line_type=dots,points(
 [[0, p0], [x0,p0],[x0,0]]),line_type=solid,key=
 "Marginal Revenue",explicit(mr(x,A0,a0),x,
 0,1.25*xe0),  ... explicit(p(x, A0, a0),
 x, 0, 1.25*xe0),  ... explicit(mc(x,C0,c0),x,0,
 1.25*xe0),title= "Unexploited Values")$
```

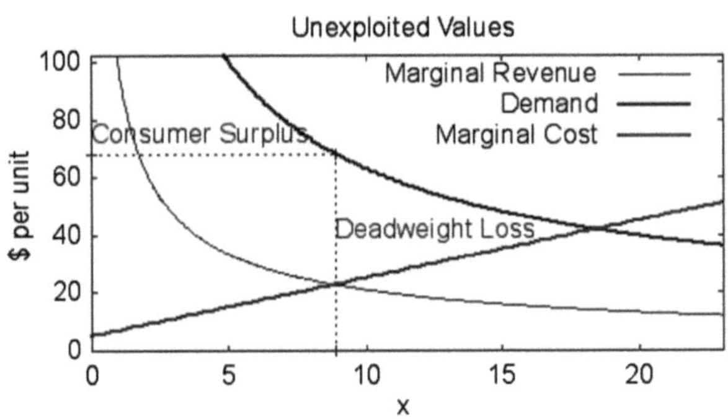

13.1.1 Pricing Mechanisms for Capturing Surplus

The remainder of this chapter shows how a producer might use more complex pricing schemes to appropriate some of these unexploited surpluses. We consider the implications of these schemes for the seller's profits and for the impacts on buyers. Use of these schemes depends on market segmentation or nonlinear pricing (or a combination of the two).

- *Market segmentation.* If markets are segments (or can be segmented by the seller's actions), price differences among the segments can increase profits. Markets are often segmented by geography. Also, sellers can often find ways to segment markets by buyer characteristics: commercial buyers often pay different prices than their noncommercial counterparts, seniors and students often pay lower prices, careful shoppers gain discounts by clipping coupons or by shopping during sales periods.

- *Nonlinear pricing*. The linear pricing scheme yields this revenue function of the quantity sold to any single buyer: $TR = p \cdot x$. The following *nonlinear* revenue function might be feasible: $TR = \texttt{R0} + (\texttt{p0} \cdot x) + (\Delta p \cdot (x - \texttt{x0}))$. If $\texttt{R0} > 0$, then buyers pay for the option to buy any units of the product, as well as paying a per-unit price. Amusement parks often charge a per-day fee and then a per-ride price. If $\Delta p < 0$, a buyer pays p0 per-unit for any number of units up to x0 and a lower price $(\texttt{p0} + \Delta p)$ for all subsequent units.

13.1.2 Conditions for Capturing Surplus

Three conditions must be met for a firm to add to its profits by using either of these mechanisms.[1] First, the firm must be able to set its price within a range dictated by its cost and buyers' willingness to pay. A price-taking firm sells at a price dictated by the market; a price-searching firm can select the price or set of prices for which it will sell its product.

The second condition is twofold. First, differences must exist in buyers' willingness to pay. These differences may be across buyers or there may be differences in the values that a single buyer places on the unit(s) being sold. Second, the seller must have some information about these differences, so that the seller can devise a pricing mechanism that will add to its profits.

The final condition is that the product must be sold to the appropriate buyer, and resale must be precluded. The firm must be able to identify buyers by type or to induce buyers to self-identify. Also, buyers of low-price items must not be able to resell those items. That is, arbitrage must not take place. This is one reason that many examples of deviations from linear pricing, like movie ticket pricing and airline fares, involve services.

The existence of some arbitrage activities does not preclude gains from nonlinear pricing. With online buying, for example, it becomes easier for buyers of textbooks to find books that were sold for a low price in a foreign market and to import those books. Such opportunities reduce the advantages of deviating from a simple single-price approach but do not remove them.

13.1.3 Market Segmentation

A seller might segment its markets by identifiable characteristics of either the product or the buyer. Traditionally, pricing schemes based on segmentation are called *price discrimination*. Three types of price discrimination are considered.

[1]Similar conditions must hold for other pricing strategies. These strategies include tying or bundling and the use of quality differences to facilitate market segmentation.

- *First-degree price discrimination* occurs when the seller places a separate price on each unit sold, whether multiple units are sold to individual buyers or each of many individual buyers purchases a single unit.
- *Second-degree price discrimination* occurs when the seller knows that more than one class of buyer exists but lacks information about individuals or lacks the ability to use such information, perhaps due to legal restrictions. The seller offers all buyers the same pricing schedule but designs the schedule in a way that encourages different consumers to opt for different menu selections.
- *Third-degree price discrimination* occurs when the markets for the seller's product are separated by factors like geography or time of the day or year, or can be separated by actions that the seller takes. In these cases, the segments are identifiable: the seller knows whether the unit is being sold in Albania or Bulgaria, or whether it is being sold at lunch time or in the evening, or whether the buyer is a senior citizen. Third-degree price discrimination adds to profits when price elasticities of demand differ across segments.

13.2 First-Degree Price Discrimination and Two-Part Pricing

The cell below shows willingness to pay for each of ten units of a good. For our purpose, these values could pertain to a single individual who might buy up to ten units (rides at an amusement park) or to ten individuals, each of whom might buy one unit (a new automobile). The analysis begins with linear pricing: the seller finds the single price that is to be charged for each unit.

The functional form for the willingness-to-pay (`wtp`) function exhibits constant elasticity. Both marginal revenue (under the single-price assumption) and marginal cost are determined by taking differences rather than by applying continuous formulas. For this example $mr > mc$ up to the fifth unit, so $x = 5$ yields maximum profit. The price is 22.361.

```
xlist: makelist(i,i,1,10)$ wtplist:50.0/xlist^0.5$
trlist:xlist*wtplist$  tclist:7*xlist+0.5*xlist^2$
makelist(trlist[i]-trlist[i-1],i,2,10)$   mrlist:
 append(["---"],%)$ makelist(tclist[i]-tclist[i-1],
 i,2,10)$     mclist: append([tclist[1]],%)$
 Plist: trlist-tclist$
```

```
names:matrix(["x","wtp","tr","tc","profit", "mr",
 "mc"])$ numbers:matrix(xlist,wtplist,trlist,tclist,
 Plist,mrlist,mclist)$ addrow(names,numbers);
```

x	wtp	tr	tc	profit	mr	mc
1	50.0	50.0	7.5	42.5	– – –	7.5
2	35.355	70.711	16.0	54.711	20.711	8.5
3	28.868	86.603	25.5	61.103	15.892	9.5
4	25.0	100.0	36.0	64.0	13.397	10.5
5	22.361	111.8	47.5	64.303	11.803	11.5
6	20.412	122.47	60.0	62.474	10.671	12.5
7	18.898	132.29	73.5	58.788	9.8131	13.5
8	17.678	141.42	88.0	53.421	9.1338	14.5
9	16.667	150.0	103.5	46.5	8.5786	15.5
10	15.811	158.11	120.0	38.114	8.1139	16.5

We determine the values of the two unexploited surpluses. Consumer surplus equals the sum of willingness to pay for each of the first five units *less* the price, which equals the willingness to pay for the fifth unit.

The efficient number of units is $x = 9$, because for each of the first nine units, willingness to pay exceeds marginal cost. Therefore, the deadweight loss equals the sum of the difference between willingness to pay and marginal cost for the sixth through ninth units.[2]

```
P0:(wtplist[5]*5-tclist[5])$ cs0:sum(wtplist[i]
 - wtplist[5],i,1,5)$ dwl0:sum(wtplist[i]-
 mclist[i],i,6,9)$ matrix(["Profit, single price",
 "Consumer Surplus","DWL"],[P0,cs0,dwl0]);
```

Profit, single price	Consumer Surplus	DWL
64.303	49.78	17.655

Suppose that the seller sells each unit for a price that is infinitesimally less than the buyer's (or buyers') willingness to pay for that unit. In this case the seller appropriates almost all of the consumer surplus. Moreover, the price paid for each unit equals the marginal revenue for that unit, because the seller no longer reduces prices on inframarginal units to sell additional units. Therefore, the seller produces nine units, thereby capturing the deadweight loss as well as the consumer surplus.

The cell below confirms that the firm's profits increase when it engages in first-degree price discrimination. It also confirms that the profit under this pricing mechanism is the sum of the profit, consumer surplus, and deadweight loss that occur under single-price regime.

---

[2] An exercise: Assume that the good can be sold in arbitrarily small units so that the equations for the demand, marginal revenue, and marginal cost curves can be used to determine all values. Determine the profit-maximizing quantity, the efficient quantity, and the values of the two unexploited surpluses. Why are these values larger than the ones reported in the text? Use a graph to illustrate your answer.

Perfect first-degree price discrimination removes the inefficiency that single-price monopoly causes. It does so, however, by adding both the deadweight loss and consumer surplus to the seller's profit. That is, all value created by the sale of the good accrues to the seller, none to the buyer(s).

```
[P1:sum((wtplist[i]-mclist[i]),i,1,9),
P0+cs0+dwl0];
```

$[131.74, 131.74]$

Perfect first-degree price discrimination is an ideal that cannot be realized. Nonetheless, analyzing this ideal highlights some important issues. First, deviations from simple linear pricing can increase economic efficiency. Second, the advantages to sellers of approximating this ideal are obvious.

Consider some ways that this ideal might be approximated. Many car dealers train their sales staff to elicit information about buyers' willingness to pay and use that information to charge different prices, depending on the information gathered. Online sellers allegedly engage in practices such as offering discounts to shoppers who delay hitting the "add to cart" button and charging higher prices when the buyer is using an expensive computer. Colleges often state a high "sticker price" and then offer rebates that depend on students' income, or that of their parents (which serves as a proxy for ability to pay).[3]

An alternative to selling each unit to a seller at a different price is to use a two-part pricing scheme. A *two-part tariff* is one in which the buyer is charged one price for the privilege of buying the product and then is allowed to decide how many units to buy at a seller-determined price. In the illustration above, the seller would receive maximum profit by setting the price equal to 16.667, which is just above the marginal cost of the ninth unit. As the next cell shows, the entry fee, `A0`, is 85.239, and the profit level is the same as determined in the preceding cell.

```
A0:sum(wtplist[i]-wtplist[9],i,1,9)$
P2:A0+sum(wtplist[9]-mclist[i],i,1,9)$
matrix(["Entry Fee","Profit"],[A0,P2]);
```

$$\begin{bmatrix} \text{Entry Fee} & \text{Profit} \\ 85.239 & 131.74 \end{bmatrix}$$

13.2.1 A Special Case: Two-Part Pricing, Identical Buyers

Suppose a seller's product faces the demand curves of 12 identical buyers (any number would do equally well). The next cell shows the market for buyers whose demand curves exhibit a constant elasticity of -2.0. Each buyer's quantity is x, and total quantity is X.

[3]Lest one appear cynical and say that this implies that colleges are engaging in profit-maximizing behavior realize that managing a limited budget for granting rebates efficiently requires the same sort of behavior that one would observe if profit maximization were the goal.

```
kill(x,X,pd,mr,mc,X0,p0,Xe0,pe0)$
x:1000*p^-2.0$ X:12*x$  matrix(
["Individual Demand","Market Demand"],[x,X]);
```

$$\begin{bmatrix} \text{Individual Demand} & \text{Market Demand} \\ \frac{1000}{p^{2.0}} & \frac{12000}{p^{2.0}} \end{bmatrix}$$

Next we create expressions for inverse market demand and marginal revenue, along with an expression for a linear marginal cost curve. Then we determine the profit-maximizing quantity, the efficient quantity, and the corresponding prices.

```
pd:(Xs/12000)^-0.5$   mr:.5*pd$  mc:2+0.5*Xs$
X0: find_root(mr - mc, Xs, 0.01, 200)$   p0:
 subst(Xs=X0,pd)$  Xe0:find_root(pd-mc,Xs,0.01,200)$
pe0:subst(Xs=Xe0,pd)$     matrix(["Demand","mr","mc",
"X0","p0","Xe0","pe0"],[pd,mr,mc,X0,p0,Xe0,pe0]);
```

$$\begin{bmatrix} \text{Demand} & \text{mr} & \text{mc} & \text{X0} & \text{p0} & \text{Xe0} & \text{pe0} \\ \frac{109.54}{Xs^{0.5}} & \frac{54.772}{Xs^{0.5}} & 0.5\,Xs+2 & 20.308 & 24.308 & 33.726 & 18.863 \end{bmatrix}$$

The graph below shows the market demand curve, marginal revenue curve, and marginal cost curve (thin lines) along with an individual demand curve (thick line).

```
wxdraw2d( ... Some commands omitted ...
 points([[0,pe0],[Xe0,pe0],[Xe0,0]]),points([[X0,0],
 [X0,p0],[0,p0]]),... explicit(pd,Xs,0.01,1.1*Xe0),
 ... explicit(mr,Xs,0.01,1.1*Xe0),explicit(mc,Xs,
 0.01,1.1*Xe0),line_width=2,key="Individual Demand",
 explicit((xd/1000)^-0.5,xd,0.01,1.1*Xe0))$
```

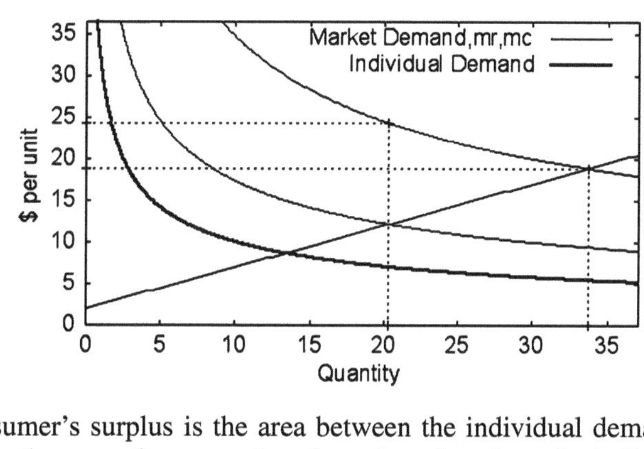

Each consumer's surplus is the area between the individual demand curve and the price, over the range from $x = 0$ to the value of x where the individual demand curve intersects the price line. Each individual can be charged an entry fee equal to that area. The area itself depends on the price that the seller establishes.

To see how two-part pricing can increase profits, calculate profits with a single price for all units and no entry fee, and then with a fee. Assuming zero fixed costs, profits equal the integral of the difference between marginal revenue and marginal cost up to the quantity at which the two equal (that is, profits are the same as producer surplus). In this illustration, the profit level is `P0` = 349.94.

To determine the entry fee for each of the 12 buyers, integrate the difference between the individual inverse demand curve and the price at which the market demand curve intersects the marginal cost curve. Each consumer can be charged (slightly less than) 53.014. Now profits equal this fee *times* 12, *plus* the area between `P0` and marginal cost up to X = Xe0. The resulting profit, `P1` is more than twice `P0`.

```
P0:float(integrate(mr-mc,Xs,10^-20,X0))$    A0:float(
 integrate((xd/1000)^-0.5-pe0,xd, 10^-20,Xe0/12))$
P1:float(12*A0+ integrate(pe0-mc,Xs,10^-20,X0))$
matrix(["P0","Fee","P1"],[P0,A0,P1]);
```

$$\begin{bmatrix} \text{P0} & \text{Fee} & \text{P1} \\ 349.94 & 53.014 & 875.52 \end{bmatrix}$$

13.2.2 Heterogeneous Buyers and Pricing Schedules

The calculation above suggests that firms can gain from two-part pricing if their buyers are relatively homogeneous. Even an approximately optimal combination of entry fee and price can add appreciably to profits. Consider this illustration, with two types of buyers. Let `m1` represent the fraction of consumers of Type A (so that 1 − `m1` is the fraction of consumers of Type B).

```
[xa(pa,a):=a-pa, pa(xa,a)  := a-xa,
mra(xa,a):=a-2*xa]; [xb(pb,b):=b-pb,pb(xb,b):=b-xb,
  mrb(xb,b):=b-2*xb];      mc:1$  P(xa,xb,A,m,a,b):=
  m*(pa(xa,a)-mc)*xa  +(1-m)*(pb(xb,b)-mc)*xb+A ;
```

$$[\mathrm{xa}(pa,a) := a - pa, \mathrm{pa}(xa,a) := a - xa, \mathrm{mra}(xa,a) := a - 2\,xa]$$
$$[\mathrm{xb}(pb,b) := b - pb, \mathrm{pb}(xb,b) := b - xb, \mathrm{mrb}(xb,b) := b - 2\,xb]$$
$$\mathrm{P}(xa, xb, A, m, a, b) :=$$
$$m\ (\mathrm{pa}(xa,a) - mc)\ xa + (1-m)\ (\mathrm{pb}(xb,b) - mc)\ xb + A$$

```
[a0,b0]:[10,20]$  wxdraw2d( ... Some commands
omitted ... explicit(pa(xa,a0),xa,0,b0+1),  ...
explicit(pb(xb,b0),xb,0,b0+1),
color=gray, key="mc",explicit(mc,x,0,b0+1)   )$
```

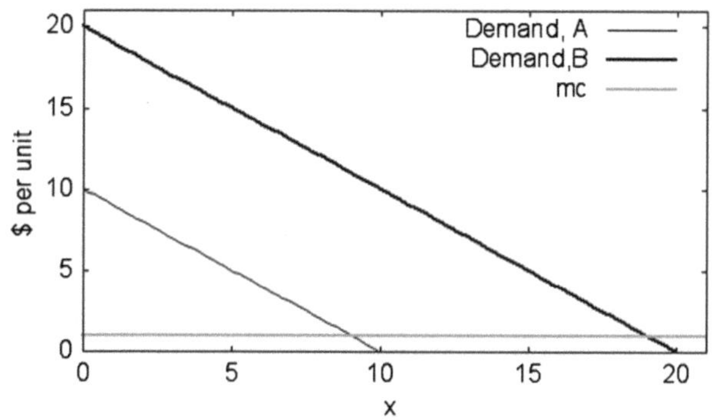

Suppose that the seller could separate consumers into two markets, but it must charge a single price in each market. Then the profit-maximizing prices would be 4.5 and 9.5, with associated quantities of $5.5 \cdot m$ and $10.5 \cdot (1 - m)$ units. The profit levels in the two markets would be $20.25 \cdot m$ and $90.25 \cdot (1 - m)$. For the moment, however, we assume that these markets are not separable.

```
matrix([ ["pa0","pb0"] ,["xa0","xb0"],"Profit"],
 [[xa0, xb0]:[find_root(mra(xa,a0)-mc,xa,0,a0),
  find_root(mrb(xb, b0)-mc,xb,0,b0)],  [pa0, pb0]:
   [pa(qa0,a0),pb(qb0,b0)],
   P0:P(xa0,xb0,0.0,m,a0,b0)]);
```

$$\begin{bmatrix} [\text{pa0,pb0}] & [\text{xa0,xb0}] & \text{Profit} \\ [4.5, \; 9.5] & [10 - qa0, \; 20 - qb0] & 20.25\, m + 90.25\, (1 - m) \end{bmatrix}$$

If the seller can both separate the markets and charge the ideal (from the seller's viewpoint) entry fee in each, then it would appropriate (almost all of) the consumers' surplus by setting the price at $1 and charging entry fees (just below) 40.5 and 180.5 (why?), so that its profit is approximately $40.5 \cdot m + 180.5 \cdot (1 - m)$.

```
40.5*m + 180.5*(1-m);
```

$$40.5\, m + 180.5\, (1 - m)$$

Suppose that the firm cannot separate its markets, but can engage in two-part pricing. Specifically, it can set an entry fee and a constant per-unit price. If it sets the price equal to marginal cost and sets the entry fee equal to the consumer surplus for Type A consumers, the firm gets $m \cdot 40.5$ from Type A consumers, and $(1-m) \cdot 40.5$ from Type B consumers, for a total profit of 40.5. One alternative is to set the entry fee equal to the consumer surplus for Type B consumers. Doing so generates $P = 180.5 \cdot (1 - m)$, because the Type A consumers buy none of the good. This will be the better of these two options if $m < 0.776$. That is, unless 77.6 % of the buyers are of Type A, the firm will not sell to Type A buyers.

```
[xae, xbe] : [xa(mc, a0), xb(mc, b0)]$
A0: float(integrate(pa(xa, a0)-mc, xa, 0, xae))$
B0: float(integrate(pb(xb, b0) -mc, xb, 0, xbe))$
m0: find_root(A0 - (1-m)*B0, m, 0, 1)$ matrix(
 ["Quantities","Per-unit profits and Critical m"],
[ [xae,xbe], [A0,B0,m0]]);
```

$$\begin{bmatrix} \text{Quantities} & \text{Per-unit profits and Critical m} \\ [9,19] & [40.5,180.5,0.776] \end{bmatrix}$$

Suppose that m is large enough that it is worth the firm's while to serve both markets. In this case, a higher price results in more revenue from the sale to high-demand Type B buyers but reduces the per-period fee that Type A buyers will pay. We can determine the price that maximizes profits by first determining the value of A as a function of the price (recall that A is the entry fee). The first input line below does that. The second line places the resulting expression for A into the profit function. The result is an expression in terms of $p2$, the per-unit price, and m. Setting the first derivative of that expression equal to zero and solving yields the optimal $p2$ as a function of m. The value of m must be at least 0.776, as seen above. For that value, the per-unit price is 3.24. If $m > 0.776$, the reduction in A due to any price increase becomes larger, so the price is closer to the marginal cost. As $m \to 1$, $p2 \to 1$, the firm's marginal cost.

```
A2:integrate(pa(xa,a0)-p2,xa,0,10-p2);
P(xa(p2,a0),xb(p2,b0),A2,m,a0,b0);
diff(%,p2);  subst(0.776,m,solve(%,p2));
p2max:rhs(%[1])$ A2min:subst(p2max,p2,A2);
```

$$\frac{p2^2-20\,p2+100}{2}$$

$$\frac{p2^2-20\,p2+100}{2}+(1-m)\,(20-p2)\,(p2-1)+m\,(10-p2)\,(p2-1)$$

$$\frac{2\,p2-20}{2}-m\,(p2-1)-(1-m)\,(p2-1)+(1-m)\,(20-p2)+m\,(10-p2)$$

$$[p2=3.24] \qquad 22.849$$

Yet another possibility is that the firm cannot charge an entry fee but can establish a price schedule that allows for quantity discounts. We continue with the same demand functions as above, but the profit function now changes. The firm will set an initial, small-quantity price that is the only price that Type A buyers pay. Type B buyers, on the other hand, pay the initial price for part of the units that they buy but pay a lower price per remaining unit.

In our illustrative case, where the distribution function consists of two types of buyers and the associated relative numbers of those buyers, the profit function is relatively simple. The partial derivatives of the profit function with respect to the price consist of a pair of linear equations in three variables, only two of which are chosen by the firm.

```
P2p:(p1-mc)*(m*xa(p1, a0)+(1-m)*xb(p1,b0))+(p2-mc)*
  (1-m)*(xb(p2,b0) - xb(p1,b0)); [dP2p1,dP2p2]:
  [ratsimp(diff(P2p, p1)),ratsimp(diff(P2p, p2))];
```

$$(1-m)\ (p1-p2)\ (p2-1)+((1-m)\ (20-p1)+m\ (10-p1))\ (p1-1)$$

$$[(1-m)\ p2-2\,p1-9m+20,(2\,m-2)\ p2+(1-m)\ p1-m+1]$$

The next cell shows the solution of the system consisting of the two first-order conditions. The profit-maximizing prices are determined as functions of m. Values are determined for $m = 0.8$. The profit-maximizing prices are 6.7895 for $x \le 13.211$ and 3.8947 for $x > 13.211$. Each Type A buyer purchases $xa = 3.2105$ units at the higher price. Each Type B buyer purchases $xb = 16.106$ units, 13.211 at the higher price and 2.8947 at the lower price.

```
soln: ratsimp(solve([dP2p1, dP2p2],[p1, p2]));
[phigh, plow] : [rhs(soln[1][1]), rhs(soln[1][2])]$
P2pmax:ratsimp(phigh*(m*xa(phigh,a0)+(1-m)*
  xb(phigh,b0))+(1-m)*plow*(xb(plow,b0)-
  xb(phigh,b0))-mc*(xb(plow,b0))); m1: 0.8$
[ph1, pl1,xa1,xbhigh1,xblow1]:subst(m=m1,
 [phigh,plow,xa(phigh,a0),xb(phigh,b0),
 xb(plow,b0)-xb(phigh,b0)])$ numbers:
 subst(m=m1,[phigh, plow,P2pmax,xa1,xbhigh1,
 xblow1])$ matrix(["phigh","plow","maximum P",
 "xa","xbhigh","xblow"],numbers);
```

$$[[p1=-\frac{19\,m-41}{m+3},p2=-\frac{9\,m-22}{m+3}]]$$

$$\frac{100\,m^2-429\,m+361}{m+3}$$

$$\begin{bmatrix} \text{phigh} & \text{plow} & \text{maximum P} & \text{xa} & \text{xbhigh} & \text{xblow} \\ 6.7895 & 3.8947 & 21.526 & 3.2105 & 13.211 & 2.8947 \end{bmatrix}$$

The graph below shows the pair of prices that generates maximum profit. The upper dashed line shows the price for lower quantities. The lower, broader dashed line shows the price for quantities above 13.211 units.

```
wxdraw2d( ... Some commands omitted ...,
 explicit(mc,x,0,b0+1),line_width=2,key=
 "Type A Demand",explicit(pa(xa,a0),xa,0,b0+1),
 color=gray,key="Type B Demand",explicit(pb(xb,b0),
 xb,0,b0+1),color=black,line_width=1,
 line_type=dots, ...key="High Price",points([[0,ph1],
 [xbhigh1,ph1]]),line_width=2,key="Low Price",
 points([[xbhigh1,pl1],[xb(pl1,b0),pl1]]) )$
```

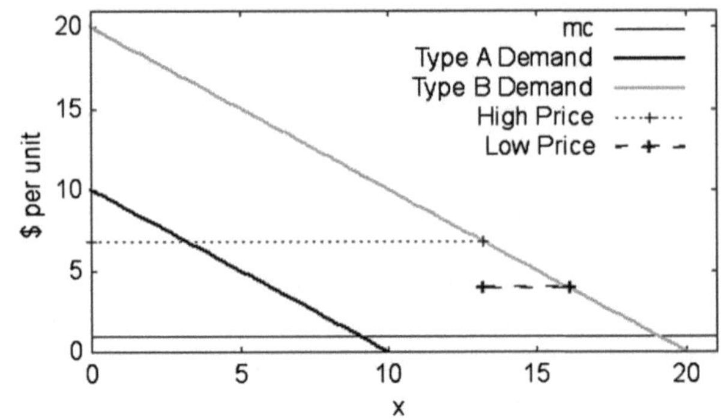

This illustration is limited to two prices and two classes of buyers. The model can be generalized to demand profiles with many types of buyers. See [1], which shows that even when the number of buyer types is quite large, a relatively simple multipart pricing scheme can closely approximate a more general scheme where each type of buyer pays a different marginal price.

13.3 Price Discrimination in Segmented Markets

The preceding section shows that firms can gain revenue by using prices to cause consumers to sort themselves. In the case of third-degree price discrimination, segmentation is much simpler, at least conceptually. An obvious example is that of geographic separation: unless buyers can cross-ship its products, the firm can treat each segment as a separate market. This section deals with this polar case, in which the quantity demanded in any one market segment depends only on the price in that segment and not on prices elsewhere. The next section relaxes this strong assumption.

Constant-elasticity demand curves illustrate this case. We proceed as follows. First, we look at the simplest case, with price discrimination and constant marginal cost (throughout, we assume that the marginal cost function is the same in both markets). The price in each market is as follows: $pi = c/(1 + 1/Ei)$, where pi is the price in the ith market, Ei is the price elasticity in that market, and c is the (common) marginal cost.[4]

Most of the analysis in this section uses a linear marginal cost function rather than the special case of constant marginal cost. This analysis proceeds in two steps. The profit-maximizing prices and quantities are determined, as are the implied profit level and consumer surplus levels. Then the firm is denied the option of selling at two different prices and the new values of all variables are determined.

[4]The analysis changes only slightly if the marginal cost differs across markets.

The cell below shows the parameter values that are used for the remainder of this section. All are counterparts to parameters used earlier: `A1`, `a1`, `B1`, and `b1` define the demand curves; and `C1`, `c1`, and `d1` define the cost curves. The parameter values are entered into a list of equalities that allow evaluation of subsequent expressions.

The terms A and B determine the location of the demand curves, while a and b are the price elasticities of demand. When marginal cost is a constant, the total cost function is $TC = C + (c \cdot (xa + xb))$. When marginal cost is a function of output, the total cost function is $TC = C + (\frac{d}{2} \cdot (xa + xb)^2)$.

```
pa:(xa/A)^(1/a)$  pb:(xb/B)^(1/b)$ [A1,a1,B1,
  b1,C1,c1,d1]:[2000,-1.5,3000,-2.5,100,5,0.4]$
params1:[A=A1,a=a1,B=B1,b=b1,C=C1,c=c1,d=d1]$
```

13.3.1 Constant Marginal Cost

This is the simplest case. The marginal revenue curves are defined as indicated above, so the prices are easily derived. Once the prices are determined, so are the quantities.

The following notation is used in this section. The price and quantity in market A are `paCCD1` and `xaCCD1`, where the uppercase letters refer to "constant-cost, discrimination." The counterparts for the case of increasing cost and discrimination are `paICD1` and `xaICD1`. When discrimination is not possible, the counterparts become `paICN1` and `xaICN1`. Notation for market B is similar. Uppercase P refers to the profit level and uppercase S refers to consumer surplus. We do not consider the case of non-discrimination when marginal cost is a constant.

The cell below shows the equilibrium values of the prices and quantities. As noted, the values for price are determined quite directly. Given the price in market A, this inverse demand expression is solved in order to determine quantity:

$$\left(\frac{xa}{A}\right)^{\frac{1}{a}} = \texttt{paCCD1}.$$

Values for market B are determined in like fashion, and all values are listed below.

```
[paCCD1,pbCCD1]:[c1/(1+1/a1),c1/(1+1/b1)]$ paCCD:
 subst(params1,pa)$  float(solve(%-paCCD1, xa))$
xaCCD1:rhs(%[1])$   pbCCD:subst(params1,pb)$
 float(solve(%-pbCCD1, xb))$ xbCCD1:rhs(%[1])$
XCCD1:xaCCD1+xbCCD1$       matrix(["paCCD1","xaCCD1",
 "pbCCD1", "xbCCD1", "XCCD1"],
    [paCCD1,xaCCD1,pbCCD1,xbCCD1,XCCD1]);
```

$$\begin{bmatrix} \text{paCCD1} & \text{xaCCD1} & \text{pbCCD1} & \text{xbCCD1} & \text{XCCD1} \\ 15.0 & 34.427 & 8.3333 & 14.965 & 49.391 \end{bmatrix}$$

13.3.2 Variable-Cost Price Discrimination

When the marginal cost depends on the total quantity sold in the two markets, the analysis becomes more complex. The firm must simultaneously allocate output between the two markets in quantities that yield equal marginal revenue in the markets, and it must determine the total quantity at which this joint marginal revenue equals marginal cost.

Mathematically, the problem is relatively simple. The next cell defines the firm's profit function `PICD` and the two relevant first derivatives of that function given the system's parameters. The `mnewton` numerical method yields the values of xa and xb that are consistent with profit maximization. Given these values the prices and profit level are determined directly.

The table below collects the values that are determined in the cell. Just above the table is an expression for the profit function given the system's parameters and that the firm allocates its output between the markets in a profit-maximizing manner.

```
PICD: subst(params1,pa*xa+pb*xb-(C+d*(xa+xb)^2/2));
diffPICDa:diff(PICD,xa)$  diffPICDb:diff(PICD,xb)$
[diffPICDa1,diffPICDb1]:subst(params1,
   [diffPICDa,diffPICDb])$        load(mnewton)$
mnewton([diffPICDa1,diffPICDb1],[xa,xb],[5,5])$
[xaICD1, xbICD1]:[rhs(%[1][1]),rhs(%[1][2])]$
[paICD1, pbICD1]:[pa,pb],params1,xa=xaICD1,xb=xbICD1$
PICD1: PICD, params1, xa=xaICD1, xb=xbICD1$
matrix(["paICD1","xaICD1","pbICD1","xbICD1",
"xICD1","PICD1"],[paICD1,xaICD1,pbICD1,xbICD1,
  xaICD1+xbICD1, PICD1]);
```

$$-0.2\,(xb+xa)^2+24.595\,xb^{0.6}+158.74\,xa^{0.333}-100$$

$$\begin{bmatrix} \text{paICD1} & \text{xaICD1} & \text{pbICD1} & \text{xbICD1} & \text{xICD1} & \text{PICD1} \\ 24.704 & 16.288 & 13.725 & 4.299 & 20.587 & 276.62 \end{bmatrix}$$

The output below shows the expression for the marginal cost curve. It also shows that the values above generate the condition that marginal revenue in market A equals marginal revenue in market B, and that both equal marginal cost. The `cons` command constructs a new list by adding the definition of `X` to the values in the list `params1`.[5]

```
MC: diff(C+d*X^2/2,X)$ print("MC = ",MC)$ matrix(
["MC","MRa","MRb"], subst(cons(X=xaICD1+xbICD1,
  params1),[MC,paICD1*(1+1/a),pbICD1*(1+1/b)]));
```

[5]The syntax of `subst` calls for a single list of substitutions. Because `param` is such a list, we need a way to add to this list rather than inserting a second list. The `cons` command accomplishes this task.

$$MC = d\,X \qquad \begin{bmatrix} MC & MRa & MRb \\ 8.2348 & 8.2348 & 8.2348 \end{bmatrix}$$

The figure below shows these results graphically. The thin dotted lines show the prices and quantities. The heavier dotted line shows the (common) marginal revenue value, the quantity sold in each market, and the total quantity.

```
[pa1,pb1]:subst(params1,[pa,pb])$ xmax:1.1*(xaICD1+
xbICD1)$   wxdraw2d( ... Some commands omitted ...
points([[xaICD1,0],[xaICD1,paICD1],[0,paICD1]]),
points([[xbICD1,0],[xbICD1,pbICD1],[0, pbICD1]]),
... points([[0,paICD1*(1+1/a1)],[xaICD1+xbICD1,
d1*(xaICD1+xbICD1)],[xaICD1+xbICD1,0]]), ...
explicit(pa1,xa,1,xmax), ... explicit(
pa1*(1 + 1/a1),xa,1,xmax), ... explicit(pb1,xb,1,
xmax), ... explicit(pb1*(1+1/b1),xb,1,xmax), ...
explicit(d1*x, x, 1, xmax)   )$
```

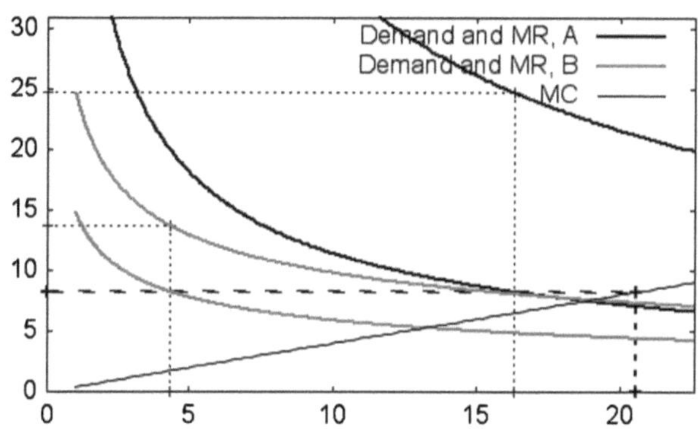

13.3.2.1 Price and Quantity Without Discrimination

Suppose that the firm cannot engage in price discrimination. The two demand curves combine to form a single demand curve (horizontal summation). The firm's task is to determine the quantity at which the resulting marginal revenue equals its marginal cost.

We determine the resulting solution by imposing the equality of the two prices. The result is an expression of the quantity of one good in terms of the other. We express xb as a function of xa. We use the first expression in the `soln` list, the one that yields positive values for both variables.

```
assume(xa>0)$ soln:float(solve(pa1=pb1,xb));
```

$$[xb = 0.00945\,xa^{\frac{5}{3}},\quad xb^{\frac{1}{5}} = -0.394\,xa^{\frac{1}{3}}]$$

The firm's profit function with nondiscriminatory pricing, PICN, is now a function of the single quantity xa. The find_root command yields the value of xa, from which other values are determined. These are reported in the table below. Inspection shows that buyers in market A gain from the removal of the firm's ability to discriminate. Buyers in market B lose; they pay a higher price for a lower quantity. The firm's profits fall. In this example, removing the ability to discriminate reduces total output, but this result cannot be generalized.

```
PICN:subst(cons(soln[1],params1),pa*xa+pb*xb-
   (C+d*(xa+xb)^2/2));   diffPICN1 : diff(PICN, xa)$
xaICN1:find_root(diffPICN1,xa,0.1,100)$  xbICN1:
 float(subst(xa=xaICN1,rhs(soln[1])))$
xICN1:xaICN1+xbICN1$  [paICN1,pbICN1]:subst(
[xa=xaICN1,xb=xbICN1],[pa1,pb1])$
PICN1:float(subst([xa=xaICN1,xb=xbICN1],PICN))$
matrix(["paICN1","xaICN1","pbICN1","xbICN1",
 "xICN1","PICN1"],[paICN1,xaICN1,pbICN1,xbICN1,
 xICN1,PICN1]);
```

$$-0.2\left(0.00945\,xa^{\frac{5}{3}}+xa\right)^2+1.5\,xa+158.74\,xa^{0.333}-100$$

$$\begin{bmatrix} \text{paICN1} & \text{xaICN1} & \text{pbICN1} & \text{xbICN1} & \text{xICN1} & \text{PICN1} \\ 22.383 & 18.886 & 22.383 & 1.2657 & 20.152 & 269.84 \end{bmatrix}$$

The cell below reports the values for consumer surplus under each of the two regimes. It also reports the total surpluses—consumer surplus combined with seller profits. Comparing the latter (SICD with SICN) shows that in this example price discrimination reduces total surplus. That is, the seller's gain from practicing price discrimination is less than the reduction in combined consumer surplus.

The output shows that the transfer caused by price discrimination is not simply a transfer from buyers to the seller. Consumers in market A suffer a reduction in consumer surplus due to price discrimination, but those in market B realize a gain in consumer surplus.

```
CSaICD1:float(integrate(pa1-paICD1, xa, 0, xaICD1))$
CSaICN1:float(integrate(pa1-paICN1, xa, 0, xaICN1))$
CSbICD1:float(integrate(pb1-pbICD1, xb, 0, xbICD1))$
CSbICN1:float(integrate(pb1-pbICN1, xb, 0, xbICN1))$
CSICD1:CSaICD1 + CSbICD1$  CSICN1:CSaICN1+CSbICN1$
SICD1:CSICD1+PICD1$  SICN1:CSICN1+PICN1$    matrix(
["CSaICD","CSaICN","CSbICD","CSbICN","SICD","SICN"],
[CSaICD1, CSaICN1, CSbICD1, CSbICN1, SICD1, SICN1]);
```

$$\begin{bmatrix} \text{CSaICD} & \text{CSaICN} & \text{CSbICD} & \text{CSbICN} & \text{SICD} & \text{SICN} \\ 804.77 & 845.47 & 39.335 & 18.886 & 1120.7 & 1134.2 \end{bmatrix}$$

13.4 Imperfect Price Discrimination and Product Line Pricing

The preceding section addresses the pricing of a good that is sold in two separate markets. Often, however, the markets are not completely separate. In such cases, the firm must take into account the effect of the price it charges in one market on the demand for its product in the other market (or markets).

This section begins with the analysis of a case in which two markets for a good are perfectly separated, as in the preceding section. Then it extends the analysis to allow for interdependence between the markets. Finally it addresses a closely related issue, the pricing of a product line.

The cell below shows the functional relationships that are used throughout the section. They consist of demand functions for market (or good) A and market (or good) B and the firm's profit function. A simplifying assumption of constant marginal cost is built into the model.

This section illustrates the principles involved with linear demand functions. This functional form is used rather than the constant elasticity form because the latter must be solved numerically and the numerical method `mnewton` has difficulty in converging on solutions when the constant elasticity functional form is used, unless the elasticities are quite low (roughly, own-price elasticities between -1.0 and -2.0 and cross elasticities below 0.1).

Even though the functional form used here is linear, we do specify a set of elasticities, from which commands derive the requisite slope values. We use this approach because elasticities, being free of units, are easier than slopes to interpret.

The own-price elasticity values are negative (demand curves slope downward), and the cross-price elasticity values are positive (a higher price in market A causes an increased demand in market B and vice versa). In effect, the good sold in market A and the same good sold in market B are substitutes. Complementary goods are considered below.

This last command in this cell provides a logic check. If $Eab0$ and $Eba0$ do not have the same sign (positive for substitutes, negative for complements), then a warning is posted.

```
kill(all)$  [xa0,xb0,pa0,pb0]:[100,100,10,10]$
[Eaa0,Eab0,Eba0,Ebb0]:[-2.5,0,0.0,-4.0];
[a1,a2,b1,b2]:[Eaa0*xa0/pa0,Eab0*xa0/pb0,
     Eba0*xb0/pa0, Ebb0*xb0/pa0];
if sign(Eab0) = sign(Eba0) then
"OK: Eab0 and Eba0 have the same sign" else
"Error: Eab0 and Eab0 must have the same sign";
```

$[-2.5,\ 0,\ 0.0,\ -4.0]\qquad [-25.\ 0, 0,\ 0.0,\ -40.0]$

OK: Eab0 and Eba0 have the same sign

13.4.1 Separate Markets and Price Discrimination

In the cell above, the two cross-price elasticities are $Eab = Eba = 0$. That is the demand for the good in market A is not affected by the price in market B, and vice versa. With this restriction, we review the case of price discrimination using linear representations, to complement the analysis of the preceding section.

The price and quantity values above, coupled with the slope values that the elasticities imply, generate a pair of intercept terms for the demand functions. We arbitrarily set the (constant) marginal cost values equal to the minimum of the two price-axis intercept values. In the first two parts of this section we maintain $ca = cb$. We relax this imposition in the third part.

We now have two demand curves and a profit function, with profit, P, stated as a function of the two prices.

```
a0:xa0-a1*pa0-a2*pb0$   b0: xb0-b1*pa0-b2*pb0$
[ca0, cb0]  : min(-a0/a1, -b0/b2)/2$
xa  : a0 + a1*pa + a2*pb;   xb: b0 + b1*pa + b2*pb;
P  : xa*(pa  -  ca0)  + xb*(pb-cb0);
```

$$350.0 - 25.0\,pa \qquad 500.0 - 40.0\,pb$$
$$(500.0 - 40.0\,pb)\,(pb - 6.25) + (350.0 - 25.0\,pa)\,(pa - 6.25)$$

Generating the two first partial derivatives of the profit function with respect to prices and setting these functions equal to zero yields the profit-maximizing prices.

```
P; diff(P,pb);
```

$$(500.0 - 40.0\,pb)\,(pb - 6.25) + (350.0 - 25.0\,pa)\,(pa - 6.25)$$
$$-40.0\,pb - 40.0\,(pb - 6.25) + 500.0$$

```
soln1:float(solve([diff(P,pa),
diff(P,pb)],[pa,pb]));
```

$$[[pa = 10.125, \quad pb = 9.375]]$$

These prices imply the profit-maximizing quantities, the price elasticities at these values, the marginal revenue in each market, and the profit level. We confirm that the price is lower in the market with more elastic demand, that the marginal revenue is the same in each market, and that this marginal revenue value equals marginal cost.

```
[pa1, pb1]:[rhs(soln1[1][1]), rhs(soln1[1][2])]$
[xa1, xb1]:[xa, xb],soln1[1]$
P1:subst([pa=pa1,pb=pb1],P)$  [Eaa1, Ebb1]:
 [a1*pa1/xa1, b2*pb1/xb1]$   [mra1, mrb1]:
 [pa1*(1+1/Eaa1), pb1*(1+1/Ebb1)]$ matrix(
 ["pa","xa","Eaa","mra"],[pa1, xa1,Eaa1,mra1]);
matrix(["pb","xb","Ebb","mrb"],[pb1,xb1,Ebb1,
 mrb1]);   matrix(["mc","P"],[ca0, P1] );
```

$$\begin{bmatrix} pa & xa & Eaa & mra \\ 10.125 & 96.875 & -2.6129 & 6.25 \end{bmatrix}$$
$$\begin{bmatrix} pb & xb & Ebb & mrb \\ 9.375 & 125.0 & -3.0 & 6.25 \end{bmatrix} \qquad \begin{bmatrix} mc & P \\ 6.25 & 766.02 \end{bmatrix}$$

13.4.2 Imperfect Market Segmentation

If the two markets are not completely separable (some who reside in area A can buy in area B, or some who have Type A characteristics can disguise themselves and buy at the price set for Type B buyers or vice versa), then the cross-price elasticities are positive.

```
kill(all)$  [xa0,xb0,pa0,pb0]:[100,100,10,10]$
[Eaa0,Eab0,Eba0,Ebb0]:[-2.5,0.5,0.25,-4.0];
[a1,a2,b1,b2]:[Eaa0*xa0/pa0, Eab0*xa0/pb0,
 Eba0*xb0/pa0,Ebb0*xb0/pa0];
if sign(Eab0) = sign(Eba0) then
 "OK: Eab0 and Eba0 have the same sign" else
 "Error: Eab0 and Eab0 must have the same sign";
```

$$[-2.5,\ 0.5,\ 0.25,\ -4.0] \qquad [-25.0,\ 5.0,\ 2.5,\ -40.0]$$
OK: Eab0 and Eba0 have the same sign

```
a0:xa0-a1*pa0-a2*pb0$  b0:xb0-b1*pa0-b2*pb0$
[ca0, cb0] : min(-a0/a1, -b0/b2)/2$
xa:a0+a1*pa+a2*pb$    xb:b0+b1*pa+b2*pb$
P : xa*(pa - ca0) + xb*(pb-cb0)$   soln1:
float(solve([diff(P,pa),diff(P,pb)],[pa,pb]));
```

$$[[pa = 10.215,\ \ pb = 9.2963]]$$

The analysis here is much like that above. Both examples share one characteristic: a more elastic demand in a market results in a lower price in that market. The two marginal revenue values are no longer the same.

The major difference between the two cases is that in both markets, the firm sells less than the amount that results in the equality of marginal revenue and marginal cost. The reason is that the firm must recognize that selling additional units in either market now has two effects. First, it reduces the price in that market, as is the case for any seller whose product faces a downward-sloping demand curve. In addition, though, this lower price shifts the demand curve in the other market to the left. Therefore, increasing the quantity in either market places downward pressure on the prices in *both* markets.

```
[pa1,pb1]:[rhs(soln1[1][1]), rhs(soln1[1][2])]$
[xa1,xb1]:[xa, xb],soln1[1]$ P1:subst([pa=pa1,
pb=pb1],P)$  [Eaa1, Ebb1]:[a1*pa1/xa1, b2*pb1/xb1]$
[mra1, mrb1]:[pa1*(1+1/Eaa1), pb1*(1+1/Ebb1)]$
matrix(["pa","xa","Eaa","mra"],[pa1, xa1,Eaa1,mra1]);
matrix(["pb","xb","Ebb","mrb"],[pb1,xb1,Ebb1,mrb1]);
matrix(["mc", "P"],[ca0, P1]);
```

$$\begin{bmatrix} pa & xa & Eaa & mra \\ 10.094 & 95.055 & -2.6548 & 6.2919 \end{bmatrix}$$

$$\begin{bmatrix} pb & xb & Ebb & mrb \\ 9.4815 & 120.98 & -3.135 & 6.4571 \end{bmatrix} \quad \begin{bmatrix} mc & P \\ 5.9375 & 823.84 \end{bmatrix}$$

The results above are phrased in terms of two imperfectly separated markets for the same good. The model could as easily have been interpreted as analyzing pricing of a line of two substitute goods offered by the same seller. The next part of this section changes the signs of the cross-price elasticities, so that it analyzes the pricing of a pair of complementary goods offered by the same seller.

13.4.3 Pricing Complementary Goods

Our third example considers a firm that sells two complementary goods, goods with negative cross-price effects. We keep the same parameter values as above except for changing the sign of the cross-price elasticities. We continue to use $ca = cb$, but this restriction can be relaxed.

```
[xa0,xb0,pa0,pb0]:[100,100,10,10]$
[Eaa0,Eab0,Eba0,Ebb0]:[-2.5,-0.5,-0.25,-4.0];
[a1, a2, b1, b2]: [Eaa0*xa0/pa0, Eab0*xa0/pb0,
  Eba0*xb0/pa0, Ebb0*xb0/pa0];
if sign(Eab0)=sign(Eba0) then
  "OK: Eab0 and Eba0 have the same sign" else
  "Error: Eab0 and Eab0 must have the same sign";
```

$$[-2.5, -0.5, -0.25, -4.0] \quad [-25.0, -5.0, -2.5, -40.0]$$

OK: Eab0 and Eba0 have the same sign

```
a0:xa0-a1*pa0-a2*pb0$  b0:xb0-b1*pa0-b2*pb0$
[ca0, cb0] : min(-a0/a1, -b0/b2)/2$
xa:a0+a1*pa+a2*pb$      xb:b0+b1*pa+b2*pb$
P:xa*(pa - ca0) + xb*(pb-cb0)$    soln1:
 float(solve([diff(P,pa),diff(P,pb)],[pa,pb]));
```

$$[[pa = 10.215, \quad pb = 9.2963]]$$

Note the similarity of imperfect market separation and the more general selling of substitute goods. As before, the markup is higher for good A, the good with the lower (in absolute value) own-price elasticity.

With the two complementary goods, the firm sells a quantity of each good such that $mr < mc$. That is, it sells more than the quantity that would maximize profits if only its good-specific marginal revenue and marginal cost were taken into account. In this case selling more of good A increases the demand for good B and vice versa.

```
[pa1,pb1]:[rhs(soln1[1][1]),rhs(soln1[1][2])]$
[xa1,xb1]:[xa, xb],soln1[1]$  P1:
subst([pa=pa1,pb=pb1],P)$   [Eaa1,Ebb1]:
[a1*pa1/xa1, b2*pb1/xb1]$   [mra1,mrb1]:
[pa1*(1 + 1/Eaa1), pb1*(1+1/Ebb1)]$  matrix(
["pa","xa","Eaa","mra"],[pa1, xa1,Eaa1,mra1]);
matrix(["pb","xb","Ebb","mrb"],[pb1,xb1,Ebb1,
 mrb1]);    matrix(["mc", "P"],[ca0, P1] );
```

$$\begin{bmatrix} pa & xa & Eaa & mra \\ 10.215 & 98.145 & -2.602 & 6.2891 \end{bmatrix}$$

$$\begin{bmatrix} pb & xb & Ebb & mrb \\ 9.2963 & 127.61 & -2.9139 & 6.1059 \end{bmatrix} \quad \begin{bmatrix} mc & P \\ 6.5625 & 707.33 \end{bmatrix}$$

Reference

1. Wilson RB (1997) Nonlinear pricing. Oxford University Press, New York

Further Readings

The table in the preface lists chapters in microeconomics textbooks that relate to this chapter. The following can also be of value:

Anderson ET, Dana JD (2009) When is price discrimination profitable? Manage Sci 55:980–989. The paper develops a framework for analyzing many important types of price discrimination: intertemporal price discrimination, damaged goods, advance purchase discounts, coupons (rebates), and information goods (list taken from the paper).

McAfee RP (2007) Pricing damaged goods. Economics Discussion Papers, No 2007-2, Kiel Institute for the World Economy. http://www.economics-ejournal.org/economics/discussionpapers/2007-2. This paper shows how firms can segment markets by impairing some units its product and selling them at a reduced price. It presents numerous examples. A paper by Deneckere and McAfee (1996), cited in this paper, contains additional examples.

Chapter 14
Oligopoly

Abstract Chapter 12 points out that, alone among industry structures, oligopoly (few sellers) forces strategic reasoning on member firms. This chapter explores some of the implications of this aspect of oligopoly and some of the models that have been used to gain insights into the behavior of oligopolists. In doing so, it deals tangentially with elements of game theory.

14.1 Introduction and Overview

We begin by offering a more precise definition of oligopoly. We use the term "noncooperative oligopoly" to emphasize that each firm makes its owns decisions. We briefly consider cooperation later in the chapter.[1]

Noncooperative oligopoly is a market where a small number of firms act independently but in which each firm is aware of the effects of its actions on other firms' environment and of their reactions to these effects.

The following assumptions typically underlie models of oligopolistic behavior.

- Each member Firm maximizes its profits based on its expectations about actions of the other firms in the industry.
- Each consumer is a price taker. If a few consumers were also price searchers, then the models would become even more complex, for consumers would also have incentives to behave strategically.
- All firms produce homogeneous products. We relax this aspect of the model during the analysis.
- Neither entry into or exit from the industry occurs. That is, the number of firms is fixed for analytical purposes.

---

[1]This chapter follows the notes of [1]. The illustrations that follow are adapted from this source.

M.R. Hammock and J.W. Mixon, *Microeconomic Theory and Computation: Applying the Maxima Open-Source Computer Algebra System*,
DOI 10.1007/978-1-4614-9417-1_14,

- Firms collectively face a single downward-sloping demand curve. One implication of this assumption is that the price can exceed the marginal cost of member firms. This assumption is relaxed when heterogeneous products are introduced.
- Each firm sets only its price or output. This does not rule out collaboration in other areas like advertizing and lobbying.

Looking ahead, we summarize some of the main results of the analysis that follows.

- The equilibrium price(s) lies (lie) between the monopoly and perfect competition prices.
- Each firm's expected profits are maximized when its (conjectural) marginal revenue equals marginal cost.
- The marginal revenue for a firm depends on its residual demand curve (market demand minus the output supplied by other firms).

14.1.1 Game Theory

We have stressed that oligopoly inherently involves conjectural decision-making, so that firms must behave strategically, or "play games." Game theory is a systematic approach to such situations. In our terminology, a *game* formally represents a situation in which a number of decision makers (players) interact in a setting of strategic interdependence. As we have noted, which action that any individual player thinks is best depends on what the player's conjecture about other players' actions.

Within game theory, *Nash equilibrium* holds a prominent position. A Nash equilibrium consists of a set of strategies. Specifically, the set of strategies constitutes a Nash equilibrium if, holding other players' strategies constant, none of the players can obtain a higher payoff by choosing a different strategy. In a Nash equilibrium, no player wants to change its strategy.[2]

14.1.2 Games of Oligopoly Behavior

Game theoretic models of oligopoly have the following characteristics:

- Two or more firms constitute the industry.
- The number of firms is small enough that each individual firm's action has an appreciable effect on the other firms. Each firm has a few rivals.

[2]This does not imply that the players are satisfied, or that they view the Nash equilibrium as a good outcome. It is possible, for example, that every player wishes that all players could agree on a collective set of strategies that would improve every player's lot, but which the nature of the game precludes.

- Each firm attempts to maximize its expected profit (payoff).
- Each firm knows the effects of other firm's actions on its profit, but it does not know what those actions will be.
- Equilibrium payoffs are determined by the number of firms, the rules of the games and the length of the game.

We briefly consider the last of these items, and specifically the length of the game. A game might consist of a single interaction, a set of a known and finite number of sequential interactions, or an indefinite number of sequential interactions. The formal models of oligopoly that follow are of the first type. Even so, they provide insights into possible behavior in the second and third types.

We consider three models of strategical oligopolist behavior. The first is the Cournot model, in which each player takes other players' output levels as given. Second is the Bertrand model, in which each player takes the other players' prices as given. Finally, we consider the Stackelberg model, in which one player leads and the others follow. Historically, each of these models was developed and presented as having players interact through a series of actions and reactions until an equilibrium is reached. Game theory collapses these hypothetical back and forth interactions into a single round. That is, it treats the players as "meeting" and acting once. Thus, they do not study each other's behavior and adjust their behavioral assumptions accordingly in subsequent rounds.

14.2 The Cournot Model

The first model of noncooperative oligopoly behavior was published in 1838. The French mathematician Augustin Cournot stated the logic in terms of a duopoly, two sellers, but it extends directly to any number. His heuristic version of the model's behavior begins with one incumbent with constant per-unit cost. That firm produces as a monopolist until one rival firm enters the market. That firm knows the market demand curve and how much the incumbent firm produces and takes that output level as given. It produces a quantity that maximizes its profits given its perceived situation. Thus ends the first "day" of activity.

In the next "day" the first firm perceives that the market is smaller, by the amount that the rival firm produces, and it reduces its quantity accordingly. In response, the second firm increases its output. Thus begins a series of adjustments that results in a determinate amount of output by each firm (equal amounts if their costs are identical).

We can capture the essence of this process by developing *reaction functions*. These functions collapse Cournot's serial representation into a single interaction between the two firms. For our purposes, we assume that the market demand curve is linear and that each of the two firms has a linear marginal cost function (these may or may not be identical).

More formally, our assumptions are these:

- Two firms (no additional entry).
- Homogeneous product. This assumption allows us to define industry output as the sum of the firms' output: $X = x1 + x2$.
- Single period of production and sales.
- The linear inverse demand curve $p = A + B \cdot X = A + B \cdot (x1 + x2)$.
- Each firm has a linear marginal cost curve: $mc1 = b1 + c1 \cdot x1$ and $mc2 = b2 + c2 \cdot x2$.
- Each firm determines its quantity and the market determines the price.

Often illustrations of Cournot duopoly specify $c1 = c2 = 0$, and they often begin with the added assumption that $b1 = b2$. We treat these special cases.

```
p(x1,x2,A,B):=A+B*(x1+x2); [mc1(x1,b1,c1):=
    b1+c1*x1,mc2(x2,b2,c2):= b2+c2*x2];
```

$$\mathrm{p}\,(x1, x2, A, B) := A + B\ (x1 + x2)$$
$$[\mathrm{mc1}\,(x1, b1, c1) := b1 + c1\,x1, \mathrm{mc2}\,(x2, b2, c2) := b2 + c2\,x2]$$

We specify a set of parameter values.

```
[A0,B0,b10,c10,b20,c20] : [100, -2, 5, 2, 10, 1]$
[p(x1,x2,A0,B0),mc1(x1,b10,c10),mc2(x2,b20,c20)];
```

$$[100 - 2\ (x2 + x1)\,, \quad 2\,x1 + 5, \quad x2 + 10]$$

We can define each firm's reaction function in terms of the other firm's quantity produced. To do this we first define the two firms' profit functions, given the assumed behavior. The firms' fixed costs are `F1` and `F2`.

```
ratsimp(p(x1,x2,A,B)*x1-(F1+integrate(
mc1(x1,b1,c1),x1))); P1(x1,x2,A,B,b1,c1):= ''%$
ratsimp(p(x1,x2,A,B)*x2-(F2+integrate(
mc2(x2,b2,c2),x2))); P2(x1,x2,A,B,b1,c1):= ''%$
```

$$-\frac{2\,F1+\left(-2\,x1\,x2-2\,x1^2\right)B-2\,x1\,A+c1\,x1^2+2\,b1\,x1}{2}$$
$$-\frac{2\,F2+\left(-2\,x2^2-2\,x1\,x2\right)B-2\,x2\,A+c2\,x2^2+2\,b2\,x2}{2}$$

14.2.1 Firms' Optimal Reactions

Next take derivatives with respect to each firm's quantity and determine the quantity that generates maximum profit, given the other firm's product. The results indicate that Firm 1 reduces its output by $\frac{B}{(2 \cdot B - c1)}$ units per one-unit increase in $x2$. If $c1 = 0$, then the per-unit change is $1/2$; a one-unit change in $x2$ causes $x1$ to change by 1/2 units in the opposite direction.

The second command in each pair of commands below rephrases this result as a functional expression. The final command generates expressions for the reaction curves, given the parameters listed above. Inspection shows that $x1 = 0$ when $x2 = \frac{95}{2}$ and that $x = 0$ when $x1 = 45$. It also shows that the slope of Firm 1's reaction curve is $\frac{dx1}{dx2} = \frac{1}{3}$ and that Firm 2's reaction curve has a slope $\frac{dx2}{dx1} = \frac{2}{5}$.

```
solve(diff(P1(x1,x2,A,B,b1,c1),x1),x1);
 R1(x2,A,B,b1,c1):=''(rhs(%[1]))$
solve(diff(P2(x1,x2,A,B,b2,c2),x2),x2);
 R2(x1,A,B,b2,c2):=''(rhs(%[1]))$  radcan(
 [R1(x2,A0,B0,b10,c10),R2(x1,A0,B0,b20,c20)]);
```

$$[x1 = -\frac{x2\,B+A-b1}{2\,B-c1}] \quad [x2 = -\frac{x1\,B+A-b2}{2\,B-c2}] \quad [-\frac{2\,x2-95}{6}, -\frac{2\,x1-90}{5}]$$

Each reaction curve is conjectural in that $x1$ can be any of a range of values depending on $x2$, and vice versa. For both firms to be at equilibrium, the conjectures must be consistent with actual behavior. To determine the Nash equilibrium is, therefore, a direct process in the Cournot case. We solve the two reaction curve functions for $x1$ and $x2$. The resulting expressions are labeled `x1N` and `x2N`. Both of these expressions are numerical values determined by the system's parameters.

```
solnNash:solve([R1(x2,A,B,b1,c1)=x1,
  R2(x1,A,B,b2,c2)=x2],   [x1,x2]);     [x1N,x2N]:
  [rhs(solnNash[1][1]),rhs(solnNash[1][2])]$
```

$$[[x1 = -\frac{(A+b2-2\,b1)\,B-c2\,A+b1\,c2}{3\,B^2+(-2\,c2-2\,c1)\,B+c1\,c2}, \quad x2 = -\frac{(A-2\,b2+b1)\,B-c1\,A+b2\,c1}{3\,B^2+(-2\,c2-2\,c1)\,B+c1\,c2}]]$$

A graph of the two reaction curves can be useful. The graph below shows the two reaction curves. (For graphing Firm 2's reaction curve, we use the `implicit` command because it is of the form $x1 = f(x2)$.) The expressions names `x2max` and `x1max` and the values that result from our parameters are used to guide the graphing.

The graph shows that if Firm 2 were to produce no output then Firm 1 (a monopolist) would produce $x1 = \frac{47.5}{3} = 15\frac{5}{7}$. Each one-unit change in $x2$ would reduce $x1$ by $\frac{1}{3}$, as we saw above. Likewise, if Firm 2 were a monopolist, the resulting output would be 18 units, and Firm 2's output falls by $\frac{2}{5}$ units per one-unit increase in $x1$, so that $x1 = 45$ results in $x2 = 0$.

```
[x1max,x2max]:[rhs(solve(R1(x1,A,B,b2,c2),x1)
  [1]),rhs(solve(R1(x2,A,B,b1,c1),x2)[1])]$
[x1max0,x2max0]:subst([A=A0,B=B0,b1=b10,b2=b20],
  [x1max,x2max]);   wxdraw2d(... Some commands
 omitted ... key="R1",implicit( R1(x2,A0,B0,
  b10,c10)-x1,x1,0,x1max0,x2,0,x2max0),
 ... ,key="R2",explicit(R2(x1,A0,B0,b20,c20),
  x1,0,x1max0), title="Reaction Curves")$
```

$$[45, \frac{95}{2}]$$

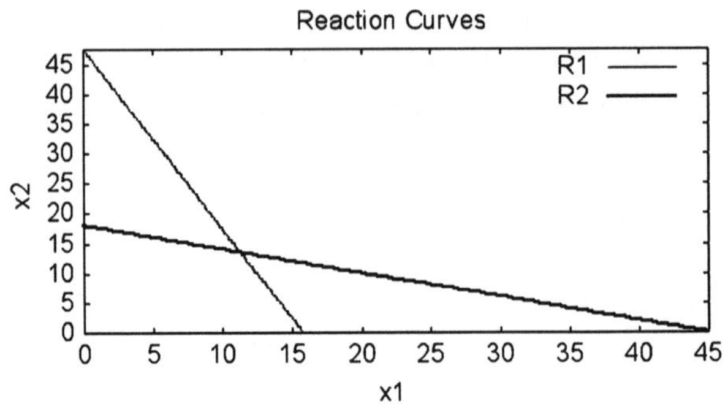

14.2.2 Solution

The intersection of these two curves represents the Cournot solution. The values appear below.

```
[x1N0,x2N0]:float(subst([A=A0,B=B0,b1=b10,b2=b20,
 c1=c10,c2=c20],[x1N,x2N])); p0: p(x1N0,x2N0,A0,B0);
```

[11.346, 13.462] 50.385

The two firms earn profits (ignoring fixed costs) of

```
P1cournot0:x1N0*p0-integrate(mc1(x1,b10,c10),x1,0,
x1N0);    P2cournot0:x2N0*p0-integrate(mc2(x2,b20,
c20),x2,0,x2N0); Pcournot0:P1cournot0+P2cournot0;
```

386.21 453.03 839.24

We consider three aspects of the Cournot solution: how it compares to a situation in which the firms are price takers, the inefficiency implicit in the Cournot solution, and how Cournot behavior compares to a situation in which the two firms collude to form a perfect cartel (monopoly).

14.2.2.1 Price Takers

Suppose that the two firms act as price takers and that the market clears. Then the horizontal sum of the firms' marginal cost curves is the industry supply curve. The first two steps in the cell below generate the inverse supply curve. The first step converts the firms' marginal cost curves to supply curves. The second step creates the industry supply curve. The third step specifies the inverse industry demand curve.

```
solve([mc1(x1,b1,c1)=p, mc2(x2,b2,c2)=p],[x1,x2]);
Xs(p,b1,c1,b2,c2):= ''(rhs(%[1][1])+rhs(%[1][2]));
Xd(p,A,B):= ''(rhs(solve(p=A + B*X,X)[1]));
```

$$[[x1 = \frac{p-b1}{c1}, x2 = \frac{p-b2}{c2}]]$$

$$\mathrm{Xs}\,(p, b1, c1, b2, c2) := \frac{p-b2}{c2} + \frac{p-b1}{c1}$$

$$\mathrm{Xd}\,(p, A, B) := -\frac{A-p}{B}$$

If the two firms were price takers, then the market would be as represented in the graph below, with $X \approx 35$ and $p \approx 30$. We determine the values below.

```
wxdraw2d( ... Some commands omitted ...
 implicit(Xs(p,b10,c10,b20,c20)-X,X,0,50,p,0,50 ),
 line_width=2,key="Demand",
 implicit(Xd(p,A0,B0)=X,X,0,50,p,0,50) )$
```

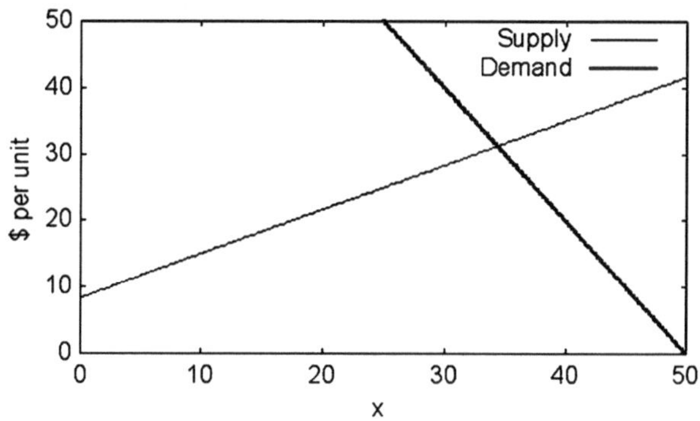

The first line below determines the competitive price (the price if firms were price takers). The second line creates a functional expression for this price as a function of the system's parameters. The next four lines bind names of the competitive price and quantities.

```
pcompexpr: solve(Xs(p,b1,c1,b2,c2)=Xd(p,A,B),p)$
pcomp(A,B,b1,c1,b2,c2):=''(rhs(%[1]))$ pcomp0:
float(pcomp(A0,B0,b10,c10,b20,c20))$  x1comp0:
float(rhs(solve(mc1(x1,b10,c10)=pcomp0,x1)[1]))$
x2comp0:float(rhs(solve(mc2(x2,b20,c20)=pcomp0,x2)
 [1]))$  Xcomp0: x1comp0+x2comp0$    matrix(
 ["Price, function","Price", "x1, comp", "x2, comp",
 "Market X"],[rhs(pcompexpr[1]),pcomp0,x1comp0,
 x2comp0,Xcomp0]);
```

Price, function	Price	x1, comp	x2, comp	Market X
$\frac{(b1\,c2+b2\,c1)\,B-c1\,c2\,A}{(c2+c1)\,B-c1\,c2}$	31.25	13.125	21.25	34.375

We can determine the profit levels (again, ignoring fixed costs) that would result if the firms were to behave as price takers. The result would be a substantial profit reduction.

```
P1comp0: pcomp0*x1comp0 -
  integrate(mc1(x1,b10,c10),x1,0,x1comp0)$
P2comp0: pcomp0*x2comp0 -
  integrate(mc2(x2,b20,c20),x2,0,x2comp0)$
Pcomp0 : P1comp0+P2comp0$  matrix(["Firm 1 Profit",
"Firm 2 Profit", "Total Profit,
 competition"],[P1comp0,P2comp0,Pcomp0]);
```

$$\begin{bmatrix} \text{Firm 1 Profit} & \text{Firm 2 Profit} & \text{Total Profit, competition} \\ 172.27 & 225.78 & 398.05 \end{bmatrix}$$

14.2.2.2 Efficiency

Consumers gain if the firms behave as price takers. The next cell shows how much consumer surplus increases. It also shows the total surplus (consumer surplus plus profits, ignoring fixed costs). The numbers show that the firms' gains from behaving as oligopolists are less than the harm done to consumers.

```
CScournot0:integrate(A0+B0*X,X,0,x1N0+x2N0)-
p0*(x1N0+x2N0);  CScomp0:integrate(
 A0+B0*X,X,0,Xcomp0)-pcomp0*Xcomp0;
TotalCournot0: Pcournot0 + CScournot0;
TotalComp0:Pcomp0+CScomp0$   matrix(["CS, Cournot",
"CS, comp","Surpluses, Cournot","Surpluses, comp"],
 [CScournot0,CScomp0,TotalCournot0,TotalComp0]);
```

$$615.42 \quad 1181.6 \quad 1454.7$$

$$\begin{bmatrix} \text{CS, Cournot} & \text{CS, comp} & \text{Surpluses, Cournot} & \text{Surpluses, comp} \\ 615.42 & 1181.6 & 1454.7 & 1579.7 \end{bmatrix}$$

14.2.2.3 Collusion

Suppose that the firms can cooperate in a way that maximizes their profits. Such cooperation requires that the firms allocate production between themselves in a way that equalizes marginal cost and that they select the quantity at which the joint marginal cost equals marginal revenue. In effect, the industry becomes a monopoly in which the single firm operates two plants.

When marginal cost curves are linear, equalizing marginal costs implies a linear relationship between $x1$ and $x2$. We assign names to the quantities that each firm would produce if they colluded to form a joint monopoly, *x1monop* and *x2monop*. The first output item shows the relationship between two firms' output values that must hold in order to maintain equal marginal costs (to keep industry total cost at its minimum value). The second output item shows the value of $x1$ as a function of total monopoly output, Xmonop, and the third item shows the corresponding value of $x2$.

The fourth item is a list of expressions given the system's parameter values: $x1 = \frac{5}{3} + \frac{Xmonop}{3}$ and $x2 = -\frac{5}{3} + \frac{2 \cdot Xmonop}{3}$. At very low values of Xmonop, only Firm 1 would have positive output.

```
solnmc: solve(mc1(x1,b1,c1)=mc2(x2,b2,c2),x2);
x1monop: rhs(solve(Xmonop=x1+rhs(solnmc[1]),x1)[1]);
x2monop: radcan(Xmonop-x1monop); [x1monop0,x2monop0]:
  subst([b1=b10,c1=c10,b2=b20,c2=c20],
  [x1monop,x2monop]);
```

$$[x2 = \frac{c1\,x1 - b2 + b1}{c2}] \qquad \frac{c2\,Xmonop + b2 - b1}{c2 + c1}$$

$$\frac{c1\,Xmonop - b2 + b1}{c2 + c1} \qquad [\frac{Xmonop + 5}{3}, \frac{2\,Xmonop - 5}{3}]$$

The output below plugs the two expressions above into the firms' marginal cost functions. The output items confirm that $mc1 = mc2$. The second list shows the marginal cost function for this cartel: $MC = \frac{25}{3} + \frac{2 \cdot Xmonop}{3}$

```
radcan([mc1(x1monop,b1,c1),mc2(x2monop,b2,c2)]);
subst([b1=b10,c1=c10,b2=b20,c2=c20],%);
```

$$[\frac{c1\,c2\,Xmonop + b1\,c2 + b2\,c1}{c2 + c1}, \frac{c1\,c2\,Xmonop + b1\,c2 + b2\,c1}{c2 + c1}]$$

$$[\frac{2\,Xmonop + 25}{3}, \frac{2\,Xmonop + 25}{3}]$$

Because output is allocated between the two firms so that $mc1 = mc2$, we can define the industry marginal cost as either. We now have marginal cost as a function of total output and the cost parameters. The second function, MCmonop0 is the functional relationship given the specific values of these parameters.

```
MCmonop(Xmonop,b1,c1,b2,c2):=
''(radcan(mc1(x1monop,b1,c1)));    MCmonop0(Xmonop):=
  ''(MCmonop(Xmonop,b10,c10,b20,c20));
```

$$\text{MCmonop}[(Xmonop, b1, c1, b2, c2) := \frac{c1\,c2\,Xmonop + b1\,c2 + b2\,c1}{c2 + c1}$$

$$\text{MCmonop0}\,(Xmonop) := \frac{2\,Xmonop + 25}{3}$$

Now we can determine the profit-maximizing quantity by setting marginal cost equal to marginal revenue.

```
solve(MCmonop(Xmonop,b1,c1,b2,c2)=A+2*B*Xmonop,
 Xmonop);   Xmonop(A,B,b1,c1,b2,c2):=''(rhs(%[1]));
```

$$[Xmonop = -\frac{(c2+c1)\,A-b1\,c2-b2\,c1}{(2\,c2+2\,c1)\,B-c1\,c2}]$$

$$\mathrm{Xmonop}\,(A,B,b1,c1,b2,c2) := -\frac{(c2+c1)\,A-b1\,c2-b2\,c1}{(2\,c2+2\,c1)\,B-c1\,c2}$$

Given this general expression for the profit-maximizing output, we can determine the value for our illustrative system. The first output item in the list below shows that $Xmonop = 19.643$. The second line, an embedded list, shows the division of this output between the two firms. The third list item shows the monopoly price.

```
[Xmonop0 : float(Xmonop(A0,B0,b10,c10,b20,c20)),
[x1monop0,x2monop0]:subst(Xmonop=Xmonop0,[x1monop0,
 x2monop0]),        pmonop0: A0+B0*Xmonop0];
```

$$[19.643, \quad [8.2143, \quad 11.429], \quad 60.714]$$

The graph below shows that same information regarding industry output, $X = 19.643$, and price, $p = \$60.714$. The marginal cost curve is constructed so that at every quantity the marginal cost is the same for each of the two firms.

```
wxdraw2d( ... Some commands omitted ...  explicit(
 MCmonop(X,b10,c10,b20,c20),X,0,50),line_width=2,
 key="Demand",explicit(A0+B0*X,X,0,50),color=gray,
 key="MR",explicit(A0+2*B0*X,  X,0,25), ...)$
```

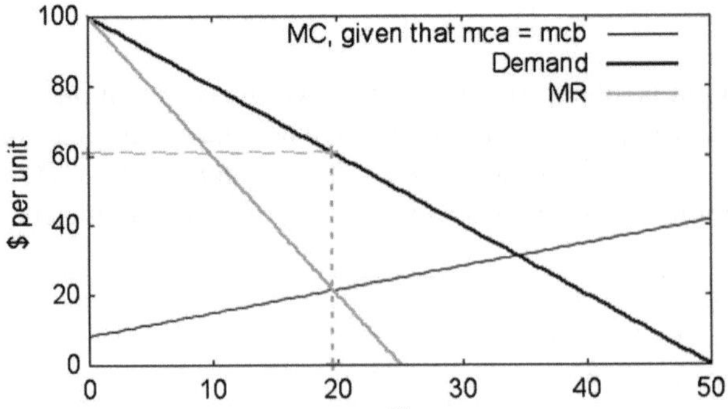

The surpluses are determined next. The table shows that the two firms gain, both jointly and individually, from colluding, given that profits are allocated according to the quantities produced individually. Consumers gain most if the firms behave as price takers and least if firms collude. As one would expect, the Cournot solution is better for consumers than monopoly but worse than competition.

```
CSmonop0:float(integrate(A0+B0*X-pmonop0,X,0,k
 Xmonop0))$  P1monop0:pmonop0*x1monop0 -
 integrate(mc1(x1,b10,c10),x1,0,x1monop0)$
P2monop0: pmonop0*x2monop0 -integrate(
 mc2(x2,b20,c20),x2,0,x2monop0)$ SurplusMonop0:
 CSmonop0+P1monop0+P2monop0$
 matrix(["","Cournot",
 "Competition","Collusion"],["CS",CScournot0,
 CScomp0,CSmonop0],["P1",P1cournot0,P1comp0,
 P1monop0],["P2",P2cournot0,P2comp0,P2monop0],
 ["Surpluses",CScournot0+P1cournot0+P2cournot0,
 CScomp0+P1comp0+P2comp0,
 CSmonop0+P1monop0+P2monop0]);
```

	Cournot	Competition	Collusion
CS	615.42	1181.6	385.84
P1	386.21	172.27	390.18
P2	453.03	225.78	514.29
Surpluses	1454.7	1579.7	1290.3

Each firm gains from collusion. If, however, Firm 1 holds the expectation that Firm 2's output is not affected by Firm 1's output, then it can do even better. Likewise for Firm 2. The graphs below show why. Each graph shows a firm's marginal cost and marginal revenue curves, given the Cournot assumption. The triangles defined by the segment of the dotted line and the segments of marginal cost and marginal revenue between the firms' allotted output and the quantities at which the lines intersect depict the added profit that each firm can gain, assuming that the other firm maintains its allotted output level.

```
Firm1: gr2d( ... Some commands omitted ... explicit(
 mc1(x,b10,c10),x,0,25),line_width=2,key="mr1",
 explicit(A0-x2monop0+2*B0*x,x,0,25), ... )$
Firm2: gr2d( ... explicit(mc2(x,b20,c20),x,0,25),
 line_width=2,key="mr2", explicit(A0-x1monop0+
 2*B0*x, x,0,25), ... )$
wxdraw(Firm1,Firm2,dimensions=[480,400])$
```

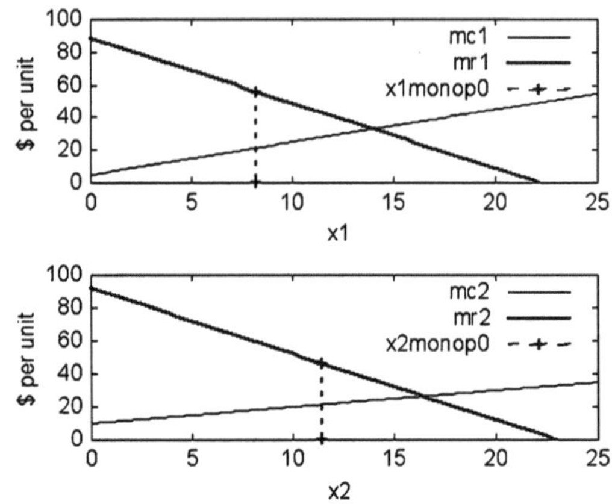

Given the attraction of colluding, one might wonder why firms do not routinely find ways to monitor each other and, thereby, reap the gains from collusion. Stigler, [3], argues that forming and maintaining cartels (formal collusions) is not as easy as it might seem (antitrust considerations aside). In particular, he shows the gain to each colluding firm from selling more than the allotted share as a function of the number of firms and the model's parameters.

The first problem is to allot each firms its share of production and its share of profits. The two are likely to be linked but not necessarily as directly as in our example. No firm knows its own marginal cost curve, and it certainly does not know the cost curve(s) of its fellow oligopolist(s). Each firm has an incentive to understate its cost in order to gain a larger share of the profits. Therefore, incentives exist that result in a cartel output that exceeds that of the ideal cartel.

Once the cartel is formed, the incentives to cheat come into play. These are likely to be especially strong if either cost curves or demand is subject to fluctuations. In such a case, the allotment will necessarily be defined in terms of market conditions, which each firm must estimate. This uncertainty gives each firm cover if it expands its production, because the other firm(s) will not be able to determine whether any sales erosion is due to changes in market conditions or to cheating by the other firm(s) in the industry.

The last observation, regarding cheating by the other firm(s) in the industry is symmetrical. Each firm must know that its output level is not easily monitored by the others and that this is true for all members of the cartel. Therefore, the Cournot assumption has some attraction, at least for small deviations from the ideal cartel output.

14.2.3 Increasing the Number of Firms

As the number of firms increases, the Cournot model yields results that approach those of the competitive model. In the special case of identical firms with constant marginal cost, the following relationship holds:

$$p = \frac{A}{N+1} + \frac{N}{N+1} \cdot c,$$

where c is the common marginal cost, A is the price intercept of the linear demand curve, and N is the number of Cournot oligopolists.[3]

We show that adding a third firm, again with a linear marginal cost curve, results in an output that approaches the competitive output.

We represent Firm 3's marginal cost as intermediate between that of the first two firms: its intercept is between `b10` and `b20`, and its slope is between `c10` and `c20`. Other values could be used.

```
p(x1,x2,x3,A,B):= A + B*(x1 + x2 + x3);
mc1(x1,b1,c1) := b1 + c1*x1; mc2(x2,b2,c2):=
 b2+c2*x2;  mc3(x3,b3,c3):= b3+c3*x3;
[A0,B0,b10,c10,b20,c20,b30,c30] :
   [100,-2,5.0,2.0,10.0,1.0,7.5,1.5]$
```

$$\mathrm{p}(x1, x2, x3, A, B) := A + B\ (x1 + x2 + x3)$$
$$\mathrm{mc1}(x1, b1, c1) := b1 + c1\, x1$$
$$\mathrm{mc2}(x2, b2, c2) := b2 + c2\, x2$$
$$\mathrm{mc3}(x3, b3, c3) := b3 + c3\, x3$$

The profit functions are defined as they were in the duopoly case.

```
p(x1,x2,x3,A,B)*x1-(F1+integrate(mc1(x1,b1,c1),
 x1));    P1(x1,x2,x3,A,B,b1,c1):= ''%$
p(x1,x2,x3,A,B)*x2 -(F2+integrate(mc2(x2,b2,c2),
 x2));    P2(x1,x2,x3,A,B,b1,c1):= ''%$
p(x1,x2,x3,A,B)*x3-(F3+integrate(mc3(x3,b3,c3),
 x3));    P3(x1,x2,x3,A,B,b1,c1):= ''%$
```

$$-F1 + x1\ ((x3 + x2 + x1)\ B + A) - \frac{c1\, x1^2}{2} - b1\, x1$$
$$- F2 + x2\ ((x3 + x2 + x1)\ B + A) - \frac{c2\, x2^2}{2} - b2\, x2$$
$$- F3 + x3\ ((x3 + x2 + x1)\ B + A) - \frac{c3\, x3^2}{2} - b3\, x3$$

We now solve for three reaction curves, each stated in terms of the variables $x1$, $x2$, and $x3$; the demand parameters, A and with B; and firm-specific marginal cost parameters. The fourth output line shows the reaction curves given the parameter values specified above.

[3]This proposition is derived in most textbooks. See Hallam, p. 14. Hallam (pp. 16–18) also treats the case in which firms have different, though still constant, marginal cost functions.

```
solve(diff(P1(x1,x2,x3,A,B,b1,c1),x1),x1)$
  R1(x2,x3,A,B,b1,c1):=''(rhs(%[1]));
solve(diff(P2(x1,x2,x3,A,B,b2,c2),x2),x2)$
  R2(x1,x3,A,B,b2,c2):=''(rhs(%[1]));
solve(diff(P3(x1,x2,x3,A,B,b3,c3),x3),x3)$
  R3(x1,x2,A,B,b3,c3):=''(rhs(%[1]));
[R10,R20,R30]:radcan([R1(x2,x3,A0,B0,b10,c10),
  R2(x1,x3,A0,B0,b20,c20),R3(x1,x2,A0,B0,b30,c30)]);
```

$$\mathrm{R1}(x2, x3, A, B, b1, c1) := -\frac{(x3+x2)\,B+A-b1}{2\,B-c1}$$

$$\mathrm{R2}(x1, x3, A, B, b2, c2) := -\frac{(x3+x1)\,B+A-b2}{2\,B-c2}$$

$$\mathrm{R3}(x1, x2, A, B, b3, c3) := -\frac{(x2+x1)\,B+A-b3}{2\,B-c3}$$

$$\left[-\frac{2\,x3+2\,x2-95}{6}, -\frac{2\,x3+2\,x1-90}{5}, -\frac{4\,x2+4\,x1-185}{11}\right]$$

The three firms' output values are determined below. Firm 1 and Firm 2 produce less than before, but total output rises.

```
soln3firms: float(solve([R10=x1,R20=x2,R30=x3],
              [x1,x2,x3])[1]);
```

$$[x1 = 9.1087, x2 = 10.478, x3 = 9.6957]$$

The appearance of a third firm, which has cost curves similar to the first two firms, increases output from just under 25 units to 29.283 units. The price falls from 50.385 to 41.435. We leave the computation of profits and consumer surplus when three firms constitute the industry as an exercise.

```
[X3:rhs(sum(soln3firms[i],i,1,3)),
P3: A0+B0*X3];
```

$$[29.283, \quad 41.435]$$

14.3 The Stackelberg Model

The Stackelberg model is like the Cournot model in that Firm 1 does not think that Firm 2 will respond to Firm 1's output decision, and vice versa. It differs from the Cournot model in that one firm acts first and then the second firm follows. The leader assumes that the follower will react to the leader's best response function. This implies that the leader determines its output, taking the follower's output into account. More specifically, the follower's reaction curve determines the demand for the leader's product.

Recall the demand and marginal cost functions that are used in the Cournot example.

```
p(x1,x2,A,B):= A+B*(x1+x2); [mc1(x1,b1,c1):=
b1+c1*x1,mc2(x2,b2,c2):=b2+c2*x2]; [A0,B0,b10,
 c10,b20,c20]:[100.0,-2.0,5.0,2.0,10.0,1.0]$
 [p(x1,x2,A0,B0),mc1(x1,b10,c10),mc2(x2,b20,c20)];
```

$$\mathrm{p}(x1,x2,A,B) := A + B\ (x1 + x2)$$
$$[\mathrm{mc1}(x1,b1,c1) := b1 + c1\,x1, \mathrm{mc2}(x2,b2,c2) := b2 + c2\,x2]$$
$$[100.0 - 2.0\ (x2 + x1), 2.0\,x1 + 5.0, 1.0\,x2 + 10.0]$$

Because Firm 1, the leader, takes Firm 2's reaction function as given, we begin by deriving that function. Here is Firm 2's profit function.

```
P2(x1,x2,A,B,b2,c2):= ''(p(x1,x2,A,B)*x2-
integrate(mc2(x2,b2,c2),x2 )- F2);
```

$$\mathrm{P2}(x1,x2,A,B,b2,c2) :=$$
$$-F2 + x2\ ((x2 + x1)\ B + A) - \frac{c2\,x2^2}{2} - b2\,x2$$

We determine the reaction function for Firm 2 by setting the derivative of the profit function P2 with respect to its output level equal to zero and solving for $x2$. We name the solution stackx2 and use it to define Firm 2's reaction function.

```
x2stack: solve(diff(P2(x1,x2,A,B,b2,c2),x2),x2);
R2stack(x1,A,B,b2,c2):=''(rhs(x2stack[1]));
```

$$[x2 = -\frac{x1\,B + A - b2}{2\,B - c2}]$$

$$\mathrm{R2stack}(x1,A,B,b2,c2) := -\frac{x1\,B + A - b2}{2\,B - c2}$$

Firm 1 knows that Firm 2 behaves this way, and so it commits to an output level that maximizes its profits. The inverse demand for Firm 1's product is given below, as its profit function.

```
p1(x1,A,B,b2,c2):=''(subst(x2stack, p(x1,x2,A,B)));
P1(x1,A,B,b1,c1,b2,c2):= ''(p1(x1,A,B,b2,c2)*x1 -
  integrate(mc1(x1,b1,c1),x1) - F1);
```

$$\mathrm{p1}(x1,A,B,b2,c2) := B\ \left(x1 - \frac{x1\,B + A - b2}{2\,B - c2}\right) + A$$
$$\mathrm{P1}(x1,A,B,b1,c1,b2,c2) :=$$
$$-F1 + x1\ \left(B\ \left(x1 - \frac{x1\,B + A - b2}{2\,B - c2}\right) + A\right) - \frac{c1\,x1^2}{2} - b1\,x1$$

The expression for Firm 1's reaction function is derived as before. The profit function's derivative with respect to its output is set equal to zero, the result is solved for $x1$. The result is a quantity, an expression in terms of the system's parameters.

```
diff(P1(x1,A,B,b1,c1,b2,c2),x1);
x1stack: rhs(solve(%,x1)[1]);
```

$$x1\,B\ \left(1 - \frac{B}{2\,B - c2}\right) + B\ \left(x1 - \frac{x1\,B + A - b2}{2\,B - c2}\right) + A - c1\,x1 - b1$$
$$-\frac{(A + b2 - 2\,b1)\ B - c2\,A + b1\,c2}{2\,B^2 + (-2\,c2 - 2\,c1)\ B + c1\,c2}$$

We insert the expression `x1stack` into the function that characterizes Firm 2's reaction. The result is a quantity expressed in terms of the system's parameters.

```
R2stack(x1stack,A,B,b2,c2);
```

$$-\frac{-\frac{B\,((A+b2-2\,b1)\,B-c2\,A+b1\,c2)}{2\,B^2+(-2\,c2-2\,c1)\,B+c1\,c2}+A-b2}{2\,B-c2}$$

We determine how much each firm produces. Compare this result with the Cournot solution. Firm 1, the leader, produces more than with Cournot reaction functions, and Firm 2 produces less. Total output is more than in the Cournot case.

Two exercises: Determine the profits of the two firms and the consumer surplus given this Stackelberg equilibrium. Then repeat this section, making Firm 2 the leader.

```
x1stack0: subst([A=A0,B=B0,b1=b10,
 c1=c10,b2=b20,c2=c20],x1stack);
R2stack(x1stack0,A0,B0,b20,c20);
```

13.409 12.636

14.4 The Bertrand Model

Both the Cournot and Stackberg models assume that each oligopolist (duopolist in our illustration) believes that its quantity decision will not affect that of its fellow oligopolist(s), and therefore treats the output of the fellow oligopolist(s) as exogenous.

Bertrand developed a model of duopoly that treats each firm as taking the other's *price* as given. That is, the conjecture of each oligopolist is that price is exogenous to its own pricing decision. Like the other models, the simplest Bertand model is one in which the two firms produce identical goods, so that a single price must pertain when the system is in equilibrium. We proceed as before, restating the demand for the product and the marginal cost of the two sellers in the market. We also specify the same set of parameters that we used in the Cournot duopoly model.

```
p(x1,x2,A,B):=A+B*(x1+x2)$   [mc1(x1,b1,c1):=
 b1+c1*x1, mc2(x2,b2,c2):=b2+c2*x2]$
[A0:100,B0:-2,b10:5,c10:2,b20:10,c20:1]$
[p(x1,x2,A0,B0),mc1(x1,b10,c10),mc2(x2,b20,c20)]$
```

The Bertrand assumptions imply that the demand and marginal revenue curves that face a seller are not continuous and differentiable functions. The functions below define the demand and marginal revenue curves of the duopolists. The inverse demand curve that faces Firm 1's product cannot be higher than the price for which Firm 2 sells. If, however, Firm 1 charges even a slightly lower price than Firm 2, then the demand for its product coincides with the demand curve. The marginal revenue function is constructed in like fashion, with the marginal revenue curve's slope being twice that of the inverse demand curve. Firm 2's functions are defined in the same way.

```
p1(x1,A,B,p2):= if A+B*x1>p2 then p2 else A+B*x1;
mr1(x1,A,B,p2):=if A+B*x1>p2 then p2 else A+2*B*x1;
p2(x2,A,B,p1):= if A+B*x2>p1 then p1 else A+B*x2;
mr2(x2,A,B,p1):=if A+B*x2>p1 then p1 else A+2*B*x2;
```

$$\text{p1}\,(x1, A, B, p2) := \text{if } A + B\,x1 > p2 \text{ then } p2 \text{ else } A + B\,x1$$
$$\text{mr1}\,(x1, A, B, p2) := \text{if } A + B\,x1 > p2 \text{ then } p2 \text{ else } A + 2\,B\,x1$$
$$\text{p2}\,(x2, A, B, p1) := \text{if } A + B\,x2 > p1 \text{ then } p1 \text{ else } A + B\,x2$$
$$\text{mr2}\,(x2, A, B, p1) := \text{if } A + B\,x2 > p1 \text{ then } p1 \text{ else } A + 2\,B\,x2$$

Each firm reacts to each of the set of possible prices that the other firm might charge. We begin by defining one end of that price range. To do that, we determine the quantity and price that would result if each firm were the only firm in the industry. Firm 1 would product $x1 = 15.833$ units and sell at a price $p = 68.333$. The corresponding values for Firm 2 are $x2 = 18.0$ and $p2 = 64.0$.

```
Firm1: ["1", x10: float(rhs(solve(
  A0+2*B0*x1=b10+c10*x1,x1)[1])), p10:A0+B0*x10]$
Firm2: ["2",x20: float(rhs(solve( A0+2*B0*x2=
 b20+c20*x2,x2)[1])),p20:A0+B0*x20]$  matrix(
 ["Firm","Monopoly x","Monopoly p"],Firm1,Firm2);
```

Firm	Monopoly x	Monopoly p
1	15.833	68.333
2	18.0	64.0

We begin to examine the price-setting process by evaluating the inverse demand and marginal revenue functions given the pair of prices from the table above.

```
p1(x1,A0,B0,p20);  mr1(x1,A0,B0,p20);
p2(x2,A0,B0,p10);  mr2(x2,A0,B0,p10);
```

$$\text{if } 100 - 2\,x1 > 64.0 \text{ then } 64.0 \text{ else } 100 - 2\,x1$$
$$\text{if } 100 - 2\,x1 > 64.0 \text{ then } 64.0 \text{ else } 100 - 4\,x1$$
$$\text{if } 100 - 2\,x2 > 68.333 \text{ then } 68.333 \text{ else } 100 - 2\,x2$$
$$\text{if } 100 - 2\,x2 > 68.333 \text{ then } 68.333 \text{ else } 100 - 4\,x2$$

The next graph shows these functions along with the two firms' marginal cost functions. The kink in Firm 1's demand curve and the discontinuity in its marginal revenue curve occur at $x1 = 15.833$. If Firm 1 charges a price above \$64, it produces nothing. If, however, it charges a price slightly lower than \$64, it can sell just over 64 units. Suppose that $p = 63.999$. Then the marginal revenue for the first 15.833 units is \$63.999, after which marginal revenue drops dramatically.

Of course, Firm 2 would lose the entire market if it continued to post a price of \$64, so it will reduce its price to a level just below \$63.999. The next question is whether this process has a logical destination. As we see below, it does.[4]

```
Firm1:gr2d( ... Some commands omitted ...
 explicit(mc1(x1,b10,c10),x1,0,-A0/B0), ...
 explicit(p1(x1,A0,B0,p20),x1,0,-A0/B0 ), ...
 explicit(mr1(x1,A0,B0,p20),x1,x20,-A0/B0 ))$
Firm2:gr2d( ... explicit(mc2(x2,b20,c20),x2,0,-A0/B0),
 ...explicit(p2(x2,A0,B0,p10),x2,0,-A0/B0 ), ...
 explicit(mr2(x1,A0,B0,p10),x1,x20,-A0/B0 ))$
wxdraw(Firm1,Firm2,columns=2,dimensions=[640,400])$
```

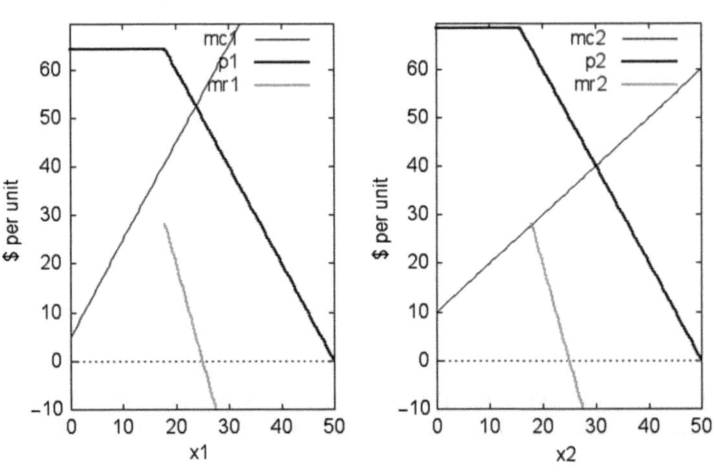

[4]We could have started our thought process with Firm 2, which would begin with the option of charging \$68.333. We know, however, that Firm 2 would sell for \$64.0 even if it were a monopolist, so the price will fall to this level, from which the process described in the text would proceed.

14.4.1 The Competitive Equilibrium

The result of the bidding process in this simple Bertrand model is the competitive outcome. To see this, we define the competitive outcome and confirm that, if this outcome pertains, neither firm gains by changing its behavior. The next cell defines the competitive supply expression, Xs.

```
solnx1: solve(mc1(x1,b1,c1)=p,x1);      solnx2:
 solve(mc2(x2,b2,c2)=p,x2);
Xs: rhs(solnx1[1])+rhs(solnx2[1]);
```

$$[x1 = \frac{p-b1}{c1}] \qquad [x2 = \frac{p-b2}{c2}] \qquad \frac{p-b2}{c2} + \frac{p-b1}{c1}$$

We derive the demand curve from the expression for the inverse demand curve and set that expression, Xd equal to Xs. For our parameters, the resulting price is \$31.25.

```
Xd: rhs(solve(A+B*X=p,X)[1]);
pcomp: float(rhs(solve(Xs=Xd,p)[1]));   pcomp0:subst(
 [A=A0,B=B0,b1=b10,c1=c10,b2=b20,c2=c20],pcomp);
```

$$-\frac{A-p}{B} \qquad \frac{(b1\,c2+b2\,c1)\,B-1.0\,c1\,c2\,A}{(c2+c1)\,B-1.0\,c1\,c2} \qquad 31.25$$

The cell below confirms that the quantity demanded equals the quantity supplied.

```
subst([A=A0,B=B0,b1=b10,c1=c10,b2=b20,
 c2=c20,p=pcomp0],     [Xd,Xs]);
```

$$[34.375, 34.375]$$

The graph below replicates the previous one, except that the part of marginal revenue that corresponds to the downward-sloping segment of each firm's demand curve is removed ($mr < 0$, and irrelevant).

At this price level, both firms produce the quantity at which $mc = mr = p$, so they have no reason to expand the quantity sold.

```
Firm1:gr2d( ... Some commands omitted ...
 explicit(mc1(x1,b10,c10),x1,0,-A0/B0), ...
 explicit(p1(x1,A0,B0,pcomp0),x1,0,-A0/B0)$
Firm2:gr2d( ... explicit(mc2(x2,b20,c20),x2,0,-A0/B0),
 ... explicit(p2(x2,A0,B0,pcomp0),x2,0,-A0/B0 ) ...
 explicit(mr2(x1,A0,B0,pcomp0),x1,x20,-A0/B0 )*/)$
wxdraw(Firm1,Firm2,columns=2,dimensions=[640,400])$
```

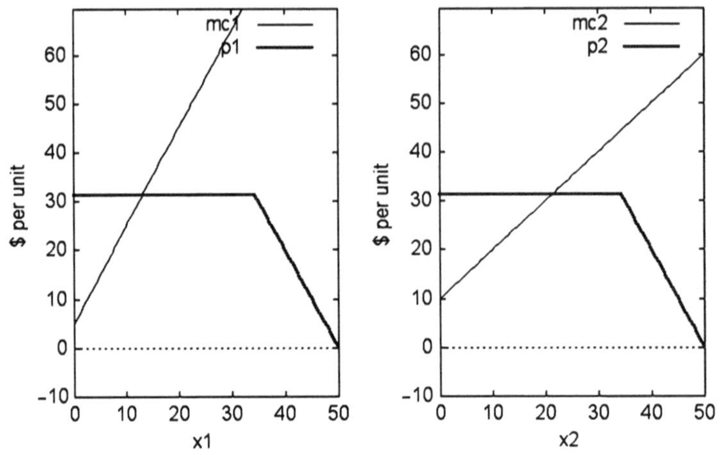

The result depicted in the graph above, and the values shown below, reflect each firm's assumption that the price is exogenous. Of all possible exogenous prices, $p = 31.25$ is the only one at which both firm's expectations are consistent with a cleared market. Given this price $x1 = 13.125$, $x2 = 21.25$, and so $X = 34.375$ is the quantity demanded and the quantity supplied.

Neither firm has any reason to change its behavior. Consider Firm 1. It will not reduce its price, because selling more than $x1 = 13.125$ results in $mr < mc$. It will not raise its price, because it assumes that Firm 2 will hold the price at $p = 31.25$ and so it will not be able to sell units for which $mr > mc$. Likewise, with Firm 2.

```
solnx1x2:float(solve([b10+c10*x1=pcomp0,b20+c20*x2=
 pcomp0],[x1,x2])[1])$    [x1comp0,
 x2comp0]:[rhs(solnx1x2[1]),rhs(solnx1x2[2])]$
Xcomp0:x1comp0+x2comp0$  Xdemand0:-(A0-pcomp0)/B0$
matrix(["x1comp0","x2comp0","Xcomp0","Xdemand0"],
 [x1comp0,x2comp0,Xcomp0,Xdemand0]);
```

$$\begin{bmatrix} \text{x1comp0} & \text{x2comp0} & \text{Xcomp0} & \text{Xdemand0} \\ 13.125 & 21.25 & 34.375 & 34.375 \end{bmatrix}$$

In Bertrand's original specification $c1 = c2 = 0$ and $b1 = b2$. Thus either of the two firms could produce a large enough quantity to meet market demand. Given this specification, the analysis above implies that $p = b$, the common marginal cost. This is Bertrand's conclusion.

Consider also the case in which $c1 = c2 = 0$, but $b1 \neq b2$. Suppose that $b1 > b2$. In this case, Firm 2 will bid a price just below $p = b1$ and will be the only seller.

Despite being the only seller, however, Firm 2 does not enjoy monopoly power (though it does earn a profit), as long as Firm 1 can resume production whenever $p \geq b1$. This case resembles *contestable markets* (see [2]).

14.4.2 An Alternative Specification

Modify Bertrand's specification as follows: Each firm expects the other to match its price. Now, the price cannot fall to the competitive level. Suppose that it does. Consider the elasticity of demand for each firm's product, given our modified assumption. If either firm raises its price, the other will follow. Thus, the firm's revenue from a higher price will be offset by two factors: reduced total quantity demanded and increased output by the other firm. The value of the first factor is $1/B$, and the value of the second is the inverse of the slope of the other firm's marginal cost curve.[5]

The implications for marginal revenue are shown below. Keep in mind that $p = mc = 31.25$ for both firms. For each firm, $mr < mc$, so both each firm would gain from raising its price, even though the other would match the price increase.

```
Ep10: (1/B0-1/c20)*pcomp0/x1comp0$ mr10:pcomp0*
   (1+1/Ep10)$        Ep20:(1/B0-1/c10)*pcomp0/x2comp0$
mr20: pcomp0*(1+1/Ep20)$      matrix(
  ["","Elasticity","mr"],["Firm 1",Ep10,mr10],
  ["Firm 2",Ep20,mr20]);
```

	Elasticity	mr
Firm 1	−3.5714	22.5
Firm 2	−1.4706	10.0

Before deriving a general solution to this model specification, consider the special case $c1 = c2 = 0$ and $b1 = b2$. In that case, this specification leads to an indeterminacy: If both firms charge the same price, neither firm's market share can be predicted.[6]

In general both firms must charge the same price, and that price must clear the market. Furthermore, each firm sets its own price (no collusion), which implies that each firm produces a quantity such that its marginal revenue equals its marginal cost. Recall that $mr = p \cdot (1 + \frac{1}{E})$, where E is the price elasticity of demand, a negative number.

[5]The expressions for marginal revenue are equivalent to those derived in Chap. 9 for the product of any member of an industry of price-taking firms.

[6]As an exercise, prove that if the firms can collude to agree on and maintain a fixed market share for each, then the result is the monopoly output.

The cell below generalizes the expressions above, defining the two price elasticities and marginal revenue values in terms of price and the system's parameters. The table shows the values given the parameter values used in this section.

```
[E1,E2] : [(1/B - 1/c2)*p/x1, (1/B - 1/c1)*p/x2]$
[mr1,mr2] : [p*(1+1/E1), p*(1 + 1/E2 )]$
[E10,E20,mr10,mr20]:subst([A=A0,B=B0,c1=c10,c2=c20],
   [E1,E2,mr1,mr2])$
matrix(["","Elasticity","Marginal Revenue"],
 ["Firm 1",E10,mr10],["Firm 2", E20,mr20]);
```

$$\begin{bmatrix} & \text{Elasticity} & \text{Marginal Revenue} \\ \text{Firm 1} & -\frac{3\,p}{2\,x1} & p\left(1-\frac{2\,x1}{3\,p}\right) \\ \text{Firm 2} & -\frac{p}{x2} & p\left(1-\frac{x2}{p}\right) \end{bmatrix}$$

The next cell solves for p, $x1$, and $x2$ in terms of the system's parameters. The expressions are quite long and not very intuitive, so they are suppressed.

```
solnx1x2p: solve([mr1=b1+c1*x1, mr2=b2+c2*x2,
   p=A+B*(x1+x2)],[p,x1,x2])[1]$
```

Substituting the parameter values into the solution defined above generates a set of prices and quantities that are consistent with equilibrium.

```
solnx1x2p0:float(subst([A=A0,B=B0,b1=b10,c1=c10,
 b2=b20,c2=c20],solnx1x2p));
[pBeq0,x1Beq0,x2Beq0] :[rhs(solnx1x2p0[1]),
rhs(solnx1x2p0[2]), rhs(solnx1x2p0[3])]$
```

$$[p = 41.364,\quad x1 = 13.636,\quad x2 = 15.682]$$

```
[E1Beq0,E2Beq0,mr10,mr20]:subst([p=pBeq0,x1=x1Beq0,
x2=x2Beq0,B=B0,c1=c10,c2=c20],[E10,E20,mr1,mr2]);
```

$$[-4.55,\quad -2.6377\ 32.273,\quad 25.682]$$

```
[mc10: b10+c10*x1Beq0, mc20: b20+c20*x2Beq0]$
matrix(["E1Beq0","E2Beq0","mr10","mc10", "mr20",
 "mc20"],[E1Beq0,E2Beq0,mr10,mc10, mr20,mc20]);
```

$$\begin{bmatrix} \text{E1Beq0} & \text{E2Beq0} & \text{mr10} & \text{mc10} & \text{mr20} & mc20 \\ -4.55 & -2.6377 & 32.273 & 32.273 & 25.682 & 25.682 \end{bmatrix}$$

```
Firm1:gr2d( ... Some commands omitted ...
explicit(mc1(x1,b10,c10),x1,0,-A0/B0), ...
explicit(p1(x1,A0,B0,pBeq0),x1,0,-A0/B0 ),...)$
Firm2:gr2d(...explicit(mc2(x2,b20,c20),x2,0,-A0/B0),
...explicit(p2(x2,A0,B0,pBeq0),x2,0,-A0/B0 )... )$
```

```
wxdraw(Firm1,Firm2,columns=2,dimensions=[640,400])$
```

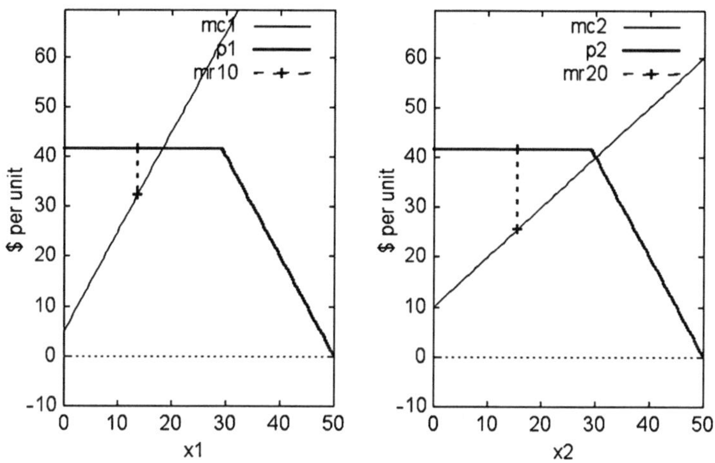

Now neither firm has an incentive to change its action: mc = mr for each of the two firms. Profits are the trapezoids between price and mc up to the dashed line segments less any fixed costs.

14.5 Nonidentical Products

This section considers cases in which a pair of duopolists sell similar but not identical products. We begin with the Bertrand model in which each duopolist takes the other's price as given. Then we consider the Cournot and Stackelberg models.

14.5.1 The Bertrand Model

This specification appears a bit more involved than the previous analysis, in which the two firms' products are identical, but the analysis is actually easier. In particular, calculus-based techniques can now be applied to the Bertrand model. The cell below contains a linear specification for demand and marginal cost functions. We change notation slightly from that used above, because two demand curves must be specified. The two firms' demand curves have intercept terms $A1$ and $A2$.

The coefficients of Firm 1's price, $p1$, are $a1$ and $a2$. The coefficients of Firm 2's price, $p2$ are $b1$ and $b2$. Each marginal cost curve includes a constant term ($C1$ or $C2$) and a slope term ($c1$ or $c2$). The named parameter list paramsEG is provided for later use. The last output line shows the two demand curves for this set of parameters. The expressions with EG appended to the name are derived by applying these parameter values.

```
[mc1:C1 + c1*x1, mc2:C2 + c2*x2];
[x1:A1 + a1*p1 + b1*p2, x2: A2 + a2*p1 + b2*p2];
paramsEG: [A1=1000,A2=1500,a1=-2,b1=1,a2=2,b2=-5,
   C1=10,c1=0.2,C2=20,c2=0.1]$
[x1EG,x2EG]:subst([paramsEG],[x1,x2]);
```

$$[C1 + c1\,x1, C2 + c2\,x2]$$
$$[A1 + b1\,p2 + a1\,p1, A2 + b2\,p2 + a2\,p1]$$
$$[p2 - 2\,p1 + 1000, -5\,p2 + 2\,p1 + 1500]$$

The analysis differs only slightly from what we have seen in the previous section, so we can sketch it quickly. The next cell shows expression $P1$, Firm 1's profit as a function of its output rate (ignoring fixed cost), and the implied reaction curve. The first input line contains two commands that determine the function for total variable cost. The second line defines Firm 1's profit. The third line creates a named expression for Firm 1's reaction curve. The appended bracketed term, [1], extracts the items from the list.

The last line shows the function given the parameter list provided above. It shows that $p1$ and $p2$ are directly related, as one would expect.

```
integrate(C1+c1*x,x)$  TVC1:subst(x1,x,%);
P1:radcan(p1*x1-TVC1);
rcBertrand1: solve(diff(P1,p1),p1)[1];
rcBertrand1EG: radcan(subst(paramsEG,%));
```

$$(A1 + b1\,p2 + a1\,p1)\,C1 + \frac{c1\,(A1+b1\,p2+a1\,p1)^2}{2}$$
$$-((2\,A1 + 2\,b1\,p2 + 2\,a1\,p1)\,C1 + c1\,A1^2$$
$$+ (2\,b1\,c1\,p2 + (2\,a1\,c1 - 2)\,p1)\,A1 + b1^2\,c1\,p2^2$$
$$+ (2\,a1\,b1\,c1 - 2\,b1)\,p1\,p2 + \left(a1^2\,c1 - 2\,a1\right)\,p1^2)/2$$
$$p1 = -\frac{a1\,C1+(a1\,c1-1)\,A1+(a1\,b1\,c1-b1)\,p2}{a1^2\,c1-2\,a1}$$
$$p1 = \frac{7\,p2+7100}{24}$$

The next cell repeats the process described above, deriving expressions that pertain to Firm 2. The two commands in the first line of the cell above are merged below.

```
TVC2:subst(x2,x,integrate(C2+c2*x,x)); P2:p2*x2-TVC2;
rcBertrand2: solve(diff(P2,p2),p2)[1];
rcBertrand2EG: radcan(subst(paramsEG,%));
```

$$(A2 + b2\,p2 + a2\,p1)\ C2 + \frac{c2\,(A2+b2\,p2+a2\,p1)^2}{2}$$

$$-(A2 + b2\,p2 + a2\,p1)\ C2 - \frac{c2\,(A2+b2\,p2+a2\,p1)^2}{2} + p2\ (A2 + b2\,p2 + a2\,p1)$$

$$p2 = -\frac{b2\,C2+(b2\,c2-1)\,A2+(a2\,b2\,c2-a2)\,p1}{b2^2\,c2-2\,b2}$$

$$p2 = \frac{6\,p1+4700}{25}$$

The next cell shows the prices, $p1 = 377.06$ and $p2 = 278.49$, that are consistent with Nash equilibrium, given the system's parameters. It also shows the implied quantity sold by each firm and the industry quantity, $x1 = 524.37$, $x2 = 861.65$, and $X = x1 + x2 = 1386.0$.

```
pricesEG: float(solve([rcBertrand1EG,
 rcBertrand2EG],[p1,p2]))[1];
 subst(pricesEG,[x1EG,x2EG]); %[1]+%[2];
```

$$[p1 = 377.06,\quad p2 = 278.49] \quad [524.37, 861.65] \quad 1386.0$$

The cell below uses the `implicit` command to graph the two reaction curves, expressed in terms of prices. The line before the `wxdraw2d` command is used to define the ranges for the graphs.

```
[p1EGmax,p2EGmax]:    [1.25*rhs(pricesEG[1]),
 1.25*rhs(pricesEG[2])]$ wxdraw2d( ... Some commands
 omitted ... title="Bertrand Reaction Curves", ...
 key="Reaction Curve 1",implicit(rcBertrand1EG,p1,0,
 p1EGmax,p2,0,p2EGmax), ... key="Reaction Curve 2",
 implicit(rcBertrand2EG,p1,0,p1EGmax,p2,0,p2EGmax))$
```

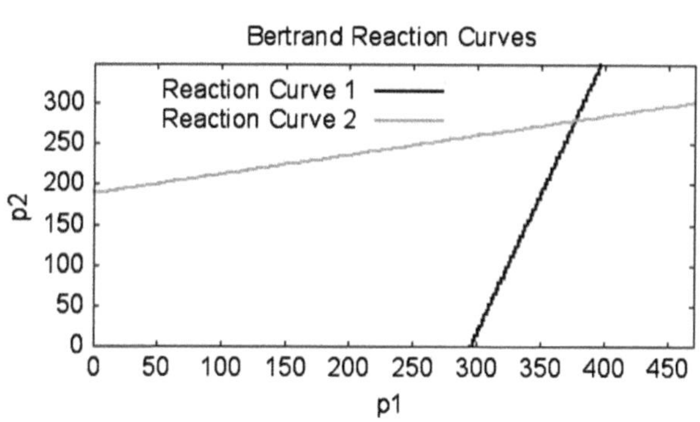

14.5.2 The Cournot Model

To analyze the Cournot and Stackelberg models we cast the analysis in terms of the inverse demand curves. The first cell derives the relevant expressions.

```
[mc1:C1+c1*x1,  mc2:C2+c2*x2]; solnDem: solve(
 [x1=A1+a1*p1+b1*p2,x2=A2+a2*p1+b2*p2],[p1,p2]);
[p1,p2]:[rhs(solnDem[1][1]),rhs(solnDem[1][2])]$
```

$$[C1+c1\,x1, C2+c2\,x2]$$

$$[[p1=\frac{b1\,(x2-A2)+b2\,A1-b2\,x1}{a2\,b1-a1\,b2},\quad p2=-\frac{a1\,(x2-A2)+a2\,A1-a2\,x1}{a2\,b1-a1\,b2}]]$$

As before we place the parameter values in a named list, to be applied throughout the remaining analysis. The expressions for inverse demand curves are $p1 = 812.5 - 0.625 \cdot x1 - 0.125 \cdot x2$ and $p2 = 625 - 0.25 \cdot x1 - 0.25 \cdot x1$.[7]

```
paramsEG: [A1=1000,A2=1500,a1=-2,b1=1,a2=2,b2=-5,
   C1=10,c1=0.2,C2=20,c2=0.1]$
[p1EG,p2EG]:radcan(subst([paramsEG],[p1,p2]));
```

$$[-\frac{x2+5\,x1-6500}{8},\quad -\frac{x2+x1-2500}{4}]$$

We next determine the profit function and the implied reaction curve expression for Firm 1.

```
TVC1:integrate(C1+c1*x1,x1); P1:radcan(p1*x1-TVC1);
rcCournot1: solve(diff(P1,x1),x1)[1];
rcCournot1EG: radcan(subst(paramsEG,%));
```

$$x1\,C1+\frac{c1\,x1^2}{2}$$

$$-\frac{(2\,a1\,b2-2\,a2\,b1)\,x1\,C1-2\,b1\,x1\,A2+2\,b2\,x1\,A1+2\,b1\,x1\,x2+((a1\,b2-a2\,b1)\,c1-2\,b2)\,x1^2}{2\,a1\,b2-2\,a2\,b1}$$

$$x1=-\frac{(a1\,b2-a2\,b1)\,C1-b1\,A2+b2\,A1+b1\,x2}{(a1\,b2-a2\,b1)\,c1-2\,b2}$$

$$x1=-\frac{5\,x2-32100}{58}$$

The counterparts for Firm 2 are determined in like fashion.

```
TVC2:integrate(C2+c2*x1,x1); P2:radcan(p2*x2-TVC2);
rcCournot2: solve(diff(P2,x2),x2)[1];
rcCournot2EG: radcan(subst(paramsEG,%));
```

[7] That the coefficient of $x1$ equals that of $x2$ in the $p2$ equation should be of no concern. The units in which $x1$ and $x2$ are expressed are not identical, so the slope coefficients cannot be compared.

$$x1\,C2 + \frac{c2\,x1^2}{2}$$

$$-\frac{(2\,a1\,b2 - 2\,a2\,b1)\,x1\,C2 + 2\,a1\,x2\,A2 - 2\,a2\,x2\,A1 - 2\,a1\,x2^2 + 2\,a2\,x1\,x2 + (a1\,b2 - a2\,b1)\,c2\,x1^2}{2\,a1\,b2 - 2\,a2\,b1}$$

$$x2 = \frac{a1\,A2 - a2\,A1 + a2\,x1}{2\,a1}$$

$$x2 = -\frac{x1 - 2500}{2}$$

Solving the system of reaction curves provides values for each firm's output. The first output line shows these values, along with the total output. The second output line shows the implied prices.

The Bertrand quantities are $x1 = 524.37$, $x2 = 861.65$, and $X = x1 + x2 = 1386.0$. The Bertrand prices are $p1 = 377.06$ and $p2 = 278.49$. We can conclude that the Bertrand solution differs from the Cournot solution. We cannot, however, make a general statement about the relative values.

```
[CournotQuantitiesEG: float(solve([rcCournot1EG,
 rcCournot2EG],   [x1,x2]))[1],
 CournotQuantitiesEG[1]+CournotQuantitiesEG[2]];
subst(CournotQuantitiesEG,[p1EG,p2EG]);
```

$$[[x1 = 465.77,\quad x2 = 1017.1],\quad x2 + x1 = 1482.9] \qquad [394.26,\quad 254.28]$$

The graph below shows the two reaction curves and the values produced above. This graph differs its Bertrand counterpart in that it is expressed in terms of quantities rather than prices.

```
temp:solve(rcCournot1EG,x2); wxdraw2d( ... Some
 commands omitted ... key= "Firm 1 Reaction Curve",
 explicit(rhs(temp[1]),x1,0,600), ... key=
 "Firm 2 Reaction Curve",
 explicit(rhs(rcCournot2EG),x1,0,600))$
```

$$[x2 = -\frac{58\,x1 - 32100}{5}]$$

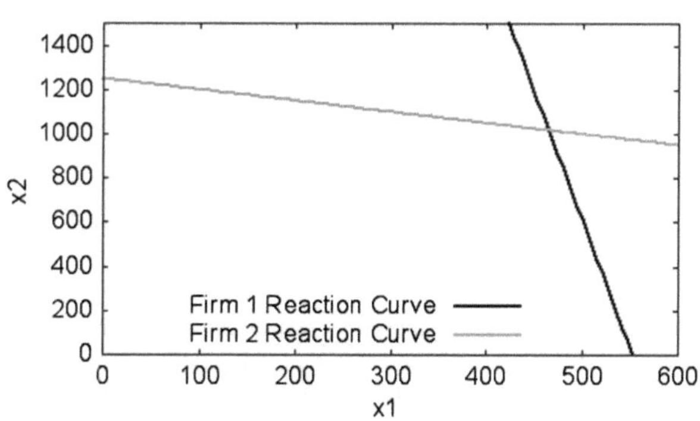

14.5.3 The Stackelberg Model

For the Stackelberg model, we insert Firm 2's Cournot reaction function into Firm 1's inverse demand curve and proceed as before. The demand curve and the associated profit function appear below.

```
p1Stack: rhs(subst(rcCournot2,solnDem[1][1]));
p1StackEG: float(subst(paramsEG,%));
P1Stack: p1Stack*x1 - TVC1;
P1StackEG: radcan(subst(paramsEG,P1Stack));
```

$$\frac{b1\left(\frac{a1\,A2-a2\,A1+a2\,x1}{2\,a1}-A2\right)+b2\,A1-b2\,x1}{a2\,b1-a1\,b2}$$

$$-0.125\,(-0.25\,(2.0\,x1-5000.0)+5.0\,x1-6500.0)$$

$$-x1\,C1+\frac{x1\left(b1\left(\frac{a1\,A2-a2\,A1+a2\,x1}{2\,a1}-A2\right)+b2\,A1-b2\,x1\right)}{a2\,b1-a1\,b2}-\frac{c1\,x1^2}{2}$$

$$-\frac{53\,x1^2-51700\,x1}{80}$$

The next cell shows the output for Firm 1, stated as a function of the system's parameters.

```
x1Stack:solve(diff(P1Stack,x1),x1);
x1StackEG:subst(paramsEG,x1Stack);
```

$$[x1=-\frac{\left(2\,a1^2\,b2-2\,a1\,a2\,b1\right)C1-a1\,b1\,A2+(2\,a1\,b2-a2\,b1)\,A1}{\left(2\,a1^2\,b2-2\,a1\,a2\,b1\right)c1-4\,a1\,b2+2\,a2\,b1}]$$

$$[x1=487.74]$$

Substituting Firm 1's output into Firm 2's reaction curve yields Firm 2's output. For the current set of parameters, we see that the results differ from both the Bertrand solution and the Cournot solution.

```
subst(x1Stack,rcCournot2);
subst(paramsEG,%);  %+x1StackEG[1];
```

$$x2=\frac{-\frac{a2\left(\left(2\,a1^2\,b2-2\,a1\,a2\,b1\right)C1-a1\,b1\,A2+(2\,a1\,b2-a2\,b1)\,A1\right)}{\left(2\,a1^2\,b2-2\,a1\,a2\,b1\right)c1-4\,a1\,b2+2\,a2\,b1}+a1\,A2-a2\,A1}{2\,a1}$$

$$x2=1006.1 \qquad x2+x1=1493.9$$

References

1. Hallam A (2004) Noncooperative oligopoly models. Online course notes at http://www2.econ.iastate.edu/classes/econ501/hallam/documents/Oligopoply.pdf
2. Martin S (2000) The theory of contestable markets. At www.krannert.purdue.edu/faculty/smartin/aie2/contestbk.pdf
3. Stigler GJ (1964) A theory of oligopoly. J Polit Econ 72:44–61

Further Readings

The table in the preface lists chapters in microeconomics textbooks that relate to this chapter. The following can also be of value:

Gibbons R (1992) Game theory for applied economists. Princeton University Press, Princeton. A general treatment of game theory is beyond the scope of this chapter. This book provides a good overview of game theory as it applies to economic decisions.

Anderson ET, Dana JD (2009) When is price discrimination profitable? Manage Sci 55:980–989. The paper develops a framework for analyzing many important types of price discrimination: intertemporal price discrimination, damaged goods, advance purchase discounts, coupons(rebates), and information goods (list taken from the paper).

McAfee RP (2007) Pricing damaged goods. Economics Discussion Papers, No. 2007-2, Kiel Institute for the World Economy. http://www.economics-ejournal.org/economics/discussionpapers/2007-2. This paper shows how firms can segment markets by impairing some units its product and selling them at a reduced price. It presents numerous examples. A paper by Deneckere and McAfee (1996), cited in this paper, contains additional examples.

Further Reading

Chapter 15
Time

Abstract This chapter deals with implications of time for economic analysis. First, it analyzes investment, the implications of which span a number of time periods. The chapter considers cases in which costs and benefits occur at discrete intervals and cases in which they occur continuously. It also considers the implications of uncertainty for evaluating investments. The chapter closes with a dynamic market model, in which the implications of a sequence of single-period actions are considered.

15.1 Introduction

Most analysis so far has been conducted in terms of a single time period. Consumers maximize utility by selecting per-period consumption bundles. Firms maximize per-period profits. The market quantity is a number of units per period, given the price or pricing schedule applied to those units.

We analyzed shifts in demand and cost curves, but these comparisons are better thought of as "with vs without" the shift, rather than "before and after" the shift. An exception is the development of long-run supply curves, but even here, time is stylized and does not relate directly to the passage of a specific number of periods. That is, the analysis consists of comparative statics analysis as opposed to dynamic analysis.

Time comes into play in either of two ways: intertemporal decisions may be analyzed, or dynamic paths may be analyzed. We are on firmer ground in the first one, in which we develop methods of reducing a series of payments or receipts to a set of values that guides current decision making.

Regarding dynamics, the best we can expect is to develop some general insights. Unlike physical scientists, who can develop dynamic models of the movements of particles and of bodies, economists are developing models of human behavior. If an analyst were to develop a model that correctly predicted the dynamic response

M.R. Hammock and J.W. Mixon, *Microeconomic Theory and Computation: Applying the Maxima Open-Source Computer Algebra System*,
DOI 10.1007/978-1-4614-9417-1_15, © Springer Science+Business Media New York 2013

to a change in the economic system, those engaged in making decisions would adjust their behavior in ways that would allow them to benefit from the model's predictions. These adjustments would change the system's dynamics.

15.2 Intertemporal Decisions

So far, decisions made by consumers and firms have not involved the passage of time except that consumption and production are quantities per unit of time. In all cases, decisions relate only to values occurring during the relevant time frame. Models that deal in snapshots, ignoring the implications of current decisions for future options, provide insights into many aspects of consumers' and firms' behavior, but they cannot analyze decisions that inherently tie the present to the future.

This chapter examines aspects of consumers' behavior and firms' behavior when today's decisions have direct implications for the future. Consumers make decisions regarding housing and education, among others, in which today's decisions can have implications for the rest of their lives (and beyond, allowing for bequests and for capital markets). Firms must choose among investment options, with each option implying a specific stream of income and costs in the future. Some firms must manage renewable resources such as trees, choosing among rotation cycle lengths.

15.2.1 Review

Following sections develop models of decision making in an intertemporal setting for consumers and for firms. This section reviews basic relationships that provide a foundation for these models. It examines the relationship between current actions (typically expenditures) and future outcomes (typically flows of net income). A central variable in all such models is the interest rate, which provides an incentive for saving and serves to ration available funds among competing uses.

15.2.1.1 The Interest Rate, Time, and Compounding

The first expression below, `F1`, shows the *future value* of P invested today with the accrued sum being compounded at the end of each period. In this expression, P is a principal that is deposited, r is the interest rate, and t is the number of periods that the principal is deposited.

The second expression, `F2`, allows for compounding, n times per period. Of course, `F1` is a special case of `F2`. The third expression shows that as n becomes large, $F2 \to P \cdot e^{r \cdot t}$, where e is the basis for the system of natural logarithms.

```
[F1(P,r,t):=P*(1 + r)^t,   F2(P,r,n,t)  :=
 P*(1 + r/n)^(n*t), limit(F2(P,r,n,t),n,inf)];
```

$$[\text{F1}\,(P, r, t) := P\,(1 + r)^t, \quad \text{F2}\,(P, r, n, t) := P\left(1 + \frac{r}{n}\right)^{n\,t}, \quad e^{r\,t}\,P]$$

The graph below shows the effects of periodic compounding. In each case, $P = 100$ and compounding occurs until $t = 50$. The focus is on the value during the last 20 years, when the effect of the difference in compounding frequency is most pronounced. The lowest line shows the future values when compounding is annual, and the highest line shows these values when compounding is instantaneous. The two intermediate lines (not labeled) show values for semiannual and quarterly compounding.

The lines show that the frequency of compounding matters increasingly as the number of periods increases. At the end of 50 years, annual compounding yields $F = 1146.7$, but instantaneous compounding yields $F = 1218.2$, about 6 % more.

```
matrix(["Annual","Biannual","Quarterly","Continuous"],
[F2(100,0.05,1,50),F2(100,0.05,2,50), F2(100,0.05,
 4,50),100*exp(0.05*50)]);          wxdraw2d(xlabel =
 "Number of periods", ylabel="$",key="Annual",
 explicit(F1(100,.05,t),t,30,50),color=gray,key="",
 explicit(F2(100,.05,2,t),t,30,50),explicit(F2(100,
 .05,4,t),t,30,50),line_width=2,key="Instantaneous",
 explicit(100*exp(.05*t),t,30,50))$
```

$$\begin{bmatrix} Annual & Biannual & Quarterly & Continuous \\ 1146.7 & 1181.4 & 1199.5 & 1218.2 \end{bmatrix}$$

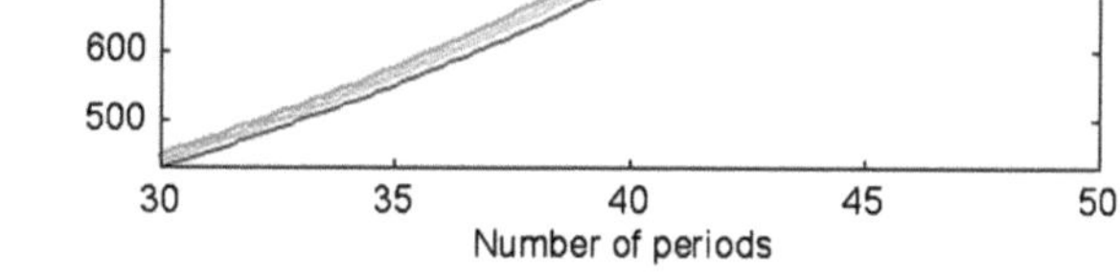

We can restate P as a present value. Suppose that an amount equal to $F1$ or $F2$ is to be received at the end of t periods. Then the present values are indicated by the expressions in the cell below. These present values are the amounts of money, that if invested at the present at the specified interest rate, would provide an amount FV (future value) at the end of t periods.[1]

[1]The equations used here must be modified slightly if the payment is to be paid at the beginning of period t, but nothing of importance changes.

Like future value, present value can allow for multiple compounding per period and for instantaneous compounding. The three solutions in the list solnPV show the expressions for present value when compounding is once per period, n times per period, and continuous. We refer to the inverse of compounding as *discounting*.

```
solnPV: [solve(F1(P,r,t)=FV,P),
 solve(F2(P,r,n,t)=FV, P),solve(P*exp(r*t)=FV,P)];
```

$$[[P = \frac{FV}{(r+1)^t}], \quad [P = \frac{FV}{(\frac{r}{n}+1)^{nt}}], \quad [P = e^{-rt}\, FV]]$$

The cell below specified these expressions as functions, with the appropriate arguments in parentheses. The output is suppressed.

```
PV1(FV,r,t):=       ''(rhs(solnPV[1][1]))$
PVn(FV,r,n,t):=     ''(rhs(solnPV[2][1]))$
PVcont(FV,r,t):=    ''(rhs(solnPV[3][1]))$
```

The cell below shows how the effect of discounting is affected by the interest rate and by the number of periods that pass before FV is to be received. Suppose that $FV = 100$ in all cases and that FV is to be received after 50 periods. The present value is $PV = 369.71$ when $r = 0.01$ but $PV = 0.0726$ when $r = 0.10$.

We leave finding the effects on these graphs of increasing compounding frequency as exercises.

```
[PV1(1000,.01,100),PV1(1000,.05,100),
 PV(1000,.10,100)];   wxdraw2d(... Some commands
 omitted ... explicit(PV1(1000,.02,t),t,1,50),
  ... explicit(PV1(1000,.05,t),t,1,50),
  ... explicit(PV1(1000,.10,t),t,1,50))$
```

$$[369.71, \quad 7.6045, \quad \mathrm{PV}(1000, 0.1, 100)]$$

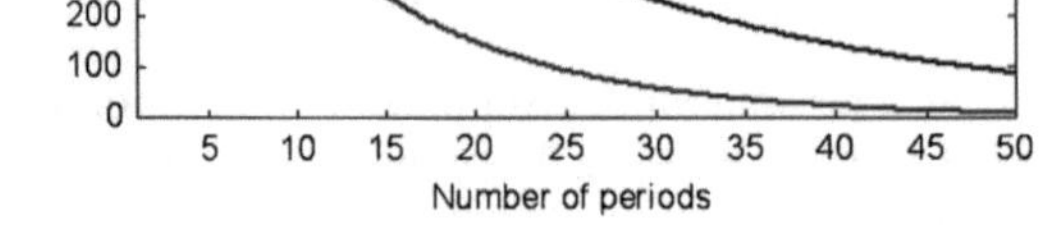

15.2.1.2 A Stream of Payments

Suppose that rather than receiving a single payment at the end of a specified number of periods, one receives a stream of payments beginning at the end of the present period and ending at the end of the specified number of periods.[2] The analysis is a direct extension of the present value calculation above: For each period, the present value of the period's contribution to the stream, Fi, where i ranges from 1 to t, the last year in the stream, is calculated. Then these individual present values are summed to give the present value of the stream, PVS:

$$PVS = \frac{FV1}{(1+r)} + \frac{FV2}{(1+r)^2} + \cdots + \frac{FVt}{(1+r)^t}.$$

The cell below illustrates this computation. Four unequal payments are to be received at the end of 1, 2, 3, and 4 years. The relevant discount (interest) rate is 5 %. The first output line shows the individual present values for the four payments. The second line shows the present value of the stream, PVS, which is the sum of the individual values in the first output line.

```
tlist:[1,2,3,4]$  FVlist:[100,125,90,150]$
PVlist: PV1(FVlist,0.05,tlist);
sum(PVlist[t],t,1,length(PVlist) );
```

$[95.238, \ 113.38, \ 77.745, \ 123.41] \qquad 409.77$

When all periodic payments are equal, as with mortgage payments, the PV equation above becomes a geometric series. Applying the formula for this series yields this expression for PVS:

$$PVS = \frac{\frac{FV}{1+r} - \frac{FV}{1+r} \cdot \left(\frac{1}{1+r}\right)^t}{1 - \frac{1}{1+r}}.$$

The expression can be simplified by multiplying the numerator and denominator by $1 + r$ and collecting terms:

$$PVS = FV \cdot \left(\frac{1-(1+r)^{-t}}{r}\right).$$

For comparison, the cell below shows the four values that sum to 354.6.

```
PVSeqn(FV,r,t):= FV*(1 - (1+r)^-t)/r;
PVSeqn(100,0.05,4); FVlist2:[100,100,100,100];
PVlist2: PV1(FVlist2,0.05,tlist);
sum(PVlist2[t],t,1,length(PVlist2) );
```

$\text{PVSeqn}(FV, r, t) := \frac{FV\left(1-(1+r)^{-t}\right)}{r} \qquad 354.6$

$[100, \ 100, \ 100, \ 100] \quad [95.238, \ 90.703, \ 86.384, \ 82.27] \qquad 354.6$

[2] For the moment, we proceed on the assumption of annual compounding. Later, we use instantaneous compounding because of its mathematical tractability.

The graph below demonstrates some important features of this function. Most obvious is that a higher interest rate reduces the present value. It also shows that the proportional reduction becomes more pronounced as the number of periods increases.

```
wxdraw2d(... Some commands omitted ... key=
 "r = 0.02",explicit(PVSeqn(100,.02,t),t,1,50),
 ... explicit(PVSeqn(100,.05,t),t,1,50),
 ... explicit(PVSeqn(100,.10,t),t,1,50))$
```

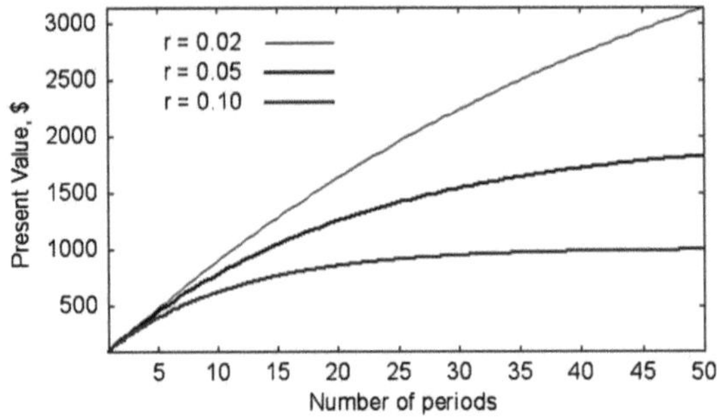

The graph suggests a second feature of the present value of a stream, that it asymptotically approaches a constant value as the number of periods increases. Thus, extending the stream beyond some number of periods adds negligible amounts to the stream's present value. When $r = 0.05$ the asymptotic value is $PVS = 2000$, and when $r = 0.10$, the asymptotic value is $PVS = 1000$. In each case, this value is F/r, which would be the value if F were to be paid at the end of each year forever. The cell below confirms this relationship.[3]

```
assume(r>0)$
limit(PVSeqn(FV, r, t), t, inf);
```

$\frac{FV}{r}$

15.2.2 Investment

The preceding section shows how to convert a flow of future earnings into a present value. This section applies the reasoning to investment. The investment may be in physical capital such as factories and equipment; it may be in financial assets such as bonds or stocks; or it may be in one's own productivity, wherein one purchases training or incurs migration costs in order to increase one's earnings. In all cases, investment involves incurring costs today, or in the near future, with the expectation of offsetting returns in the longer term. Accordingly, we revise our notation somewhat, replacing the present value of a series (PVS) with *net present value* (npv).

[3] *Maxima* must be told that $r > 0$.

15.2.2.1 Discrete Income Flow

We first analyze cases in which the investment generates a stream of discrete flows of income over a finite time period. Consider the following simple example: One can invest \$100 today in a short-lived project that generates \$25 per quarter for 1.5 years. The first output line expresses the net present value in a form much like the *pvs* expressions used above. The second output lines shows the result of an attempt to apply `solve` to determine the interest rate at which the net present value is zero. A rule for investment is to undertake all investments for which $npv > 0$. Even this relatively simple problem yields a high-order polynomial expression when we seek an analytical solution to $npv = 0$.

Irregular income flows can fail to yield any analytical solution at all or they can yield multiple solutions.[4]

The graph shows that, even though the expression for the solution to this expression is a high-order polynomial, the resulting relationship between r and npv is well behaved: Higher interest rates reduce the npv of this project. The final output line (below the graph) shows that $npv > 0$ whenever $r < 0.26$.[5]

```
npv:-100+sum(25/(1+r/2)^t,t,1,6); solve(npv,r);
wxdraw2d(xaxis=true,xlabel="r",ylabel="$",
key = "npv", explicit(npv, r, 0, 1))$
find_root(npv, r, 0, 1);
```

$$\frac{25}{\frac{r}{2}+1}+\frac{25}{\left(\frac{r}{2}+1\right)^2}+\frac{25}{\left(\frac{r}{2}+1\right)^3}+\frac{25}{\left(\frac{r}{2}+1\right)^4}+\frac{25}{\left(\frac{r}{2}+1\right)^5}+\frac{25}{\left(\frac{r}{2}+1\right)^6}-100$$

$$[0 = 2\,r^6 + 23\,r^5 + 108\,r^4 + 260\,r^3 + 320\,r^2 + 144\,r - 64]$$

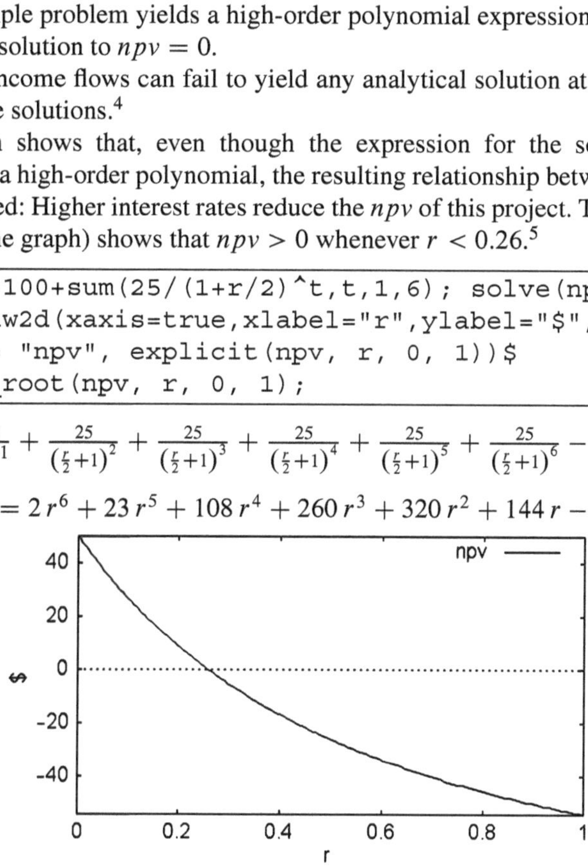

0.26

[4]An alternative method of investing is to sort potential investments according to the size of the rate at which $npv = 0$. This rate is called the investment's *internal rate of return* (IRR). This approach is flawed, however. It provides no guidance when the critical value of r cannot be calculated or it yields imaginary solutions, and only imperfect guidance when multiple real solutions occur.

[5]Confirm that applying `solve` to the formula for the present value of a finite series of equal flows, $F * (1 - (1 + r)^{-t})/r$, yields the same polynomial that appears in the cell below. Replace r with $\frac{r}{2}$.

Investments involving many more terms may be treated in a similar fashion. For example, suppose that one can invest \$10,000 today to refurbish an old house, with the result that the rental income from the house will increase by \$200 per month for 10 years. The next cell shows this example as it is cast in *Maxima*. Applying `solve` results in a polynomial expression whose order is 120.

The graph of the *npv* expression looks much like the one above and is not presented. Applying `find_root` shows that for interest rates below $r = 0.21$, this investment is profitable (it produces a positive net present value).

```
npv: -10000+sum(200/(1+r/12)^t,t,1,120)$
find_root(npv, r, 0, 1);
```

0.21

The income flows need not be constant throughout the investment period. Suppose that the stream in the first example is as indicated in the cell below, perhaps reflecting seasonality. In this example, because the larger flows occur earlier, the investment would be profitable as long as $r < 0.547$.

```
flowlist: [40,25,10,10,40,25]$
npv:sum(flowlist[t]/(1+r/4)^t,t,1,6)-100;
find_root(npv, r, 0,1);
```

$$\frac{40}{\frac{r}{4}+1}+\frac{25}{\left(\frac{r}{4}+1\right)^2}+\frac{10}{\left(\frac{r}{4}+1\right)^3}+\frac{10}{\left(\frac{r}{4}+1\right)^4}+\frac{40}{\left(\frac{r}{4}+1\right)^5}+\frac{25}{\left(\frac{r}{4}+1\right)^6}-100 \qquad 0.547$$

15.2.2.2 Continuous Income Flows

If the income flow is a continuous function of time, occurring throughout periods, then its representation becomes simpler. In the example below, we allow for different behavior over the first 5 years (perhaps before competitors can produce close substitutes for a new product) and thereafter.

```
assume(r>0)$  stream: if t<5 then 10*exp(0.1*t) else
 10*exp(0.1*5-0.05*(t-5))$   npv : -200 + integrate(
 10*exp(0.1*t - r*t),t,0,3)+integrate(10*exp(0.1*5
 -0.05*(t-5) - r*t),t,3,inf);  find_root(npv,r,0,1);
```

$$\frac{200\,e^{\frac{3}{5}-3r}}{20r+1}+10\left(\frac{10}{10r-1}-\frac{10\,e^{\frac{3}{10}-3r}}{10r-1}\right)-200 \qquad 0.0451$$

The first panel below shows the income stream for the first 50 periods. The second panel shows the net present value for interest rates between 0.01 and 0.05.

```
streamgraph: gr2d(... Some commands omitted ...
 explicit(stream, t, 0, 50) )$
npvgraph: gr2d(... ,explicit(npv,r,0.01,0.05))$
wxdraw(streamgraph,npvgraph, ... )$
```

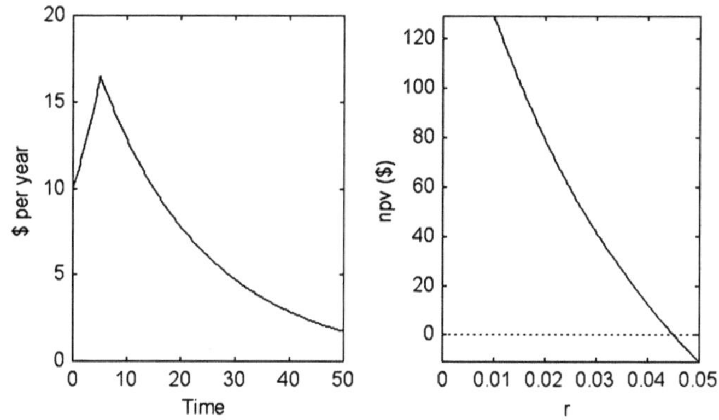

15.2.3 Risky Investment

The analysis so far has treated the stream of income flows and costs as known. Many payment streams, like mortgages or returns from high-grade bonds, are at least approximately free of risk and can, therefore, be treated as above. In many investment settings, however, the returns are not certain. This section deals with two such settings. In the first, two streams of income are listed and a probability is attached to each possible stream. (The number of possible streams and associated probabilities could be extended without affecting the substance of the analysis.) We then determine the *expected* net present value of this investment.

The second setting is one in which the investment can be undertaken either today or in the future. We find that when this latter option is added, then the relevance of the expected net present value becomes problematic.

15.2.3.1 Expected Net Present Value

The investment that we consider has these characteristics. The investor must incur an initial cost of \$100 and \$50 at the end of the first 5 years of operation, during which no revenue is accrued. Beginning at the end of year 6 and continuing for 40 years, profit contributions of \$150 per year will be realized if the project succeeds, after which profit contributions cease. If the project fails, the profit contributions will be zero.

First, determine the present value of cost, which consists of an initial cost at $t = 0$ and a stream for 5 years ($t = 1$ through $t = 5$) of \$50 per year. We use the formula developed in the preceding section for a series of constant flows into the stream.

Next, determine the present value of revenue in case of success. This involves two steps. First, use the same formula as above to discount the revenues back to the beginning of the time during which revenue flows (the beginning of period 6, which coincides with the end of period 5). Finally discount this value to the beginning of the first period (i.e., the present, $t = 0$).

In this example, any interest rate above 8.12 % results in a negative expected net present value. Given the riskiness of investments like this one, applying a rate in the range of 15 % is common. [6]

```
probsuccess: 0.25$
pvcost : 100 + (50/r)*(1 - 1/(1+r)^5)$
pvsuccess : 150*(1 - 1/(1+r)^40)*(1/(1+r)^5)/r$
npvsuccess:pvsuccess-pvcost$          Enpv:
 npvsuccess*probsuccess - pvcost*(1-probsuccess)$
r0: find_root(Enpv, r, 0.001,1)$  subst(r=r0,
 [pvcost,pvsuccess,npvsuccess,Enpv])$
numbers0:cons(r0,%)$  subst(r=.15,[pvcost,pvsuccess,
 npvsuccess,Enpv])$      numbers15:cons(0.15, %)$
numbers05:subst(r=.05,[pvcost,pvsuccess,npvsuccess,
 Enpv])$    numbers05:cons(.05,numbers05)$  names:
 ["r", "pvcost", "pvsuccess", "npvsuccess","Enpv"]$
matrix(names, numbers05, numbers0, numbers15);
```

$$\begin{bmatrix} r & \text{pvcost} & \text{pvsuccess} & \text{npvsuccess} & \text{Enpv} \\ 0.05 & 316.47 & 2016.7 & 1700.2 & 187.7 \\ 0.0812 & 299.03 & 1196.1 & 897.1 & 5.68434\,10^{-14} \\ 0.15 & 267.61 & 495.32 & 227.71 & -143.78 \end{bmatrix}$$

15.2.3.2 Timing of an Investment

The example above treats one type of risk. The firm must decide whether to undertake a project, with the prospect of one of two (or more) possible outcomes. In cases like this, the expected present value can guide decision makers.[7]

In cases like the one above, the net present value calculations address a question to which the answer is either "yes" or "no." For some investments, the choice is "yes" or "wait." Suppose that a homeowner is considering refinancing the mortgage. The NPV rule would dictate making the investment (incurring closing costs) if the result is a reduction in the present value greater than the closing cost. This rule, however, ignores another option: wait until later to see whether the mortgage rate falls even more. For simplicity, suppose that the refinancing will occur today or 1

---

[6]McAfee [2] explains why high rates apply even though the cost of capital to a firm (the interest rate at which it can borrow) is lower.

[7]Guide, but not dictate. Depending on firm's attitudes regarding risk, a firm might undertake a project with negative expected net present value, or it might reject one with positive expected net present value.

year hence. Waiting incurs a cost in that if the interest rate falls no farther, 1 year of unnecessarily high payments have been made. Worse, if the rate rises, the benefits of refinancing may be lost entirely.

Even so, the homeowner's net wealth is not likely to be maximized by refinancing now unless the net present value is some discrete positive value.[8] That is, the homeowner would not be indifferent between refinancing if the expected net present value of doing so is zero.

In a similar situation, suppose that a firm is deciding whether to begin pumping oil from a reservoir. Oil pumped and sold today is not available for sale in future periods. Situations like this are addressed by models of the *option value of investment*. Models of the option value of investment must be tailored to specific sets of conditions and can be quite arcane and mathematically challenging.

Fortunately, a relatively simple example provides insights into the implications of the option value of investment for the relevance of the simple net present value (NPV) formula. In particular, it shows that the NPV formula tends to result in too much investment at the present time and not enough waiting and investing later.

In general, determining a rule of maximizing one's expected wealth can be a difficult process. We can, however, gain significant insights into the process with a simple two-period model.[9] The firm can either make an investment of C in this period or wait until the next period to make its decision about whether to invest. The investment C could be spending on a drilling platform, or it could be the *strike price* for an option—the amount paid to purchase a share of the stock.

The actual value of the stock to be received as a result of the investment is a random variable V. We select units so that V falls between 0 and 1. Furthermore, and for simplification, we assume that V has a uniform distribution over this interval. This distribution implies that V falls between a and b (inclusive) with a probability of $b - a$. The range of V implies that $C < 1$; otherwise, the option has no positive value. The firm can make this investment at any time in the indefinite future, after waiting one period. The probability distribution function is the same in each period.

The firm must determine a *cutoff value*, which we label Vo. The NPV rule would set $Vo = C$. That is, if $V > C$, the firm makes the investment today. (It would be indifferent about whether to invest today if $V = C$.) The NPV is no longer appropriate, however. In general $Vo > C$, with the difference depending on the level of C and the discount rate r.

Denote the expected value of the option to invest as EVo. We assume that the firm wishes to undertake investments that have a positive expected value and only those investments. This assumes that the risk associated with any one investment is not correlated to that associated with other investments.

---

[8]The example of the mortgage is a complex one, in part because of correlation of interest rates across time and in part because the homeowner's options are not limited to the two periods indicated above.

[9]The development of this model follows [2], which states "This theory is developed in striking generality by Avinash Dixit and Robert Pindyck [1]."

The expression below defines EVo. The term $1 - Vo$ is the probability that $V > Vo$ in the first period (and in the second and subsequent periods). Therefore, $1 - Vo$ is the probability that the investment will be undertaken immediately. The expected value of V in the first period is the midpoint of the range $[Vo, 1]$, which equals $\frac{1+Vo}{2}$. If the opportunity occurred only once, then $EVo = (1 - Vo) * (\frac{1+Vo}{2} - C)$. If the firm can defer the decision for one period and then make a decision based on V in the second period (with Vo unchanged) and so forth into the indefinite future, then the expression below applies.[10]

```
expr:EVo=(1-Vo)*((1+Vo)/2-C)+Vo*(EVo/(1+r));
```

$$EVo = (1 - Vo)\left(\frac{Vo+1}{2} - C\right) + \frac{EVo\,Vo}{r+1}$$

We solve the expression above for EVo and then differentiate the solution with respect to Vo, the optimal value of which is to be determined. Setting the derivative equal to zero and solving for Vo yields two expressions. The second of the two implies a value $Vo > 1$ and can be ignored.[11]

```
solnEVo:ratsimp(solve(expr,EVo));  diffEVo:ratsimp(
diff(rhs(solnEVo[1]),Vo)); solndiff:solve(%,Vo);
```

$$[EVo = -\frac{((2r+2)Vo-2r-2)C+(-r-1)Vo^2+r+1}{2Vo-2r-2}]$$

$$\frac{(2r^2+2r)C+(r+1)Vo^2+(-2r^2-4r-2)Vo+r+1}{2Vo^2+(-4r-4)Vo+2r^2+4r+2}$$

$$[Vo = -\sqrt{r}\sqrt{-2C+r+2}+r+1, \quad Vo = \sqrt{r}\sqrt{-2C+r+2}+r+1]$$

The next cell extracts the relevant solution for Vo and expresses that value as a function of C and r. Each of the two panels shows the critical Vo for three values of C and for ranges of r. In the left panel, r ranges from 0 to 200 %; in the right panel it ranges from 0 to 10 %.[12] The larger r values in the first panel imply quite large discounting. Even so, $Vo > C$.

Furthermore, for smaller r values the difference is quite large. For $r = 0.10$, $Vo = 0.642$ when $C = 0$, so the NPV rule is far from appropriate. The differences are only slightly less striking for larger values of C: when $C = 0.2$, $Vo = 0.668$, and when $C = 0.4$, $Vo = 0.739$.

[10] If the firm can choose between investing today and investing after the end of the current period, then $EVo = (1 - Vo) * ((1 + Vo)/2 - C) + Vo * ((1 - Vo) * ((1 + Vo)/2 - C))/(1 + r)$.

Exercise: Confirm that this expression is correct, and confirm that it yields results that are similar to those below. The value of the option to invest in later periods falls for two reasons: (1) discounting, and (2) the decreasing size of the probability that such options will be exercised (that is, that the investment will not have been made earlier).

[11] Exercise: Derive the second derivative with respect to Vo and confirm that for plausible values of r and C, the second derivative is negative for all values of Vo.

[12] Thus, the second panel is a cutout of the upper left-hand corner of the first panel.

```
rhs(solndiff[1])$ Vo(C, r):=''%;    rmax: 2.0$
rlarge:gr2d( ... Some commands omitted ...
 grid=true,key="C = 0",explicit(Vo(0,r),r,0,rmax),
 ... explicit(Vo(0.2,r),r,0,rmax),...
 explicit(Vo(0.4,r),r,0,rmax) )$
rsmall:gr2d( ... grid=true,...  explicit(
 Vo(0,r),r,0,rmax/20), ... explicit(Vo(0.2,r),r,0,
 rmax/20), ... explicit(Vo(.4,r),r,0,rmax/20))$
wxdraw(rlarge,rsmall,columns=2 ... )$
```

$$\text{Vo}(C, r) := -\sqrt{r}\,\sqrt{-2C + r + 2} + r + 1$$

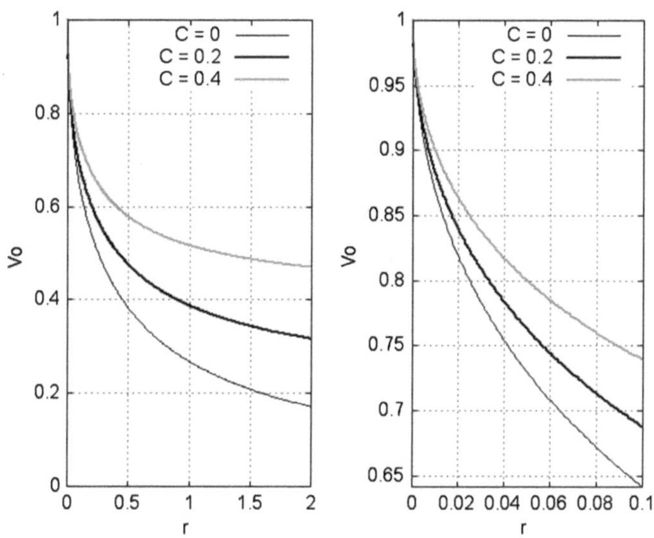

The two graphs show that at very low interest rates, Vo is quite close to 1. If $r = 0$, the firm has no incentive to invest now because it can wait until the period in which, by chance, $V = 1$ and gain a present value of 1. Higher interest rates result in larger discounting of future values, so that $Vo \to C$, but the graphs show that, especially for lower interest rates, the convergence on C can be quite slow. The NPV rule (invest whenever $V > C$) is appropriate in this environment only when future returns have no value (infinite interest rate).

15.2.4 Investments in Human Capital

The preceding section discusses criteria that firms or government agencies might use to evaluate investments in physical capital. This section applies similar reasoning to investments in one's own productivity. Gary Becker, the Nobel laureate who developed the reasoning on which the section is based, labeled this sort of investment *Human Capital*.

15.2.4.1 Continuous Income Flow

We illustrate the human capital model by specifying a functional relationship that approximates age-earnings profiles. One's earning at any time depends on the maximum earnings ($ymax$), one's initial earnings ($ymax - a$, one's age ($18 + t$ in the specification below), on the assumption that one begins to earn at age 18, when $t = 0$), and the rate at which earnings converge to $ymax$, which g determines.

```
y(ymax, a, g, t) := ymax - a*exp(-g*t)$
```

We assume that two options are available to our 18-year-old. The first option is to enter the workforce without investing in human capital with maximum earning of $ymax = 80$, $a = 30$ (initial earning of $ymax - a = 50$), and $g = 0.10$.

The second option leads to a higher maximum earning, $ymax = 100$, but with a lower initial earning reflected in $a = 70$ ($ymax - a = 30$). The result is an age-earning profile that starts lower, with $ymax-a = 30$, and which requires an explicit outlay, the present value of which is 150 (any number could be substituted). The second option has two advantages: it yields a maximum income of $ymax = 100$, and it converges to that higher value relatively rapidly, with $g = 0.15$.

```
wxdraw2d(... Some commands omitted ... line_type=
 dots,... explicit(80,t,0,45),line_width=3,
 line_type=solid, ...explicit(y(80,40,0.1,t),t,0,45),
 ... explicit(y(100,70,0.15,t),t,0,45), ...
 key="ymax=100",explicit(100,t,0,45) )$
```

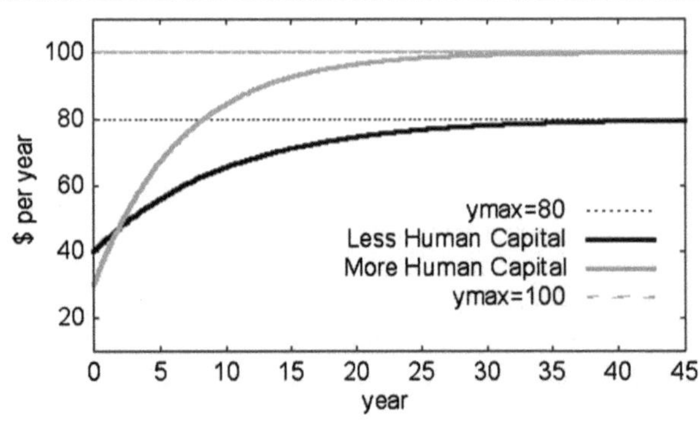

To compare these options in terms of present value, we define the function *npvdiff*, which is the difference between the present values of the two income streams *less* the present value of the explicit outlay. To determine the difference in the net present values of the two functions, we integrate the two functions from $t = 0$ ($age = 18$), to $t = 45$ ($age = 63$), at which time the individual retires.[13]

[13]Post-retirement income is assumed to be from savings during the 45 years of the person's work life and are ignored.

We do not print the expression for *npvdiff* because it is quite long and not very informative. We determine the value of r for which the expression equals zero. That value is $r = 0.0844$. For $r > 8.44\,\%$, the person's wealth is maximized by forgoing the human capital acquisition; for any lower value of r, the human capital investment increases wealth.

```
ratprint:false$ assume(r>0)$  npvdiff:
 integrate(y(100,70,0.15,t)/exp(r*t)  -
  y(80,40,0.1,t)/exp(r*t),t,0,45)-150$
r_critical: find_root(npvdiff,r,.001,1);
```

0.0844

To appreciate the importance of discounting in a setting like this, integrate the two earnings functions without discounting earnings. Total undiscounted earnings from the high human capital income profile is over 20 % higher than the counterpart for the low income profile, even though the present values of the two are equal when $r = 0.0844$ and negative for higher interest rates.

```
[totalMHC:float(integrate(y(100,70,0.15,t),t,0,45)),
totalLHC:float(integrate(y(80,40,0.1,t),t, 0, 45)),
totalMHC - totalLHC - 150];
```

[4033.9, 3204.4, 679.44]

15.2.4.2 Discrete Income Flow

We can model human capital investments in terms of a discrete flow of annual income. Doing so involves developing a list of income flows discounting these flows. The equation developed above is applied to a list of times, values of which range from 0 to 45 (the counter `i` ranges from 1 to 46, and $t = i - 1$).

The equation is phrased in terms of the difference between high human capital earnings and low human capital earnings. The third list, `disclist`, discounts these values with $r = 0.05$. Both the values of the 46 flows into the income stream and their discounted values appear below.

```
tlist:makelist(i-1,i,1,46)$ ydifflist:y(100,70,0.15,
 tlist)-y(80,40,0.10,tlist)$        discountlist:
 ydifflist/(1+r_critical)^tlist$    wxdraw2d(
 ...axis_top=false, ... points(ydifflist), ...
 key="discounted differences",points(discountlist))$
```

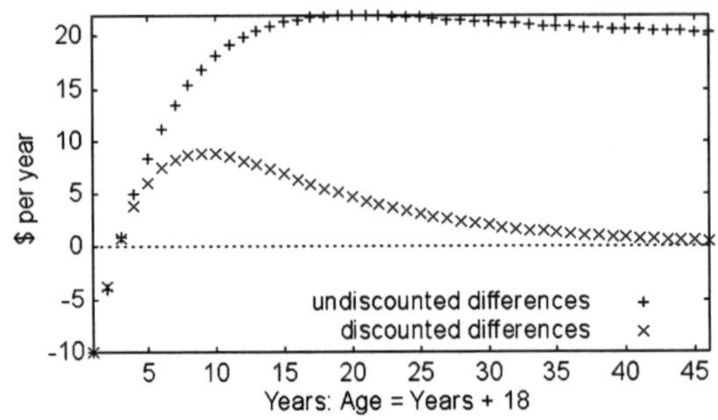

The next cell shows the undiscounted and discounted differences. The critical discount rate, determined above, is used to compute discounted values. The discounted difference is $153.0631 \approx 150$.

```
[sum(ydifflist[i],i,1,46),
sum(discountlist[i],i,1,46)];
```

[834.07, 153.06]

Determining the value of r at which the net present value equals zero proceeds as with the continuous representation. Create a list of discounted values with an unspecified value of r and sum the 46 items in that list.[14] Then apply the `find_root` command to find the r at which net present value (present value *less* the initial outlay of 150) equals zero. The result differs only slightly from the result with a continuous representation.[15] The resulting critical discount rate, $r = 0.0857$, differs only slightly from the value that applies when incomes are treated as occurring continuously.

```
discounteddifferences:(y(100,70,0.15,tlist) -
  y(80, 40,0.10,tlist))/(1+r)^tlist$
  find_root(sum(discounteddifferences[i],
  i,1,46)-150,r,0.001,1);
```

0.0857

15.2.5 *Harvest Rotation*

This section, also based on McAfee, develops a simple forest rotation model. A stand of trees is to be managed in perpetuity. In the model used here, all crops are

[14]Be aware that $t = 0$ corresponds with $i = 1$ in this list; a list cannot contain a 0th element. This resulting output, a 45th degree polynomial, is not printed.

[15]An exercise, due to McAfee: You've decided to go to university but must decide which of two to select. If you attend East State University, you pay \$4,500 per year for 4 years and earn \$50,000 per year for 40 years. If you attend North Private University, you pay \$33,000 per year for 4 years and earn \$70,000 for 40 years once you graduate. Determine the interest rate below which attending NPU yields the largest net present value. The net present value for ESU is $\sum_{t=1}^{4}\left(\frac{-4500}{(1+r)^t}\right)$ $+ \sum_{t=5}^{44}\left(\frac{-50000}{(1+r)^t}\right)$.

the same over time (inter alia, this means that land quality does not change over time, nor does the growth behavior of the trees due to hybridization or genetic manipulations) and that the price per unit of mass is constant over time and independent of when the forest is cut.

The assumption regarding price consists of two parts. First, the manager of the stand of trees is a price taker. Second, quality of the biomass produced does not change with changes in total mass. McAfee shows how to amend the model to relax the latter assumption. A third assumption, made for simplicity, is that fixed costs of cutting and replanting a stand is zero. The analysis shows the implications of selecting forest rotation based on each of two criteria. The first is to maximize the present value of the forest. The second is to maximize the output per year.

The analysis in this section is in two parts. The first part treats the general case in which the function that governs forest growth is not specified. The second part specifies a simple growth function and provides a numerical illustration of the principles developed in the first part.

A forest, to be maintained into the indefinite future, will be harvested once every T years. The biomass that is harvested, b, is a function of the length of the rotation period T. For now the functional form is not specified. The forest's owner can invest funds into an alternative investment that yields a continuously compounded rate of ρ (the Greek letter rho) per time period. Thus, the present value of a given amount of biomass, sold at the end of T years, is $p \cdot b \cdot (e^{-\rho \cdot T})$.

The revenue received T years in the future is $p \cdot b$. This revenue is discounted by the factor $e^{-\rho \cdot T}$. This process is repeated perpetually, so the present value of the mass that is produced is this:

$$p \cdot b \cdot (e^{-\rho \cdot T}) + p \cdot b \cdot (e^{-2 \cdot \rho \cdot T}) + p \cdot b \cdot (e^{-3 \cdot \rho \cdot T}) + \ldots.$$

At the end of the second period revenue of $p \cdot b$ is once more realized. This revenue must be discounted for $2T$ years; hence, the factor $e^{-2 \cdot \rho \cdot T}$. This process is repeated indefinitely, so the present value of the stream is, strictly speaking, infinite. The present values of the income in the far future are sufficiently small, however, that they can be ignored.

15.2.5.1 Analyzing the General Case

The cell below uses *Maxima*'s `nusum` command to determine the value of the series defined above. (The `sum` command is not used, because it returns a symbolic expression that cannot be assigned to a name for use in further analysis.) The first output item shows the present value as the sum of two terms. The second term becomes vanishingly small because of the term $e^{-\infty \cdot T}$. For computational purposes, this term is excised. This is accomplished by defining the present value function as the first `part` of the expression.

The resulting expression is named PV. The second entry in the `print` command calls for PV_T, but does evaluate the expression.[16]

```
nusum(p*b*%e^-(%rho*n*T),n,1,inf);   PV:part(%,1)$
print("The present value is", 'PV[T], "=", PV)$
```

$$\frac{b\,p}{e^{\rho T}-1} - \frac{b\,p\,e^{-\infty\rho T}}{e^{\rho T}-1} \text{ The present value is } PV_T = \frac{b\,p}{e^{\rho T}-1}$$

The next cell generates an expression in terms of the value of T that generates maximum PV. The `depends` command informs *Maxima* that biomass b depends on the rotation length T. With that information the `diff` command evaluates the derivative of PV with respect to T.

```
depends(b, T)$    diff(PV,T);    solve(%,T);
```

$$\frac{\left(\frac{d}{dT}b\right)p}{e^{\rho T}-1} - \frac{\rho\,b\,p\,e^{\rho T}}{\left(e^{\rho T}-1\right)^2} \quad [e^{\rho T} = \frac{\frac{d}{dT}b}{\frac{d}{dT}b-\rho\,b}]$$

The `solve` command shows the result of determining the T value that results from setting the derivative equal to zero. The output provides a mathematically correct solution to this problem, but does not phrase the solution in an economically meaningful fashion. A few manipulations are required:

$$e^{\rho\cdot T} = \frac{db}{dT} / \left(\frac{db}{dT} - (\rho \cdot b)\right),$$

so

$$e^{\rho\cdot T}\left(\frac{db}{dT} - (\rho \cdot b)\right) = \frac{db}{dT}.$$

We require an expression of the annual growth rate of the biomass in terms of the rotation length, T. Rearranging the terms in the expression above yields

$$\left(\frac{db}{dT} \cdot \frac{1}{b}\right) = \frac{\rho \cdot e^{\rho\cdot T}}{e^{\rho\cdot T} - 1} = \frac{\rho}{1 - e^{-\rho\cdot T}}.$$

This expression says that T should be selected such that the rate at which the forest is growing is somewhat faster than the value of ρ. To get a sense of what "somewhat" means, consider the case in the cell below. Let $T = 30$ (within the rotation range for many softwood species) and $\rho = 0.05$. In that case, as the output below shows, the value of T should be such that the growth rate is $\frac{db}{dT}/b = 0.0644$. Therefore, if

---

[16]The `nusum` command produces a summation of the expression. The result must be one that can be expressed in terms of integer powers, factorials, binomials, and rational functions. Our result is consistent with this requirement. The `print` command contains, in order, a text string, an unevaluated expression, a second text string, and an evaluated expression.

$T = 30$ results in a growth rate above 0.0644, then the rotation should exceed 30 years; if the growth rate is less than this value, the rotation should be less than 30 years.[17]

```
print(%rho," = 5% and T = 30 imply", %rho/
  (1-exp(-%rho*T)), "=",0.05/(1-exp(-0.05*30)))$
```

$$\rho = 5\,\% \text{ and T} = 30 \text{ imply } \frac{\rho}{1-e^{-\rho T}} = 0.0644$$

An important special case is $\rho = 0$, when no discounting is applied. In this case, Maxima applies L'Hospital's rule to determine that the limit of $\rho/(1 - e^{-\rho \cdot T})$ as $\rho \to 0$ equals $1/T$, so T should be selected such that $\frac{db}{dT}/b = 1/T$ or $\frac{db}{dT} = b/T$. This criterion says that T should be selected such that the marginal gain from increasing T by one unit equals the average volume produced per year. This equality occurs when the marginal function intersects the average function, which it does at an extreme value of the average function. Given the nature of growth functions, this extreme is the maximum value of b/T.

Thus the criterion results in maximum per-year output, labeled *maximum sustainable yield*. In general, this criterion results in a rotation cycle that is too long to yield maximum value. The next section illustrates this point.

15.2.5.2 An Illustration

The analysis above is general. To illustrate the principles, we use a *logistic growth function* to approximate the volume of biomass, b, as a function of the rotation period, T. The functional form is

$$b(T, K, b_0, r) = \frac{K \cdot b_0 \cdot e^{r \cdot T}}{K + b_0 \cdot e^{r \cdot T} - 1}.$$

This expression states that biomass is determined by rotation length T, the maximum mass that the forest can attain K, the initial biomass b_0, and the initial growth rate r. We see in the graph below that the growth rate increases over a period of time and then decreases, approaching zero as T becomes large.

The cell below contains the logistic growth function. The specification of that function is followed by a `gr2d` command that creates a scenario named `Growth` that is to be graphed below. Next, a present value function is defined, based on a specific set of parameters of the growth function: $K = 1000$, $b_0 = 1$, and $r = 0.20$.

This expression for b replaces the variable b in the expression used in the previous section. As before, `nusum` produces an expression for the present value.

---

[17]For $T = 60$, within the relevant range for many hardwoods, the corresponding value would be 0.053.

We now cast this expression as an explicit function of T, p, and ρ. Having created the equation, we create a second scenario, `PresentValue`, based a on set of parameter values, to be graphed. Finally, `wxdraw` generates the required graph.

15.2.5.3 Graphical Representation

The top panel shows the biomass, b, which asymptotically approaches 1,000 units. By $T = 60$, the growth rate is quite nearly zero. The present value reaches a maximum when the rotation period is about 40 years.[18]

```
b(T, K, r)  := K*exp(r*T)/(K + exp(r*T)  - 1); /*Growth
 Scenario*/ Growth:gr2d(... Some commands omitted ...
 ytics=[0,500,1000],explicit(b(T,1000,0.2),T,1,75))$
/*Present Value Expression*/
  nusum(p*b(T, 1000, 0.2)*%e^-(%rho*n*T),n,1,inf)$
  PV(T,  p,  %rho)  := ''(part(%,1)); /*Present Value
  Scenario*/   PresentValue: gr2d(ylabel="Dollars",
  xlabel="T",explicit(PV(T,1,0.05),T,1,75))$
wxdraw(Growth, PresentValue)$
```

$$\mathrm{b}(T,K,r) := \frac{K\exp(r\,T)}{K+\exp(r\,T)-1} \qquad \mathrm{PV}(T,p,\rho) := \frac{1000\,p\,e^{\frac{T}{5}}}{\left(e^{\frac{T}{5}}+999\right)\left(e^{\rho T}-1\right)}$$

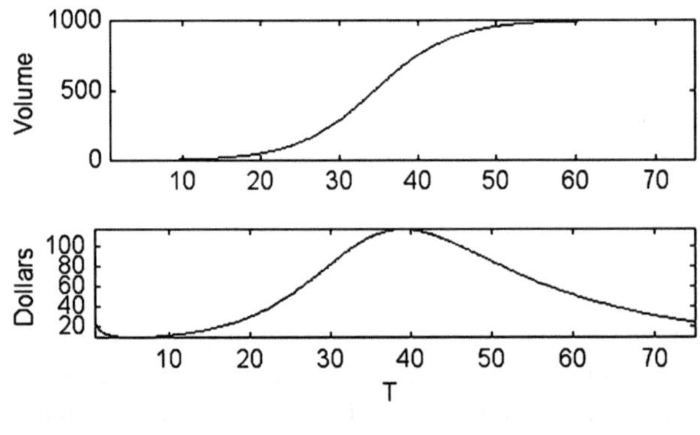

15.2.5.4 Determining the Relevant Values

The graph above shows that the approximate value of the optimal rotation period is approximately $T = 40$. The next cell uses `find_root` to compute the value.

[18]One peculiarity of the present value function is that it declines for very small values of T. This behavior represents a weakness of the logistic function in representing reality. This weakness is more apparent than real, however, for the purpose of analyzing forest rotation is to consider values of T that could apply to real trees. Typically such values are above 20 years.

The first input line instructs *Maxima* to determine the derivative of the present value function and to solve for the value of T that results in $\frac{dPV}{dT} = 0$. The first output line shows the result. The second input line defines a value Topt which is determined numerically. That value is passed to the print command that reports the optimal value $T = 39$ years.[19] The next three lines report the level of the biomass function, the annual growth rate, and the percentage growth rate at $T = 38.974$ years. The last line is a check to show that the actual growth rate (very nearly) equals the growth rate that is required in order to maximize present value. The slight difference is due to the requirement that the optimal rotation be an integer value.

```
diff(PV(T,1,0.05),T)$  solnPV: solve(%,T); Topt:
 round(find_root(rhs(solnPV[1])-lhs(solnPV[1]),
 T,10,100))$ diff(b(T,1000,0.20),T)$
 b_growth:subst(T=Topt,%)$  transpose(matrix(
 ["Optimal rotation period =",
"Total biomass as period's end =",
"Annual growth rate at Topt =",
"Percentage growth =","Percentage for optimality ="],
 [Topt, b(Topt,1000,0.20), b_growth,100*b_growth/
 b(Topt,1000,0.20),100*0.05/(1-exp(-0.05*Topt))],
["years","units","units per year",
"percent","percent"]));
```

$$[e^{\frac{T}{4}} = \frac{\left(3996\, e^{\frac{T}{20}} - 3996\right) e^{\frac{T}{5}}}{e^{\frac{T}{5}} + 999}]$$

$$\begin{bmatrix} \text{Optimal rotation period =} & 39 & \text{years} \\ \text{Total biomass as period's end =} & 709.56 & \text{units} \\ \text{Annual growth rate at Topt =} & 41.217 & \text{units per year} \\ \text{Percentage growth =} & 5.8088 & \text{percent} \\ \text{Percentage for optimality =} & 5.8294 & \text{percent} \end{bmatrix}$$

15.2.5.5 Implications of Maximum Sustainable Yield

Finally, consider the implication of imposing the maximum-sustainable-yield criterion. The first line below defines the output-per-year function, which is total biomass b divided by period length T. The second line takes the derivative of this function with respect to T. The first output line shows the result of setting that derivative equal to zero. That result is a list of two values, one of which is positive. The find_root command is invoked once more and shows that the value of $\frac{b}{T}$ is maximized at $T = 44.921$ years. With this rotation cycle, the forest returns 888.69 units per period.

---

[19] We round the solution, assuming that a planting must occur in a specific season. If planting can occur at any time, then T can be a fraction. The difference between the two is slight.

The present value of the forest's output when this criterion is used is \$105.17. In contrast, the 38.974-year cycle defined above yields a present value of \$117.70.

```
b_per_year(T,K,r):=(K*exp(r*T)/(K+exp(r*T)-1))/T$
diff_b_per_year:diff(b_per_year(T,1000,0.20),T)$
solve(%, T); Tmsy:round(find_root(%[1],T,10,100))$
transpose(matrix(["Max. sustainable yield period =",
"Total biomass as period's end =",
"PV with maximum sustainable yield =", "PV with
 optimal rotation ="],[Tmsy,b(Tmsy,1000,0.20),
 float(PV(Tmsy,1,0.05)),float(PV(Topt,1,0.05))],
 ["years","units","dollars","dollars"]));
```

$$[T = \frac{5e^{\frac{T}{5}}+4995}{999}]$$

$$\begin{bmatrix} \text{Max. sustainable yield period =} & 45 & \text{years} \\ \text{Total biomass as period's end =} & 890.24 & \text{units} \\ \text{PV with maximum sustainable yield =} & 104.89 & \text{dollars} \\ \text{PV with optimal rotation =} & 117.7 & \text{dollars} \end{bmatrix}$$

15.3 A Simple Dynamic Model

So far, the models of a competitive market have involved comparative statics analysis, in which markets are considered at the point in time at which a given condition has been established. For example, the price may have adjusted so that equilibrium has been achieved, or we may consider the effect of a mandated nonequilibrium price. These models say nothing about movement through time. That is, they do not attempt to explain the time path that leads to the values being observed.

This section considers a simple model that does take time into account. This model is often called the Cobweb Model, for reasons that will appear below. We represent the model with a linear supply curve and a linear demand curve:

$$Xs_t = A + a \cdot p_{t-1},$$

and

$$Xd_t = B + b \cdot p_t.$$

That is, the quantity supplied in the current period is a function of the price in the preceding period, and the quantity demanded is a function of the current price.

The first cell creates the functions for the supply and demand curves. The bracketed terms `[t]` and `[t + 1]` denote the time period. Moving the market forward one period in time allows the value of price in an initial period ($t = 0$) to be specified.

To solve this system, we first set the quantity demanded equal to the quantity supplied and solve for the implied time path of prices. This results in a *difference equation*, which can be phrased as follows:

$$p_{t+1} - (a/b) \cdot p_t = (A - B)/b.$$

To define a unique path for this *recursive equation* we use *Maxima*'s `solve_rec` (for solve recursive) module, which must be loaded. The final expression below defines the path, given $p0$ and the system's parameters.

15.3.1 Convergence

Before we explore a specific path, one general note is in order. For the price path to be convergent, the term (a/b) must be between 0 and -1 (b is negative). Thus the absolute value of the demand curve must exceed that of the supply curve if the system is to be stable. We assume stability for the purpose of illustration and return to the question when we generalize the model below.

```
load(solve_rec)$ Xs(p,A,a):=A+a*p[t];
Xd(p, B, b) := B + b*p[t+1];
solnp1 : solve(Xs(p,A,a)-Xd(p,B,b),p[t+1]);
solnrec : solve_rec(solnp1[1], p[t], p[0]=p0);
p(p0, t, A, a, B, b) := ''(rhs(solnrec));
```

$$\mathrm{Xs}(p, A, a) := A + a\, p_t \quad \mathrm{Xd}(p, B, b) := B + b\, p_{t+1}$$

$$[p_{t+1} = -\frac{B-A-a\,p_t}{b}] \quad p_t = \frac{\left(\frac{a}{b}\right)^t (B-A)}{b-a} - \frac{B-A}{b-a} + \left(\frac{a}{b}\right)^t p0$$

$$\mathrm{p}(p0, t, A, a, B, b) := \frac{\left(\frac{a}{b}\right)^t (B-A)}{b-a} - \frac{B-A}{b-a} + \left(\frac{a}{b}\right)^t p0$$

This output shows that the price in any period is determined by the initial price and the system's parameters. Now we examine the time path of the price for a specific set of values. First, we specify the initial price and quantity and a set of slope parameters. Then we determine the intercept terms that are consistent with that specification.

```
[X0,p00]:[1000.0,20.0]$
[a0, b0]:[40.0,-50.0]$
matrix(["A0","B0"],[A0 X0-a0*p00,
 B0:X0-b0*p00]);
```

$$\begin{bmatrix} \mathrm{A0} & \mathrm{B0} \\ 200.0 & 2000.0 \end{bmatrix}$$

Next, we create a list of 11 time periods and apply the equation for price, given the preceding parameter values. The third input line below defines the quantity demanded in each period, given the period's price.

Then we create a list of the first ten prices and determine the quantity supplied for each period. This quantity supplied must equal the quantity demanded in the next period; that is how the next-period price is determined. We create 10-period lists for the period number and for the price, by deleting the initial period ($t = 0$).

The table below shows that the quantity demanded equals the quantity supplied for each period and the price that yields this result. It also shows the lagged price, which determines the current quantity.

```
tlist:makelist(i,i,0,10)$   plist:
 p(p00,tlist,A0,a0,1.5*B0, b0)$  Xdlist:
 1.5*B0+b0*plist$  plistshort:delete(plist[11],
 plist)$   Xslist10:A0+a0*plistshort$
Xdlist10:delete(Xdlist[1],Xdlist)$  plist10:
delete(plist[1],plist)$  tlist10:delete(tlist[1],
 tlist)$ numbers:transpose(matrix(tlist10,Xdlist10,
Xslist10,plist10,plistshort))$    names:
matrix(["Time","Xdemand","Xsupply","Price",
 "LagPrice"])$      addrow(names, numbers);
```

Time	Xdemand	Xsupply	Price	LagPrice
1	1000.0	1000.0	40.0	20.0
2	1800.0	1800.0	24.0	40.0
3	1160.0	1160.0	36.8	24.0
4	1672.0	1672.0	26.56	36.8
5	1262.4	1262.4	34.752	26.56
6	1590.1	1590.1	28.198	34.752
7	1327.9	1327.9	33.441	28.198
8	1537.7	1537.7	29.247	33.441
9	1369.9	1369.9	32.602	29.247
10	1504.1	1504.1	29.918	32.602

Graphing the movement of price and quantity through time is a direct procedure, as shown below.

```
prices: gr2d(... Some commands omitted ...
 points(tlist, plist))$  quantities: gr2d( ...
 ytics=250,points(tlist10, Xslist10))$
wxdraw(prices, quantities)$
```

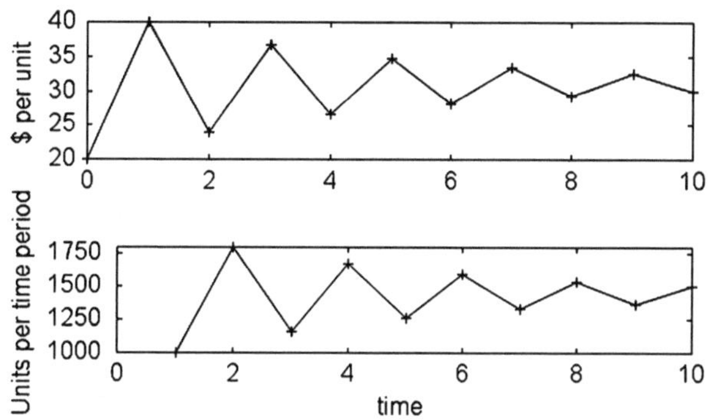

To see how the Cobweb Model gets its name, we place the values above in a traditional graph of a competitive market. To do this, we map the ten quantities supplied on the lagged price, which determines the current period's quantity. Likewise, we map the quantities demanded on the current prices.

```
Xspdata : map("[", Xslist10, plistshort)$
Xdpdata : map("[", Xdlist10, plist10)$
transpose(matrix(Xspdata, Xdpdata));
```

$$\begin{bmatrix} [1000.0, 20.0] & [1000.0, 40.0] \\ [1800.0, 40.0] & [1800.0, 24.0] \\ [1160.0, 24.0] & [1160.0, 36.8] \\ [1672.0, 36.8] & [1672.0, 26.56] \\ [1262.4, 26.56] & [1262.4, 34.752] \\ [1590.1, 34.752] & [1590.1, 28.198] \\ [1327.9, 28.198] & [1327.9, 33.441] \\ [1537.7, 33.441] & [1537.7, 29.247] \\ [1369.9, 29.247] & [1369.9, 32.602] \\ [1504.1, 32.602] & [1504.1, 29.918] \end{bmatrix}$$

The next step is not required, but it illustrates a potentially useful command, `join`. It creates a single data table that can be passed to `draw`.

```
datalist: join(Xspdata, Xdpdata)$
```

The graph below shows the data points generated above, along with the demand and supply curves. The cobweb appearance becomes apparent.[20]

[20] We have taken a labor-intensive approach, developing values point-by-point. *Maxima*'s `dynamics` module can produce representations directly. This module's `evolution` command produces a profile of prices, and its `staircase` command produces a cobweb representation.

```
Xsorted:sort(Xdlist)$  Xmax:Xsorted[11]$
Xmin: Xsorted[1]$  wxdraw2d( ... Some commands
 omitted ... key = "Initial Demand", explicit(
 (X-A0)/a0,X,.9*Xmin, 1.1*Xmax), ... key =
 "New Demand", explicit((X-1.5*B0)/b0, X,
 0.9*Xmin, 1.1*Xmax),... key = "Supply",
 explicit((X-B0)/b0,X,0.9*Xmin,1.1*Xmax), ...
 line_type=dots, key = "path", points(datalist))$
```

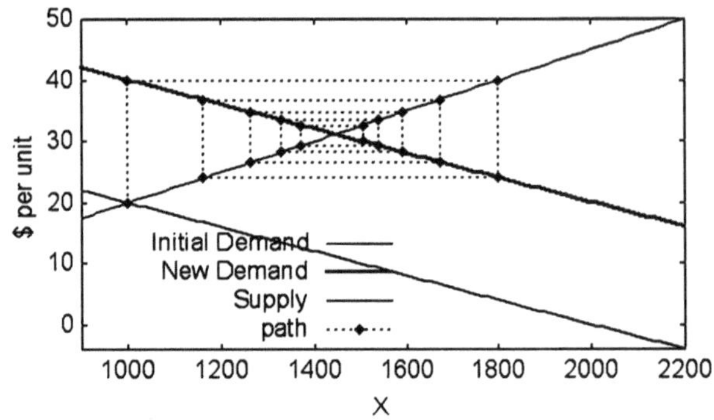

15.3.2 Adaptive Expectations

The demand and supply expressions below are the same as above, except that the price in the supply curve is now an expected price. In the simple Cobweb Model, the expectation is that the previous period's price will persist in this period, that is $pe_t = p_{t-1}$.

```
Xs(p,A,a):=A+a*pe[t];   Xd(p,B,b):=B+b*p[t];
```

$$\mathrm{Xs}(p, A, a) := A + a\ pe_t \quad \mathrm{Xd}(p, B, b) := B + b\ p_t$$

Before proceeding to determine the nature of the difference equation that this specification implies, we look at the relationship between *pe* and *p* in any period.

Market equilibrium requires that the first expression in the cell below must hold. Solving this expression provides an expression for *pe* that can be substituted as required to specify the difference equation that dictates the responses of price and quantity to demand or supply shift. Once we have a way to express *p* for any period, that period's *pe* is determined.

```
A + a*pe = B + b*p$
solve(%, pe);
```

$$[pe = \frac{B-A+b\,p}{a}]$$

To illustrate, we specify a simple *adaptive expectations* model for the determination of today's price expectations. This model treats expectations as a linear combination of expectation for the preceding period and the deviation from that expectation. The value of the coefficient e lies between 0 and 1. If $e = 1$, this model reduces to the simple model above; only the last-period price matters. If $e = 0$, then only the expectation brought into the preceding period matters; the price deviation has no effect on current expectations.[21]

```
expectations:pe[t]=pe[t-1]+e*(p[t-1]-pe[t-1]);
```

$$pe_t = pe_{t-1} + e\ (p_{t-1} - pe_{t-1})$$

Next we define the functional relationship between the price and time. The first step is to define the price that equates the quantity supplied and that demanded. Then we evaluate this expression given the definition of `expectations`. The next step is to evaluate the preceding expression given the relationship between pe and p that we derived previously.

The result of this third step is the third line of output, which expresses p_t as a difference equation. The next two lines use the `solve_rec` command to solve this difference equation, extract the expression for that solution and state price as a function with that expression on the right-hand side. The last two steps are not printed.

```
solve(Xs(p, A, a) = Xd(p, B, b), p[t]);
subst(expectations,%[1]);
subst(pe[t-1] = (B - A + b*p[t-1])/a,%);
solnrec2 : solve_rec(%, p[t], p[0] = p0);
p(t, p0, A, a, B, b, e):=''(rhs(%))$
```

$$[p_t = -\frac{B-A-a\,pe_t}{b}]$$

$$p_t = -\frac{B-A-a\left(pe_{t-1}+e\left(p_{t-1}-pe_{t-1}\right)\right)}{b}$$

$$p_t = -\frac{-a\left(e\left(p_{t-1}-\frac{B-A+b\,p_{t-1}}{a}\right)+\frac{B-A+b\,p_{t-1}}{a}\right)+B-A}{b}$$

$$p_t = \frac{\left(-\frac{(b-a)\,e-b}{b}\right)^t (B-A)}{b-a} - \frac{B-A}{b-a} + \left(\frac{a\,e}{b} - e + 1\right)^t p0$$

Consider the last output line above. If $e = 1$, then the two bracket terms equal $\frac{a}{b}$, and the expression is the same as that for the Cobweb Model. This more general model converges (is not explosive) as long as $\frac{a}{b} < \frac{-2}{e} + 1$. If $e = 1$, this reverts to the condition under which the simple model converges $\frac{a}{b} > -1$. For smaller values of e, the term on the right-hand side becomes a larger negative number, making the convergence criterion easier to meet.

To provide a numerical example, we assign values as in the case of the Cobweb Model. The only difference between this expression and the one in the preceding section is that a value of e is stipulated.

---

[21] Some manipulation shows that this model can be thought of as a weighted average of expectations and observation: $pe_t = (1-e) \cdot pe_{t-1} + e \cdot pt - 1$. Thus the weight given to expectations is $1-e$, and the weight given to the observed value in the immediately preceding period is e.

```
[X0, p00]:[1000.0,20.0]$
[a0,b0,e0]:[40.0,-50.0,0.9]$
[A0:X0-a0*p00, B0:X0-b0*p00];
```

[200.0, 2000.0]

The resulting expression results in convergence, for the term raised to the t power is between -1 and 0.

```
p(t, p00,A0,a0,B0,b0,e0);
```

$$-20.0\,(-0.62)^t + 20.0\,(-0.62)^t + 20.0$$

We shift the supply curve in this illustration. The two commands below produce the new equilibrium price and, for reference, the initial equilibrium price.

```
float([solve(.25*A0 + a0*p = B0 + b0*p, p),
solve(A0 + a0*p = B0 + b0*p, p)]);
```

$$[[p = 21.667], \quad [p = 20.0]]$$

Finally, we create three price paths from the initial price to the new one. The first path is from the Cobweb Model above. We create that list by setting $e = 1$ in the function. The next two paths show the effect of increasingly lower values of e. Path 1 is defined for $e = 0.9$. Path 2 is defined for $e = 0.4$.

```
tlist: makelist(i, i, 0, 10)$
cobweblist:  p(tlist, p00, .25*A0, a0, B0, b0, 1)$
p1list : p(tlist, p00, .25*A0, a0, B0, b0, .9)$
p2list : p(tlist,p00, .25*A0, a0, B0, b0, .4)$
names:matrix(["Time", "Cobweb Path", "Price Path 1",
"Price Path 2"])$ numbers: transpose(matrix(
 tlist,cobweblist, p1list,p2list) )$
addrow(names, numbers);
```

Time	CobwebPath	PricePath1	PricePath2
0	20.0	20.0	20.0
1	23.0	22.7	21.2
2	20.6	21.026	21.536
3	22.52	22.064	21.63
4	20.984	21.42	21.656
5	22.213	21.819	21.664
6	21.23	21.572	21.666
7	22.016	21.725	21.666
8	21.387	21.63	21.667
9	21.89	21.689	21.667
10	21.488	21.653	21.667

The table above and the graph below tell the same story. Relative to the Cobweb Model, the model with adaptive expectations shows a much smoother path from one equilibrium to the next. Even with $e = 0.9$, the price fluctuations are dampened noticeably. With $e = 0.4$, the fluctuations vanish, and the price approaches its new equilibrium value in a monotonic fashion.

```
wxdraw2d( ... Some command omitted ... key="Cobweb",
 points(tlist,cobweblist), ... line_type=dots, ...
 points(tlist,p1list), ... points(tlist, p2list)  )$
```

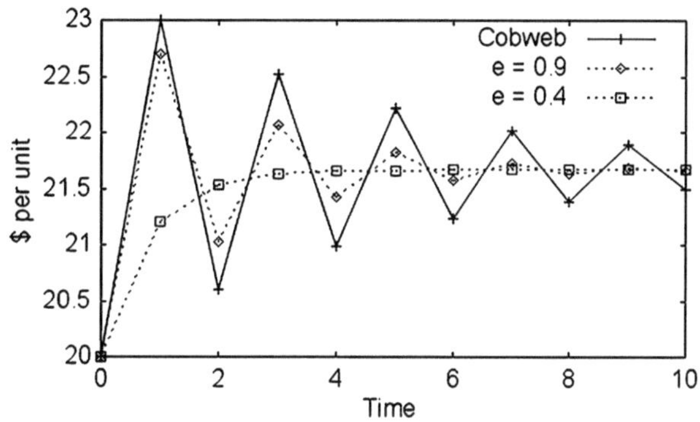

References

1. Dixit AK, Pindyck RS (1994) Investment under uncertainty. Princeton University Press, Princeton
2. McAfee RP (2006) Introduction to economic analysis, version 2.0. http://www.introecon.com

Chapter 16
Uncertainty

Abstract Most of the material covered so far deals with a world of certainty. Decision makers know the effects of consumption, employment, output, or pricing choices. This chapter considers decisions under conditions of *uncertainty*, in which economic actors cannot be sure of outcomes. *Risk* (we use the words risk and uncertainty interchangeably) relates to the existence of a range of possible outcomes that result from a decision. Specifically, the analysis treats an outcome x as the result of a gamble or lottery. The value of the outcome is random, and the probability of a particular realization of x can be described by a probability density function (or *pdf*), $f(x)$. The chapter develops a model of risk preferences that relates to the nature of a utility function. The model is applied to ways of measuring differences in risk preferences. Finally, the model is applied to choices related to insurance.

16.1 Risk and Preferences

Formal thinking about decision-making in the face of risk[1] dates back at least to Daniel Bernoulli's discussion of the *St. Petersburg Paradox* in 1738. Bernoulli proposed the following gamble. A casino offers a game in which a fair coin is tossed at each stage. The pot starts at \$1 and doubles each time a head appears. The first time a tail appears, the game ends and the player wins whatever is in the pot. Thus the player wins \$1 if a tail appears on the first toss, \$2 if a head appears on the first toss and a tail on the second, \$4 if a head appears on the first two tosses and a tail on

[1]For simplicity, we do not explore ramifications of the distinction between *risk* and *uncertainty*, as developed by Knight [2]. Knight treats risk as a situation in which the effects of decisions are not known, but the decision maker knows the *probability distributions* that pertain to the possible outcomes. True uncertainty occurs when the decision maker cannot assign probabilities to various possible outcomes. Knight argues that uncertainty but not risk is a source of profits.

M.R. Hammock and J.W. Mixon, *Microeconomic Theory and Computation: Applying the Maxima Open-Source Computer Algebra System*,
DOI 10.1007/978-1-4614-9417-1_16,

the third, \$8 if a head appears on the first three tosses and a tail on the fourth, and so forth. In summary, the player wins $2^{(i-1)}$ dollars if the first tail appears on the ith toss of the coin.

Bernoulli's question regarding what a gambler might pay to enter such a game was aimed at shedding light on the nature of preferences as regards uncertain outcomes. Specifically, the game points out that, at least in the case of this game, the *expected value* of the game would greatly exceed what gamblers would pay to play. This must be the case, because the expected value is infinite if the player is allowed an unlimited number of tosses.

Recall that the expected value of any gamble is the sum of the products of the items in a list of outcomes and the probabilities that the listed outcomes will occur. In this case, the possible outcomes are 2^i, where i is the number of flips of the coin. The first four possible payoff values are [1,2,4,8,16] where the values are associated with the following number of consecutive heads: [0,1,2,3,4].

The four values in the payoff list can be written as the first four results of applying the expression $2^{(i-1)}$, where i is the number of flips. The associated probabilities are these: [1/2,1/4, 1/8,1/16]. That is, the probability of getting two heads in the first two flips is $(1/2)^2$ and so forth. If the game were restricted to four flips of the coin, its expected value would equal $1 \cdot \frac{1}{2} + 2 \cdot \frac{1}{4} + 4 \cdot \frac{1}{8} + 8 \cdot \frac{1}{16} = 2$.

The reasoning above indicates that the expected value of this opportunity equals one half of the maximum number of coin flips that the gambler is allowed to take (not of the number of heads that actually results). That value grows without bound, as the output below shows. The first entry in each `sum` expression, 2^{i-1} is the payoff from i successive heads, and the second entry is the probability of observing that number of successive heads. The first entry and output confirm the calculation above, and the second entry and output confirm that the limiting expected value of this game is infinite.

```
[sum(2^(i-1)*(1/2)^i,i,1,4),
sum(2^(i-1)*(1/2)^i,i,1,inf)];
```

$[2, \infty]$

Why, then, would no reasonable person be willing to pay huge sums to play this game? The economist's answer (and Bernoulli's) is *risk aversion*, which is related to the *diminishing marginal utility of wealth*. Although it is possible that the gamble would reach an enormous payoff, such a huge payoff is very unlikely (one does not *expect* to observe the "expected value"), and most people are unwilling to pay a large sum for such an unlikely outcome.[2]

Put more carefully, a *risk averse* person would be unwilling to accept a *fair gamble*. That is, if offered a 50/50 chance of winning or losing a dollar, a risk averse person would decline. Similarly, a *risk preferring* or *risk loving* person would accept

[2]Determine the expected value when each of the following maximum number of tosses is allowed: 10, 50, 100. Also, determine the maximum amount that you could win in each of the three games and the probability of winning that amount. Consider how much you would pay to play each of these three games.

a fair gamble. A *risk neutral* person would be indifferent between accepting and not accepting such a lottery.[3]

We will discuss more precise ways of defining and measuring these risk preferences in subsequent sections. First we formalize our approach to risk preferences by examining a specific utility function.

16.1.1 An Example: The Exponential Utility Function

Suppose that $U1(W, R) = 1 - e^{-0.1 \cdot W/R}$ is a consumer's utility function. In this expression W is the consumer's wealth, and R is a constant. We will see that R is inversely related to measures of risk aversion. This is the *Exponential Utility Function*.[4]

To explore this function's implications, we begin with a consumer whose wealth, absent a gamble, is given. Let that value be \$25, and set $R = 2$. The graph depicts the consumer's utility function over a range of wealth values, and the utility level at this particular of wealth value.

```
U1(W,R):=1-%e^(-0.1*W/R)$  W0: 25$  R0:2$
wxdraw2d(yrange=[0,1],xlabel="Wealth ($)",
ylabel="Utility",points_joined=true,line_type=dots,
points([[0,U1(W0,R0)],[W0,U1(W0,R0)],[W0,0]]),
line_type=solid, key=concat("Utility when R= ",
string(R0)), explicit(U1(W,R0),W,0,50))$
```

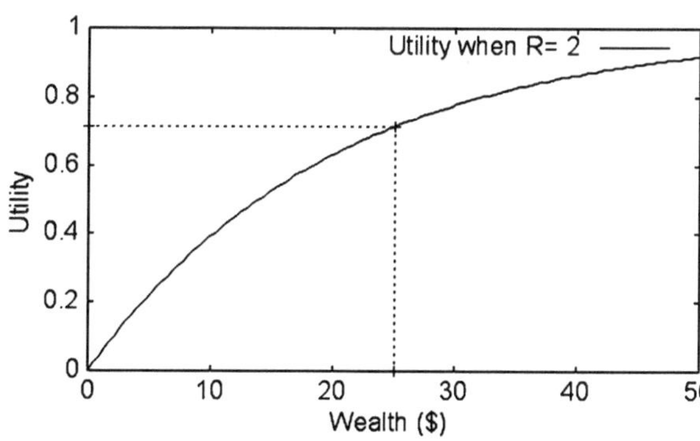

[3]Our analysis is limited to the utility that the person receives from the payoff itself. It does not take into account the pleasure from, or addiction to, the playing of the game. We use the words lottery and gamble interchangeably.

[4]Two seminal treatments of decision making under uncertainty, [1] and [4] use formulations that are consistent with this expression.

The graph shows that this utility function exhibits diminishing marginal utility of wealth. It indicates that a certain wealth level $W = 25$ generates a utility level that is just over 0.7. Specifically, the consumer's utility level is 0.713, although as is the case with all utility functions, the number is meaningless in and of itself.

```
U1(W0,R0);
```

0.713

Let this consumer be faced with a gamble. The *expected utility* of this gamble is the probability-weighted sum of the utility levels associated with the outcomes. Let p be the probability of the "bad" outcome, in which the consumer loses x, and $(1 - p)$ be the probability of the "good" outcome, in which the consumer gains y. Then expected utility is:

```
EU1(W,R,x,y,p):=p*U1(W-x,R)+(1-p)*U1(W+y,R)$
```

Suppose that $x = y = \$10$, and that outcomes x and y are equally likely. That is, the consumer faces a fair bet. In this case, expected utility is 0.677. Again, this number is, in itself, meaningless—a utility function of this type (known as a von Neumann–Morgenstern utility function) is invariant up to an affine transformation.[5] The sign of the difference between this value and that of $U1$(W0, R0), is, however, quite meaningful. See [3] and [5].

```
x0:10$ y0:10$ p0:0.5$
EU1(W0,R0,x0,y0,p0);
```

0.677

This outcome demonstrates that, even though the expected payoff of this gamble in dollar terms is \$0, the consumer's expected utility is less than the utility of getting W0 with 100% probability. This can be shown graphically.

On the graph below, the straight diagonal dotted line represents linear combinations (more specifically, weighted averages) of the utility levels associated with the two outcomes. The *expected utility* level occurs at W0, where the two weights are 0.5 and lies on this diagonal dotted line. Its value is below the $U(W0)$ due to the concavity of the utility function. This *risk averse* consumer would not choose to take this gamble.[6]

[5]For an affine transformation, all points that lie on a line initially still lie on a line after the transformation, and ratios of distances are not affected (so that, for example, the midpoint of a line segment remains the midpoint after transformation).

[6]Other points on this reflect different odds. At any point, the W value is the expected wealth level, the height of the line segment is the expected utility level, and the height of the solid curve is the utility level that would result if the expected wealth level were a certain level of wealth. Clearly, at every value, the utility level for a certain wealth level exceeds the expected utility level when wealth is not subject to risk. The graph shows, however, that for some probability values, the expected utility exceeds 0.713. We examine the implications of this fact below.

```
wxdraw2d( ... Some commands omitted ..
 points([[0,U1(W0,R0)],[W0,U1(W0,R0)],[W0,0]]),
 points([[(W0-x0),0],[(W0-x0),U1(W0-x0,R0)],
 [(W0+y0),U1(W0+y0,R0)],[W0+y0,0]]),points([
 [0,EU1(W0,R0,x0,y0,p0)],[W0,EU1(W0,R0,x0,y0,
 p0)]]),  ... explicit(U1(W,R0),W,0,50))$
```

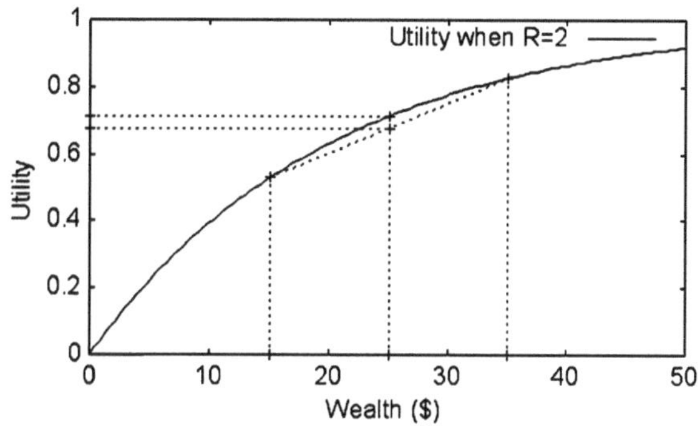

The shape of the utility function reflects risk preferences. Larger R values reduce the utility level at each wealth level, and they reduce the function's concavity. Let $R = 6$.

```
wxdraw2d( ... Some commands omitted ...
 points([[0,U1(W0,6)],[W0,U1(W0,6)],[W0,0]]),
 points([[(W0-x0),0],[(W0-x0),U1(W0-x0,6)],
 [(W0+y0),U1(W0+y0,6)],[W0+y0,0]]),      ...)$
```

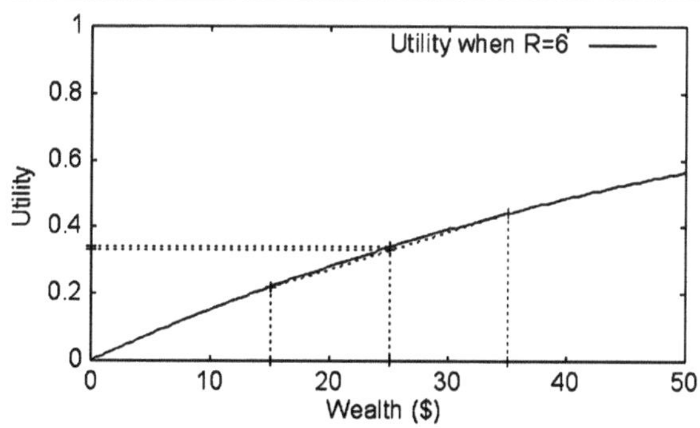

Increasing R straightens out the utility function. It also reduces the difference between the utility of a certain $25 and the gamble that results in an income of either $15 or $35. That is, as R increases, the consumer becomes less risk averse. With a convex utility function, the consumer would be a risk seeker.[7]

[7]Exercise: Given the utility function $U3(W, R) = [(W - x)/(y - x)]^R$, with $R > 1$, how will the expected value of this gamble compared to the certain $25? Note that R here no longer has the same interpretation as above.

16.1.2 Inducing Risk Taking Behavior

Even if $R = 2$, it is possible to alter the lottery to induce this risk averse consumer to accept it. This could be done in several ways. One way is to adjust the payoff probabilities. Allow p to fall (and $1 - p$ to rise) until the consumer becomes indifferent between the certain \$25 and the lottery. Call the resulting probability $p1$.

The first line below solves for the value of p that sets the utility of a certain \$25, 0.713, equal to the expected utility of the lottery. Expected post-lottery wealth now exceeds \$25, as it must if we are to induce the risk averse consumer to be indifferent between a certain outcome and a risky one.

The implied value of p is approximately 0.378. The table confirms that this value results in the same expected utility level as the certainty that $W = \$25$. It also shows that the expected wealth level, given this probability level, is \$27.449.

```
float(solve(U1(W0,R0)-EU1(W0,R0,x0,y0,p),p))$
p1: rhs(%[1])$  matrix(["W0 ","U(W0,R0) ",
 "p1 ","EU1(W0,R0,x0,y0,p1)","Expected W"],
 [W0, U1(W0,R0), p1, EU1(W0,R0,x0,y0,p1),
 p1*(W0-x0)+(1-p1)*(W0+y0 ) ] );
```

$$\begin{bmatrix} \text{W0} & U(\text{W0,R0}) & \text{p1} & EU1(\text{W0,R0,x0,y0,p1}) & \text{Expected W} \\ 25 & 0.713 & 0.378 & 0.713 & 27.449 \end{bmatrix}$$

The graph below shows expected income levels for which $EU1(W0, R0, x0, p) > U(W0, R0)$. The range between \$27.449 and \$35 corresponds to $p < 0.378$, given that $x = y = \$10$.[8]

```
wxdraw2d(xtics_rotate=true, ...Some commands
 omitted ... xtics={W0-x0,W0,p1*(W0-x0)+(1-p1)*
 (W0+y0),W0+y0}, ... points([[0,U1(W0,R0)],
 [W0,U1(W0,R0)],[W0,0]]),points([[(W0-x0),0],
 [(W0-x0),U1(W0-x0,R0)],[(W0+y0),U1(W0+y0,R0)],
 [W0+y0,0]]), points([[0,EU1(W0,R0,x0,y0,0.378)],
 [(.3780*(W0-x0)+(1-.3780)*(W0+y0)),EU1(W0,R0,x0,
 y0,0.378)],[(.3780*(W0-x0)+(1-.3780)*(W0+y0)),
 0]]), ...  explicit(U1(W,R0),W,0,50))$
```

[8] Exercise: Determine what happens to the difference between the utility of the certain \$25 and the expected utility from the lottery as R increases.

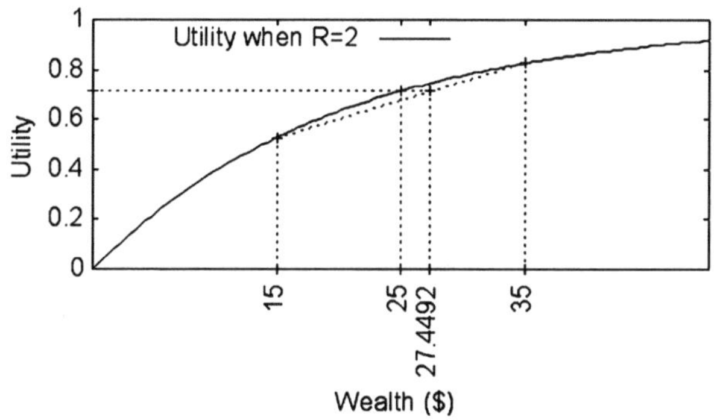

16.2 Measuring Risk Aversion

There are several ways to describe risk preferences. The analysis above hints at two, the *certainty equivalent* and the *risk premium*. These are simple ways to determine the strength of risk preferences for a particular lottery. Two more general ways of describing risk preferences, which also allow comparisons among lotteries, are also available. These are *absolute risk aversion* and *relative risk aversion*. This section examines these options.

16.2.1 Certainty Equivalent and Risk Premium

Return to the previous example, with $p = 0.5$. Recall that the expected utility of the lottery is 0.677. The *certainty equivalent* is the amount of wealth that would yield the same utility as the lottery. That is, the consumer would be indifferent between the lottery and the certainty equivalent wealth level.

```
float(solve(EU1(W0,R0,x0,y0,p0)-U1(CertEq,R0),
 CertEq));        CertEq0: rhs(%[1]);
```

$$[CertEq = 22.598] \quad 22.598$$

In our example, a certain wealth of \$22.598 is considered by the consumer to be as attractive as a gamble that leaves the consumer with an *expected* wealth of \$25. The *risk premium* is the difference between the certainty equivalent and expected outcome of the lottery. That is, the risk premium represents the wealth (in expectation) that the consumer is willing to give up in order to have certainty.

The cell below shows that, given the utility function, the risk premium is a function of the probability of a bad outcome. More specifically, it decreases as p increases. The rate of decrease for this utility function and initial wealth level is −$200 per 1% point increase in p. When $p = 0.5$, this expression's value is RiskPr0, $2.402, which is $(25 − 22.598).

```
RiskPr: (p*(W0-x0)+(1-p)*(W0+y0))-CertEq0$
diff(RiskPr,p)$   RiskPr0:subst(p=p0,RiskPr)$
matrix(["Risk premium expression",
 "Expression Slope", "Expression Value"],
 [RiskPr, diff(RiskPr,p),RiskPr0]);
```

$$\begin{bmatrix} \text{Risk premium expression} & \text{Expression Slope} & \text{Expression Value} \\ 15\,p + 35\,(1-p) - 22.598 & -20 & 2.4023 \end{bmatrix}$$

Graphically, we can see the certainty equivalent and the risk premium as distances on the horizontal axis. The graph below shows the certainty equivalent value that is derived above. The graph has been truncated in order to focus on the relevant value ranges.

```
wxdraw2d(... Some commands omitted ... xtics=
 {12.5,W0-x0,CertEq0,W0, W0+x0}, ... points(
 [[12.5,(EU1(W0,R0,x0,y0,p0))],[CertEq0,
 (EU1(W0,R0,x0,y0,p0))],[CertEq0,0]]),
 points([[(W0-x0),0],[(W0-x0),U1(W0-x0,R0)],
 [(W0+y0),U1(W0+y0,R0)],[W0+y0,0]]),
 points([[12.5,EU1(W0,R0,x0,y0,p0)],[(p0*(W0-x0)+
 (1-p0)*(W0+y0)), EU1(W0,R0,x0,y0,p0)],
 [(p0*(W0-x0)+(1-p0)*(W0+y0)),0]]),
 ... explicit(U1(W,R0),W,12.5,37.5) )$
```

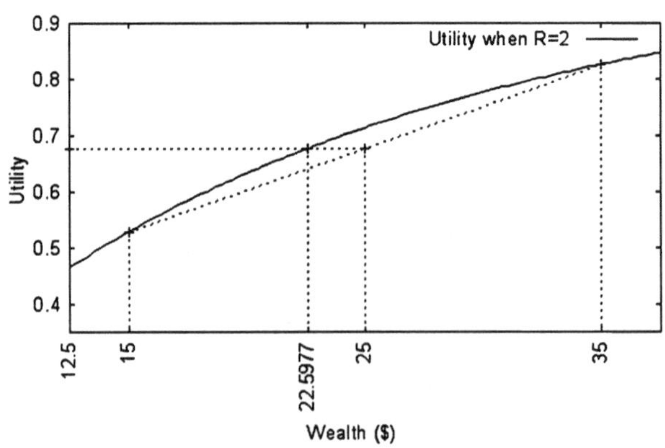

If we allow R to change, both the risk premium and the certainty equivalent change as well. The table below shows the certainty equivalent of an expected wealth of $25 when $p = 0.5$.

```
Rlist: makelist(-1+1.5*i,i,1,8)$ CertEq(R):=
  ''(rhs((solve(EU1(W0,R,x0,y0,p0)=U1(W,R),W))[1]))$
CertEqList:float(CertEq(Rlist))$
RiskPr(R):=''(p0*(W0-x0)+(1-p0)*(W0+y0)-CertEq(R))$
RiskPrList:float(RiskPr(Rlist))$   names:matrix(
  ["R","Certainty Equivalent","Risk Premium"])$
numbers:transpose(matrix(Rlist, CertEqList,
RiskPrList))$        addrow(names,numbers);
```

(%o32)

R	Certainty Equivalent	Risk Premium
0.5	18.375	6.625
2.0	22.598	2.4023
3.5	23.59	1.4095
5.0	24.007	0.993
6.5	24.234	0.766
8.0	24.377	0.623
9.5	24.475	0.525
11.0	24.546	0.454

Larger values of R correspond to utility functions for increasingly less risk averse consumers. Such consumers are willing to accept an increasingly small reduction of wealth in exchange for certainty. The risk premium is a measure of this decreasing risk aversion.

Although the certainty equivalent and risk premium are useful for examining how different risk preferences dictate different responses to a given lottery, they do not provide an adequate basis for making broader statements about risk preferences across different lotteries. Two alternatives have been developed: Absolute Risk Aversion and Relative Risk Aversion.

16.2.2 Absolute Risk Aversion

The first measure we will examine for more general comparisons of risk preferences across consumers is the Arrow-Pratt measure of risk aversion, the Absolute Risk Aversion measure, henceforth referred to as $R(W)$.

$$R(W) = -\frac{U''(W)}{U'W}$$

That is, the $R(W)$ measures the ratio of the second derivative of utility with respect to wealth and the first derivative of the function. In the case of our exponential utility function, R(W) would be:

```
R1(W):='' (-1*(diff(U1(W,R),W,2)/
diff(U1(W,R),W)));
```

$$\mathrm{R1}(W) := \frac{0.1}{R}$$

Two aspects of this expression warrant attention. First, $R1(W)$ is decreasing in R, the risk aversion parameter. This makes sense; we saw before that as R increased, the utility function became straighter. This means that the consumer was becoming *less* risk averse. Conversely, if R decreases, $R1(W)$ increases, so the consumer is becoming *more* risk averse.

The second thing worth noting about $R1(W)$ is that it degenerates to a constant (because R is a constant for a particular consumer). The exponential utility function displays Constant Absolute Risk Aversion, or CARA. The level of risk aversion does not depend on the consumer's wealth.

Utility functions can display a wide range of behavior with regard to $R(W)$. For example, consider the following utility function:

```
[U2(W,R):=(W^(1-R))/(1-R), R2(W):=
  ''(-1*(diff(diff(U2(W,R),W),W)/diff(U2(W,R),W)))];
```

$$[\mathrm{U2}(W,R) := \frac{W^{1-R}}{1-R}, \mathrm{R2}(W) := \frac{R}{W}]$$

This consumer's risk preferences depend on the level of wealth. For a given R, risk aversion falls as the consumer becomes wealthier. This utility function generates Decreasing Absolute Risk Aversion (DARA). Increasing Absolute Risk Aversion is also possible.

Finally, consider the following utility function[9]:

```
[U3(W,R):=(R*W-1)^3+1,        R3(W):=
  ''(-1*(diff(diff(U3(W,R),W),W)/diff(U3(W,R),W)))];
```

$$[\mathrm{U3}(W,R) := (R\,W-1)^3+1, \quad \mathrm{R3}(W) := -\frac{2\,R}{R\,W-1}]$$

This utility function generates different risk behavior as wealth varies. For low wealth levels ($RW < 1$, or $W < 1/R$, whatever the wealth units might be) the consumer is risk averse. For high wealth levels ($RW > 1$) the consumer is risk seeking. Such a person might reason that with a sufficiently high level of wealth, he or she can better afford to gamble a given dollar amount without endangering his or her lifestyle. Put another way, for this consumer, wealth has increasing marginal utility past a certain level.

The graph below shows the utility functions for two different individuals (so the utility levels cannot be compared). The gray diamonds indicate the wealth values (we use \$100,000 as the unit) at which each person ceases being risk averse; that is they identify inflection points on the utility functions.

[9]The constant term here makes the graph look more conventional, with utility going through the origin, but it is not necessary. Remember, expected utility functions are invariant to affine transformations.

```
wxdraw2d(... Some commands omitted ... key="R = 2.5",
 explicit(U3(W,2.5),W,0,1), ... key="R1.25",
 explicit(U3(W,1.25),W,0,1.5), ...
 key="",point_type=13,points([[1/2.5,U3(1/2.5,2.5)]]),
 points([[1/1.25,U3(1/1.25,1.25)]]) )$
```

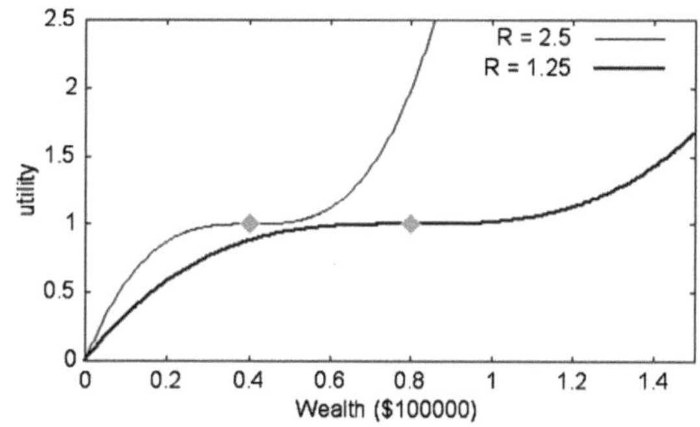

16.2.3 Relative Risk Aversion

The Absolute Risk Aversion measure holds the size of the gamble constant in dollar terms. It is sometimes more natural to think of gambles relative to one's wealth. $R(W)$ can be counterintuitive in that it describes risk preferences over gambles of a fixed dollar amount. An alternative measure of risk preferences is Relative Risk Aversion, here denoted $RR(W)$.

$$RR(W) = -W \cdot \frac{U''(W)}{U'(W)}$$

For the exponential utility function $U1(W, R)$ the $RR(W)$ would be as follows:

```
RR1(W):=''(-W*(diff(U1(W,R),W,2)/
 diff(U1(W,R),W)));
```

$$\mathrm{RR1}\,(W) := \frac{0.1\,W}{R}$$

The exponential utility function displays constant *absolute* risk aversion, but increasing *relative* risk aversion. That is, as this consumer becomes wealthier, the marginal utility of wealth falls, and the consumer will accept only those gambles that make up a smaller fraction of income.

To put it another way, constant absolute risk aversion means that this consumer would accept the same \$10 gamble when wealth is \$25 as when wealth is \$250. Increasing relative risk aversion implies that this consumer might accept a gamble for 10% of a low level of wealth, but might be willing to gamble only 8% of a larger wealth level.[10]

[10]Exercise: Derive the relative risk aversion measures for $U2(W, R)$ and $U3(W, R)$ and compare them to their absolute risk aversion counterparts.

16.3 Risk and Insurance

Risk averse consumers often have the option of reducing risk by purchasing insurance. This option makes the two possible states of the world, certainty and uncertainty, more alike. The insurer collects a payment (the insurance premium) from the consumer, and the consumer is paid if the harmful event occurs. The loss of the insurance premium in "good" state of the world, along with the payment in the "bad" state of the world, causes the two states to be more similar to each other.

16.3.1 A Given Probability

To examine this option more systematically, we again use the exponential utility function, but change the gamble. Suppose $p1 = 0.1$ is the probability of a bad outcome, the loss of $x1 = \$10$ (due to some kind of accident or health problem). The probability of a good outcome, $y1 = 0$, is $1 - p1 = 0.9$. We assume income is still \$25.[11]

In this case, the risk-free utility (which is unavailable to the consumer) is still 0.713, and the expected utility with risk is 0.695. As a result of the lottery, the consumer has expected wealth of \$24.

```
p1:0.1$ x1:10$ y1:0$    print("Expected Utility = ",
EU1(W0,R0,x1,y1,p1)," and Expected Wealth = ",
p1*(W0-x1)+(1-p1)*(W0+y1))$
```

Expected Utility = 0.695 and Expected Wealth = 24.0

The following figure depicts the lower expected utility that the gamble imposes. We truncate the graph to emphasize the difference between utility and expected utility.

```
wxdraw2d( ... Some commands omitted ...
 points([[0,U1(W0,R0)],[W0,U1(W0,R0)],[W0,0]]),
 points([[(W0-x1),0],[(W0-x1),U1(W0-x1,R0)],
 [(W0+y1),U1(W0+y1,R0)],[W0+y1,0]]),
 points([[0,EU1(W0,R0,x1,y1,p1)],
 [p1*(W0-x1)+(1-p1)*(W0+y1),EU1(W0,R0,x1,y1,p1)],
 [p1*(W0-x1)+(1-p1)*(W0+y1),0]]),  ...
 explicit(U1(W,R0),W,0,30))$
```

[11]More realistically, a consumer faces a range of possible negative outcomes with different probabilities. We use the two-outcome model for simplicity.

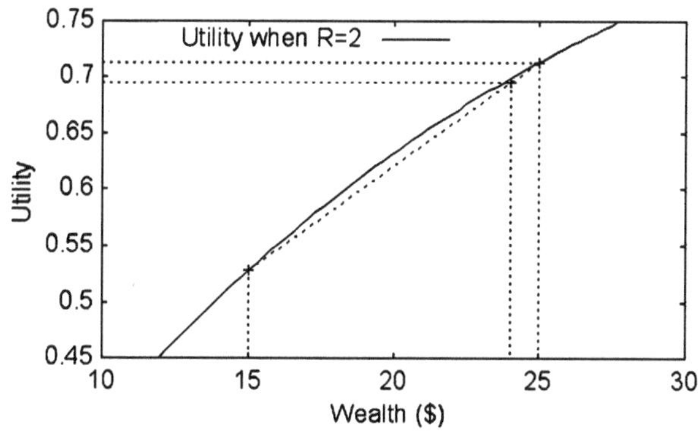

Suppose an insurer were considering offering insurance against this gamble. An *actuarially fair* insurance premium would be equal to the expected loss. In this case, the expected loss is \$1. If the insurer had no costs (aside from paying claims), it would expect to break even if it charged the consumer \$1, and paid the consumer \$10 if the bad event were to occur. If the insurer has many consumers with an identical risk, then the outcome that was unpredictable for any single consumer becomes predictable to the insurer—about 10% of consumers will suffer an accident—and the insurer can expect to break even.

Of course, real insurance companies do have labor and capital costs, aside from the costs of paying claims. We will not model the operation of a competitive insurance market (the supply side of the market), but we can easily examine how much above the actuarially fair rate the consumer would be willing to pay.

Without insurance, the consumer's expected utility is 0.695. With insurance, expected utility will be $U1(W0 - Pr, R0)$, where Pr is the insurance premium. Set that expression equal to 0.695, the consumer's expected utility when facing the gamble without insurance. The consumer would be willing to pay for insurance up to the point at which his or her expected utility is no higher with insurance than without.

```
EU1(W0,R0,x1,y1,p1);
```

0.695

```
float(solve(((U1(W0-Pr,R0))-
EU1(W0,R0,x1,y1,p1)),Pr));
```

$[Pr = 1.2571]$

The consumer would be willing to pay up to \$0.2571 *above* the actuarially fair \$1 premium.

16.3.2 A Continuous Probability Distribution Function

This consumer would pay up to \$1.2571 for insurance that would remove the financial implication of the adverse event which has a known impact on wealth.

In many cases, however, the impact of the event cannot be known, but is a random value. One way to address the implication of this source of uncertainty is to treat the implication of a negative event as having possible values that are generated by a *probability distribution function*, or pdf.

Before specifying the pdf that we use to illustrate this issue, we consider some aspects of how to apply *Maxima*. *Maxima*'s module, `distrib`, deals with probability distributions, including continuous probability distributions. To access and use *Maxima*'s built-in distributions, we first use the command `load(distrib)`.

Before loading and using `distrib`, however, we address one technical issue that pertains to our illustration. *Maxima* can have difficulty with the scale of some calculations. In our example, preliminary analysis revealed that *Maxima* could not process the command with units such that `W0` and related values are expressed in units that result in large numbers, like $25,000, so we recast the expression in money units such that $W = 25$. This *scaling* has no impact on the formal analysis, and it facilitates computation.

We now load the `distrib` module and apply it. We assume that losses follow a Chi-squared distribution that has a mean value of $10. The Chi-squared distribution can be accessed using `pdf_chi2(x,a)`, where x is the variable, a is its degrees of freedom (and its mean). The `pdf` part of the command calls for the probability distribution function.

The graph below shows that, for $a = 10$, the distribution is skewed. The probability that x will take a value as large as $10 is slightly greater than 0.5. The value of the *cumulative distribution function* for $x = 10$ is the area under the pdf to the left of $x = 10$, which is approximately 0.56.

```
load(distrib)$    wxdraw2d( ... Some commands
 omitted ... explicit(pdf_chi2(x,10),x,0,30),
 ... points([[10,0],[10,pdf_chi2(10,10)]]))$
print("cdf_chi2(10,10) = ",float(cdf_chi2(10,10)))$
```

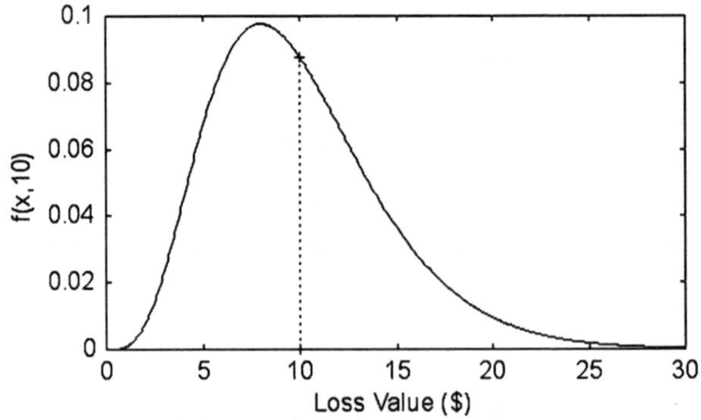

cdf_chi2(10,10) = 0.56

We calculate expected utility for this consumer by applying definite integration. We integrate over the *product* of the value of the distribution that attaches to a given

x value and the utility level that attaches to that value of x. This person's expected utility level, given the probability distribution of possible losses, is 0.515.

```
assume(x>0)$
float(integrate(((pdf_chi2(x,10))*
      U1(W0-x,R0)),x,0,inf));
```

0.515

The expected loss is $10, so the actuarially fair insurance premium is also $10. We can once again solve for the maximum amount the consumer would be willing to pay.

```
float(solve(((U1(W0- Pr,R0))-
float(integrate(((pdf_chi2(x,10))*
U1(W0-x,R0)),x,0,inf))),Pr));
```

$[Pr = 10.536]$

With this particular distribution function and set of parameters, the consumer is willing to pay $0.536 above the actuarially fair premium.[12]

References

1. Arrow KJ (1971) The theory of risk aversion. Reprinted in: Essays in the theory of risk bearing. Markham, Chicago
2. Knight FH (1921) Risk, uncertainty, and profit. Available at http://www.econlib.org/library/Knight/knRUP.html
3. Nicholson W, Snyder C (2008) Microeconomic theory: basic principles and extensions, 11th edn. South-Western, Mason
4. Pratt JW (1964) Risk aversion in the small and in the large. Econometrica 32:122–136
5. Varian HR (1992) Microeconomic analysis, 3rd edn. Norton, New York

[12]Remember units. If this person's wealth is measured in 1,000 dollar units, then the person is willing to pay $536 "regular" dollars above the actuarially fair premium.

Exercise: Examine the effect of changing R on the consumer's willingness to pay above the actuarially fair premium.

Index

M.R. Hammock and J.W. Mixon, *Microeconomic Theory and Computation: Applying the Maxima Open-Source Computer Algebra System*,
DOI 10.1007/978-1-4614-9417-1, © Springer Science+Business Media New York 2013

MIX
Papier aus verantwortungsvollen Quellen
Paper from responsible sources
FSC® C105338

If you have any concerns about our products,
you can contact us on
ProductSafety@springernature.com

In case Publisher is established outside the EU,
the EU authorized representative is:
Springer Nature Customer Service Center GmbH
Europaplatz 3, 69115 Heidelberg, Germany

Printed by Libri Plureos GmbH
in Hamburg, Germany